Lecture Notes in Computer Science 13911

The series Lecture Notes in Computer Science (LNCS), including its subseries Lecture Notes in Artificial Intelligence (LNAI) and Lecture Notes in Bioinformatics (LNBI), has established itself as a medium for the publication of new developments in computer science and information technology research, teaching, and education.

LNCS enjoys close cooperation with the computer science R & D community, the series counts many renowned academics among its volume editors and paper authors, and collaborates with prestigious societies. Its mission is to serve this international community by providing an invaluable service, mainly focused on the publication of conference and workshop proceedings and postproceedings. LNCS commenced publication in 1973.

Frank Drewes · Mikhail Volkov
Editors

Developments in Language Theory

27th International Conference, DLT 2023
Umeå, Sweden, June 12–16, 2023
Proceedings

Editors
Frank Drewes
Umeå University
Umeå, Sweden

Mikhail Volkov
Ural Federal University
Ekaterinburg, Russia

ISSN 0302-9743 ISSN 1611-3349 (electronic)
Lecture Notes in Computer Science
ISBN 978-3-031-33263-0 ISBN 978-3-031-33264-7 (eBook)
https://doi.org/10.1007/978-3-031-33264-7

This Springer imprint is published by the registered company Springer Nature Switzerland AG
The registered company address is: Gewerbestrasse 11, 6330 Cham, Switzerland

Preface

The 27th International Conference on Developments in Language Theory (DLT 2023) was held from June 12 to June 16 in Umeå, Sweden. It was organised by the Department of Computing Science at Umeå University, and co-located with the International Conference WORDS 2023 held at the same time. The organization of last year's DLT 2022 was all of a sudden interrupted by the shocking events in Ukraine, which jeopardized the conference. However, back then it was decided that not to continue our work would have been the wrong answer to the terrible events. Today, one year later, the war is still ongoing with no end in sight, and both soldiers and civilians are suffering to an extent an outsider simply cannot imagine. The world is divided more than ever since the cold war, and insane amounts of money that could be used to further science and the wellbeing of humanity are spent on killing machines instead. As scientists who are used to collaborate across the borders of different beliefs, political systems and cultural backgrounds, we condemn this development and call instead for an end to the suffering of humans caused by humans.

The DLT conference series provides a forum for presenting current developments in formal languages and automata. Its scope is very general and includes, among others, the following topics and areas: grammars, acceptors and transducers for words, trees and graphs; algebraic theories of automata; algorithmic, combinatorial and algebraic properties of words and languages; relations between formal languages and artificial neural networks; variable length codes; symbolic dynamics; cellular automata; groups and semigroups generated by automata; polyominoes and multi-dimensional patterns; decidability questions; image manipulation and compression; efficient text algorithms; relationships to cryptography, concurrency, complexity theory and logic; bio-inspired computing; and quantum computing.

Since its establishment by Grzegorz Rozenberg and Arto Salomaa in Turku (1993), a DLT conference was held every other year in Magdeburg (1995), Thessaloniki (1997), Aachen (1999), and Vienna (2001). Since 2001, a DLT conference takes place in Europe every odd year and outside Europe every even year. The locations of DLT conferences since 2002 were: Kyoto (2002), Szeged (2003), Auckland (2004), Palermo (2005), Santa Barbara (2006), Turku (2007), Kyoto (2008), Stuttgart (2009), London (2010), Milano (2011), Taipei (2012), Marne-la-Vallée (2013), Ekaterinburg (2014), Liverpool (2015), Montréal (2016), Liège (2017), Tokyo (2018), Warsaw (2019), Porto (2021) and Tampa (2022). The missing DLT 2020 was planned to be the one that would be held in Tampa, Florida, but it was cancelled due to the COVID-19 pandemic. The accepted papers were nevertheless published in volume 12086 of Lecture Notes in Computer Science, and the authors of these papers could present their work at DLT 2021 in Porto. With DLT 2023 in Umeå, the DLT conference series has now finally gone back to its established rhythm.

In 2018, the DLT conference series instituted the Salomaa Prize to honour the work of Arto Salomaa, and to increase the visibility of research on automata and formal language theory. The prize is funded by the University of Turku. The ceremony for the Salomaa

Prize 2023 took place in Umeå on June 14, 2023, as part of the combined program of DLT 2023 and WORDS 2023, and we congratulate the winner Moshe Y. Vardi, who was presented with the prize by Wolfgang Thomas, the Chair of the Prize Committee of 2023.

This volume contains invited contributions as well as the accepted papers of DLT 2023. There were 31 submissions by 66 authors from 21 countries: Canada, Chile, Czechia, Finland, France, Germany, India, Israel, Italy, Japan, The Netherlands, Norway, Poland, Portugal, Russia, South Africa, South Korea, Sweden, Taiwan, the UK and the USA. Each submission was single-blind reviewed by three experts in the field. All submissions were thoroughly discussed by the Program Committee who decided to accept 19 of the 31 submitted papers for presentation at the conference. We would like to thank the members of the Program Committee, and all external reviewers, for their excellent work in evaluating the papers and for valuable comments that led to the selection of the contributed papers.

There were five invited talks, which were presented by:

- Émilie Charlier (University of Liège, Belgium):
  Alternate Base Numeration Systems
- Ismael Jecker (University of Warsaw, Poland):
  Transducers and the Power of Delay
- Jussi Karlgren (Silo AI, Finland):
  When the Map is More Exact than the Terrain
- Will Merrill (New York University, USA):
  Formal Languages and the NLP Black Box
- Markus Schmid (Humboldt-Universität zu Berlin, Germany):
  On Structural Tractability Parameters for Hard String Problems

We would like to express our sincere thanks to the invited speakers and all authors of submitted papers. Without them, there would not have been a DLT 2023, much less a successful one.

The EasyChair conference system provided excellent support in the selection of the papers and the preparation of these proceedings. Special thanks are due to the Lecture Notes in Computer Science team at Springer for having granted us the opportunity to publish these proceedings in the series, and also for their help during the process. We are also grateful for the kind support received from the journal Algorithms from MDPI that made it possible to award prizes during the conference.

We are grateful to the members of the Organizing Committee: Martin Berglund, Johanna Björklund and Lena Strobl.

DLT 2023 was financially supported by

- Department of Computing Science, Umeå University
- Umeå municipality, Region Västerbotten and Umeå University
- MDPI (Multidisciplinary Digital Publishing Institute)

June 2023

Frank Drewes
Mikhail Volkov

Organization

Program Committee Chairs

Frank Drewes	Umeå University, Sweden
Mikhail Volkov	Ural Federal University, Russia

Steering Committee

Marie-Pierre Béal	Gustave Eiffel University, France
Cristian S. Calude	University of Auckland, New Zealand
Volker Diekert	University of Stuttgart, Germany
Yo-Sub Han	Yonsei University, Republic of Korea
Juraj Hromkovic	ETH Zürich, Switzerland
Oscar H. Ibarra	University of California, Santa Barbara, USA
Nataša Jonoska	University of South Florida, USA
Juhani Karhumäki (Chair 2010–2020)	University of Turku, Finland
Martin Kutrib	Justus-Liebig-Universität Gießen, Germany
Giovanni Pighizzini (Chair 2020–present)	University of Milan, Italy
Michel Rigo	Université de Liège, Belgium
Antonio Restivo	Università di Palermo, Italy
Wojciech Rytter	Warsaw University, Poland
Kai Salomaa	Queen's University, Canada
Shinnosuke Seki	University of Electro-Communications, Japan
Mikhail Volkov	Ural Federal University, Russia
Takashi Yokomori	Waseda University, Japan

Honorary Members

Grzegorz Rozenberg	Leiden University, Netherlands
Arto Salomaa	University of Turku, Finland

Program Committee

Marie-Pierre Béal	Université Gustave Eiffel, France
Frank Drewes (Chair)	Umeå University, Sweden
Yo-Sub Han	Yonsei University, South Korea
Galina Jirásková	Slovak Academy of Sciences, Slovakia
Jarkko Kari	University of Turku, Finland
Manfred Kufleitner	Universität Stuttgart, Germany
Nutan Limaye	IT University of Copenhagen, Denmark
Andreas Maletti	Universität Leipzig, Germany
Ian McQuillan	University of Saskatchewan, Canada
Alexander Okhotin	St. Petersburg State University, Russia
Thomas Place	Université de Bordeaux, France
Marek Szykuła	University of Wrocław, Poland
Mikhail Volkov (Chair)	Ural Federal University, Russia
Petra Wolf	University of Bergen, Norway
Tomoyuki Yamakami	University of Fukui, Japan
Hsu-Chun Yen	National Taiwan University, Taiwan

Additional Reviewers

Emmanuel Arrighi
Corentin Barloy
Martin Berglund
Henrik Björklund
Michaël Cadilhac
Arnaud Carayol
Antonio Casares
Andrei Draghici
Szilard Zsolt Fazekas
Fan Feng
Pierre Guillon
Mark Kambites
Christos Kapoutsis
Sergey Kitaev
Johan Kopra
Maria Kosche
Marina Maslennikova
Carl-Fredrik Nyberg Brodda
Erik Paul
Sylvain Perifel
Ramchandra Phawade
Elena Pribavkina
Karin Quaas
Michael Rao
Igor Rystsov
Lena Katharina Schiffer
A. V. Sreejith
Wolfgang Steiner
Tachio Terauchi
John Torr
Farhad Vadiee
Thomas Zeume

Invited Papers

Transducers and the Power of Delay

Ismaël Jecker

University of Warsaw, Poland

Abstract. Transducers are theoretical machines computing functions: they read input words and answer with output words. Several variations of the standard transducer model have been developed to compute different classes of functions. This talk surveys three fundamental classes:

- Sequential functions, the basic class computed by the simplest model;
- Regular functions, a versatile class recognized by a variety of models;
- Polyregular functions, the most expressive but most complex class.

The central theme of the talk is the notion of delay, a powerful tool that reduces problems about transducers to problems about automata.

1 Sequential Functions

Finite state transducers [12, 19] form the simplest model for computing functions, and have been studied extensively. Like finite state automata, they are machines with a finite number of states that read an input letter on each transition. However, finite state transducers also produce an output word on each transition. We consider here only machines that process the input letters deterministically, thus compute functions from input words to output words.[1] Functions computed by (deterministic) finite state transducers are called *sequential* functions.

Example: Sequential functions can be used to perform a variety of simple tasks. For example, the *double* function duplicates each letter of the input, the *erase* function removes all occurrences of specified letters from the alphabet, and the *swap* function exchanges each letter at an odd position with the following letter:

Double :	DLT2023	$\mapsto$	DDLLTT22002233
Erase numerals :	DLT2023	$\mapsto$	DLT
Swap :	DLT2023	$\mapsto$	LD2T203

In addition, the composition of two sequential functions is also sequential. Building a finite state transducer that recognises the composition of two sequential functions is straightforward: it is achieved through a simple Cartesian product.

[1] If non-determinism is allowed, the machines recognize binary relations between input words and output words, where one input can be mapped to multiple outputs.

While finite state transducers share similarities with finite state automata, they are considerably more complex to handle. For instance it is impossible to decide whether two finite state transducers map at least one input word to the same output, due to the potential for the two transducers to produce the same output at vastly different paces. To address this issue, the concept of *delay* was introduced. The delay assigns a numerical value to each pair of transducer executions, indicating their proximity. The power of the delay lies in its ability to reason about a transducer by examining the regular language of its executions, thereby reducing the study of transducers to that of automata. The notion of delay is central to many key results related to finite state transducers, including the ability to decide transducer equivalence (even beyond deterministic transducers [15]), to minimize transducers [8], and to transform a non-deterministic transducer into a deterministic one (when possible) [7].

2 Regular Functions

The beauty of regular functions lies in their versatility: they are recognised by multiple formalisms which appear vastly different at first glance. The oldest model is the *two-way transducer*, mentioned by Shepherdson in [20]. Such a machine is obtained by allowing a finite state transducer to move forward and backward along the input word. A broader interest towards this class began to surface when Engelfriet and Hoogeboom established in [13] that the functions defined by two-way transducers match a logical formalism known as *monadic second order transductions*. Finally, the introduction of the *streaming string transducer* (SST) model by Alur and Cerný in [1] solidified this class as a fundamental class of functions, and gave rise to a myriad of equivalent models, such as regular combinators (a parallel to regular expressions [3]), DREX (a declarative programming language [2]), and regular list functions (a calculus of functions [6]).

Example. On top of sequential functions, regular functions contain the *duplicate* function that doubles the input word and the *mirror* function that reverses it:

$$\begin{array}{llcl} \text{Duplicate}: & \text{DLT2023} & \mapsto & \text{DLT2023DLT2023} \\ \text{Mirror}: & \text{DLT2023} & \mapsto & \text{3202TLD} \end{array}$$

Moreover, the composition of two regular functions is also regular. For most abstract machine models composition is complex and leads to an (at least) exponential increase of the size of the state space [1, 9]. The exception to this rule is the model of *reversible two-way transducers*, a restricted form of two-way transducers. These transducers can be composed in polynomial time, yet remain expressive enough to recognise all regular functions [10].

Regular functions enjoy many good algorithmic properties : for instance, their equivalence is decidable. Recently, a notion of delay adapted to the setting of regular functions has been introduced in the hope of achieving the same success as in the case of finite state transducers, as described in [16]. While this new concept has proven useful in addressing some open problems, a number of important questions remain unanswered. For instance, there is currently no well-defined notion of minimisation for any of the models, and there is no canonical object that can be used to recognise a given regular function.

3 Polyregular Functions

The *growth rate* of a regular function f is at most linear: there exists an integer N such that for every input word u the length of $f(u)$ is at most $N(|u| + 1)$. The class of polyregular functions aims to push this limitation further: the growth rate of these functions can be a polynomial of higher degree.

The interest for polyregular functions is relatively recent. The earliest model studied is the pebble transducer [14], which extends the two-way transducer model by providing access to a fixed number of *pebbles*. These pebbles can be dropped during the execution to mark specific positions of the input word. Interest towards polyregular functions has grown rapidly following Bojańczyk's work [4] which showed that, while polyregular functions are strictly more expressive than regular functions, they are similarly versatile and can be recognized by several different formalisms, from logic to programming languages.

Example: On top of regular functions, polyregular functions contain the *squaring function* that maps each word u to the word $u^{|u|}$:

$$\text{DLT2023} \mapsto \text{DLT2023DLT2023DLT2023DLT2023DLT2023DLT2023DLT2023}$$

Moreover, the composition of two polyregular functions is also polyregular.

The study of polyregular functions is relatively young, and numerous important questions remain open. One of the key unresolved problems is whether or not equivalence of polyregular functions is decidable. While this issue is still an open question in general, recent studies have established that equivalence is decidable when the output alphabet consists of a single letter. This result was proven by reducing pebble transducers to weighted automata, as outlined in [11].

Acknowledgements. This extended abstract was inspired by several excellent surveys on the subject [5, 17, 18]. If you are interested in delving deeper into the various classes of transducers and their properties, we highly recommend referring to these resources for further reading.

References

1. Alur, R., Cerný, P.: Expressiveness of streaming string transducers. In: FSTTCS 2010 (2010). https://doi.org/10.4230/LIPIcs.FSTTCS.2010.1
2. Alur, R., D'Antoni, L., Raghothaman, M.: Drex: A declarative language for efficiently evaluating regular string transformations. In: POPL 2015 (2015). https://doi.org/10.1145/2676726.2676981
3. Alur, R., Freilich, A., Raghothaman, M.: Regular combinators for string transformations. In: CSL-LICS 2014 (2014). https://doi.org/10.1145/2603088.2603151

4. Bojanczyk, M.: Polyregular functions. CoRR abs/1810.08760 (2018). http://arxiv.org/abs/1810.08760
5. Bojanczyk, M.: Transducers of polynomial growth. In: LICS 2022 (2022). https://doi.org/10.1145/3531130.3533326
6. Bojanczyk, M., Daviaud, L., Krishna, S.N.: Regular and first-order list functions. In: LICS 2018 (2018). https://doi.org/10.1145/3209108.3209163
7. Choffrut, C.: Une caracterisation des fonctions sequentielles et des fonctions sous-sequentielles en tant que relations rationnelles. Theor. Comput. Sci. **5**(3), 325–337 (1977). https://doi.org/10.1016/0304-3975(77)90049-4
8. Choffrut, C.: Minimizing subsequential transducers: a survey. Theor. Comput. Sci. **292**(1), 131–143 (2003). https://doi.org/10.1016/S0304-3975(01)00219-5
9. Chytil, M., Jákl, V.: Serial composition of 2-way finite-state transducers and simple programs on strings. In: Automata, Languages and Programming, Fourth Colloquium (1977). https://doi.org/10.1007/3-540-08342-1_11
10. Dartois, L., Fournier, P., Jecker, I., Lhote, N.: On reversible transducers. In: ICALP 2017 (2017). https://doi.org/10.4230/LIPIcs.ICALP.2017.113
11. Douéneau-Tabot, G.: Pebble transducers with unary output. In: MFCS 2021 (2021). https://doi.org/10.4230/LIPIcs.MFCS.2021.40
12. Elgot, C.C., Mezei, J.E.: On relations defined by generalized finite automata. IBM J. Res. Dev. **9**(1), 47–68 (1965). https://doi.org/10.1147/rd.91.0047
13. Engelfriet, J., Hoogeboom, H.J.: MSO definable string transductions and two-way finite-state transducers. ACM Trans. Comput. Log. **2**(2), 216–254 (2001). https://doi.org/10.1145/371316.371512
14. Engelfriet, J., Maneth, S.: Two-way finite state transducers with nested pebbles. In: MFCS 2002 (2002). https://doi.org/10.1007/3-540-45687-2_19
15. Filiot, E., Jecker, I., Löding, C., Winter, S.: On equivalence and uniformisation problems for finite transducers. In: ICALP 2016 (2016). https://doi.org/10.4230/LIPIcs.ICALP.2016.125
16. Filiot, E., Jecker, I., Löding, C., Winter, S.: A regular and complete notion of delay for streaming string transducers. In: STACS 2023 (2023). https://doi.org/10.4230/LIPIcs.STACS.2023.32
17. Filiot, E., Reynier, P.: Transducers, logic and algebra for functions of finite words. ACM SIGLOG News **3**(3), 4–19 (2016). https://doi.org/10.1145/2984450.2984453
18. Muscholl, A., Puppis, G.: The many facets of string transducers (invited talk). In: STACS 2019 (2019). https://doi.org/10.4230/LIPIcs.STACS.2019.2
19. Schützenberger, M.P.: A remark on finite transducers. Inf. Control. **4**(2–3), 185–196 (1961). https://doi.org/10.1016/S0019-9958(61)80006-5
20. Shepherdson, J.C.: The reduction of two-way automata to one-way automata. IBM J. Res. Dev. **3**(2) (1959). https://doi.org/10.1147/rd.32.0198

When the Map is More Exact than the Terrain

Jussi Karlgren

Silo AI, Finland

Computational models of the world are often designed to conform to both our intuitions about important qualities of the world and to computational convenience. These two design principles can be at cross purposed. Hypotheses about e.g. human information processing, based on observations of effective and efficient human behaviour can be quite useful as an inspiration to how a computational model should be put together. Those intuitions and guiding principles and the metaphors they provide can later turn out to limit innovation and development, especially if they are catchy and easily communicated to the outside world. This talk will give examples related to neural processing models, and specifically vector space models of the semantics of human language.

On Structural Tractability Parameters for Hard String Problems

Markus L. Schmid

Institut für Informatik, Humboldt-Universität zu Berlin, 10099 Berlin, Germany

Abstract. For intractable graph problems, many structural "tractability parameters" exist (like, e. g., treewidth or other width-parameters), which, if bounded by a constant, yield polynomial-time solvable cases or even fixed-parameter tractability. Indeed, consulting the well-known "Information System on Graph Classes and their Inclusions" (https://www.graphclasses.org/), the knowledge on graph parameters and their corresponding graph classes (both algorithmically as well as combinatorially) seems overwhelming, especially to a non-expert in algorithmic graph theory.

For (intractable) problems on strings, the situation is quite different. Most parameters that naturally come to mind have a non-structural flavour like alphabet size, number of input strings, size of language descriptors, etc. In any case, structural tractability parameters for strings seem to be highly dependent on the actual problem at hand: some structural restriction of strings may substantially decrease the complexity of one specific problem, while it has virtually no impact for any other string problem.

In this talk, we will survey some results that stem from the task of discovering structural parameters for strings that can make hard problem instances tractable. We will focus on the approach to translate string problems into graph problems in order to make use of algorithmic metatheorems in algorithmic graph theory.

Contents

Formal Languages and the NLP Black Box

William Merrill(✉)

New York University, New York, NY 10011, USA
willm@nyu.edu

1 Introduction

The field of natural language processing (NLP) has been transformed in two related ways in recent years. First, the field moved towards using neural network architectures like LSTMs[1] [12] and transformers [31], in contrast to approaches that explicitly represent grammatical rules. Another innovation was a move towards *semi-supervised learning* [7,24,26]: language models have been used in various ways to solve downstream NLP tasks that previously would have required large labeled datasets. Transformer-based language models in particular have been remarkably empirically successful across a range of NLP tasks [28], and making the models and datasets bigger tends to not only improve performance on benchmarks [13] and linguistic generalization [30,32], but can also lead to the emergence of new algorithmic behavior, such as arithmetic and logical reasoning [33].

I will argue that there are many intriguing mysteries about these empirical results that formal language theory can clarify. It seems as if large transformer language models are implicitly learning some aspects of natural language grammar [30]. If so, it seems useful to understand what kinds of formal grammars such networks can simulate, and how grammatical dependencies are represented within them. For comparing and extending neural network architectures, it would also be useful to have a theory of how different types of neural networks compare in expressive power to one another.

Rather than just focusing on grammatical dependencies, we may also adopt a similar formal language theoretic perspective for understanding reasoning in transformer language models. Can we characterize the kinds of computational problems transformers can solve? Can we use this theory to extract algorithmic behavior from a transformer into a discrete, human-readable program? Can we predict the amount of language modeling data needed to solve various reasoning problems, or find problems that transformers can *never* solve, even at massive scale?

I will survey recent work that provides some insight on the computational model of RNNs and transformers. I will start by discussing an older line of work analyzing the power of different RNN variants in relation to one another and the Chomsky hierarchy. I will then discuss a newer line of work that analyzes the computational power of transformers using circuit complexity theory.

[1] An LSTM is a special kind of recurrent neural network [8,9].

F. Drewes and M. Volkov (Eds.): DLT 2023, LNCS 13911, pp. 1–8, 2023.
https://doi.org/10.1007/978-3-031-33264-7_1

While the techniques used in these two lines of work may be different, some unified insights emerge for understanding the capabilities and inner workings of both RNNs and transformers. In particular, *counting* seems to be a central computational ability to the kinds of processing possible in both LSTMs and transformers. Yet the benefit of counting is not the same for each architecture: transformers—but not LSTMs—can use counting to recognize k-Dyck for any k, a capability often taken to embody sensitivity to hierarchical structure [2]. On the other hand, a fundamental weakness of transformers compared to LSTMs is parallelism: while RNNs can simulate certain computation graphs linear in the size of the input sequence, transformers have a constant-depth computation graph, and thus must do much more processing in parallel.

2 RNNs

There are deep connections between neural networks with recurrence and automata: historically, one motivation for developing finite automata was to model the computation of networks of biological neurons with binary activation patterns [15]. The 1990s saw the analysis of RNN models with linear-thresholded activations, which, with infinite precision and run time, are Turing-complete and thus significantly more expressive than finite automata [27]. See [17] for further discussion of RNN results with infinite precision.

But the infinite-precision, infinite-runtime model [27] does not capture the type of RNNs used in modern deep learning, which are typically unrolled with one step per input token ("real-time"), and suffer from practical precision constraints that prevent storing an unbounded Turing machine tape in a finite number of numbers in $[-1, 1]$.[2] How then should we understand the set of languages that RNNs can learn to recognize in practice then? A central finding here is that LSTMs, one RNN extension, can recognize languages requiring counting, whereas most other RNNs cannot [34]. [16] then proposed saturated RNNs as a simplified theoretical model of bounded-precision RNNs, and showed that the expressive power of saturated RNNs often predicts the empirical abilities of RNNs.

2.1 LSTMs Can Count, Other RNNs Cannot

Aiming to understand the practical power of RNNs, [34] empirically evaluate the ability of basic RNNs and their variants, LSTMs [12] and GRUs [5], to recognize the formal language $a^n b^n$. Empirically, they show that LSTMs can recognize $a^n b^n$, while RNNs and GRUs cannot. Moreover, they show the LSTM

[2] There are at least two reasons to view this practical precision setting as more realistic. First, hardware imposes a maximum precision on each number in the RNN, which may reasonably be considered to be finite or logarithmic in the input sequence length n. Second, RNNs are trained by gradient descent, and we would like to understand the class of languages that can be learned by an RNN. Intuitively, constructions that are sensitive to low-order imprecision may be hard to learn by gradient descent [34].

achieves this by "counting": using a memory cell to track the difference between the number of a's and b's in the input. Similar results were then observed for other languages requiring counting, such as 1-Dyck or shuffled Dyck [29]. In contrast, LSTMs were not able to reliably learn 2-Dyck, which requires a stack as opposed to just counting [29].[3]

2.2 Saturated RNNs as a Model of Practical RNNs

[16,22] analyze the expressive power of *saturated* RNNs as a proxy for what unsaturated RNNs can learn by gradient descent. This technique is motivated by the hypothesis that networks requiring bounded parameter norm are unlikely to be acquired by a training process where the parameter norm is growing consistently over the course of training.[4] Given a network $f(x;\theta)$, a saturated network is the function $f'(x;\theta)$ obtained by making the parameters θ large:

$$f'(x;\theta) = \lim_{\rho \to \infty} f(x;\rho\theta).$$

The effect of making the weights large in this way is to convert the activation functions in all parts of the network to step functions. [16] place saturated RNNs, GRUs, and LSTMS in the Chomsky hierarchy. Saturated RNNs and GRUs are equivalent to finite automata [16], whereas saturated LSTMs can simulate a constrained class of counter automata that can recognize 1-Dyck, $a^n b^n$, or $a^n b^n c^n d^n$ but do not have enough memory to recognize 2-Dyck. Thus, LSTMs can be understood to cross-cut the conventional Chomsky hierarchy: able to recognize some context-sensitive languages, but unable to simulate a stack or process arbitrary hierarchical structure. This analysis of saturated networks places different types of RNNs in different relations to the Chomsky hierarchy, and these predictions largely match the type of languages that unsaturated RNNs can learn to recognize in practice.

2.3 Summary and Open Questions

Both empirical results and the theoretical model of saturated networks suggests that counting is a key capability of LSTMs that RNNs and GRUs do not have. While counting enables LSTMs to recognize some languages like $a^n b^n$ and 1-Dyck, it does not allow LSTMs to process arbitrary hierarchical or context-free structure (e.g., 2-Dyck or palindromes).

3 Transformers

Over the last 5 years, transformers have largely replaced RNNs as the backbone of neural NLP systems. A difficulty in extending automata-based analysis of

[3] See also [6] for more recent, but similar, empirical results on the ability of different RNN variants to recognize formal languages.

[4] See [18] for thorough empirical exploration of norm growth and saturation during the training of large transformer language models.

RNNs to transformers is that the transformer neural network architecture lacks autoregressive structure. Instead, recent work has made progress understanding the power of transformers by relating transformers to formal language classes defined by circuit families and logics.

3.1 Transformers with Hard Attention

[10] prove that the transformers with hard attention cannot recognize even simple formal languages like parity or 1-Dyck. [11] extend this result to prove that hard-attention transformers can be simulated[5] by constant-depth, poly-size circuit families, which recognize the formal language class AC^0. This implies [10]'s results, as well as demonstrating new languages that hard-attention transformers cannot recognize, such as majority (taking the majority vote of a sequence of bits).

3.2 Transformers with Soft Attention

[3] show that, like LSTMs, transformers have the ability to count, and can use this to recognize 1-Dyck and other related formal languages. [21,25] show that transformers can also use counting to recognize majority, implying that soft attention is stronger than hard attention.

[21] then analyze *saturated* transformers, which have simplified attention patterns compared to soft attention, but can still count. They find that saturated transformers over a floating-point data type can be simulated by constant-depth, poly-size *threshold* circuit families (i.e., the complexity class TC^0). Intuitively, counting is one of the key capabilities achievable in TC^0 but not AC^0, suggesting that counting is a good way to understand the gain in power that saturated attention grants relative to hard attention.

[19] then extend [21]'s result to show that arbitrary soft-attention transformers with precision logarithmic in the input length can be simulated in the tighter class log-space-uniform TC^0. Log-space-uniform TC^0 is conjectured to be separated from other complexity classes like L, NL, or P, which would imply that transformers cannot solve complete problems for these classes. Thus, accepting these separation conjectures, transformers cannot compute connectivity in directed or undirected graphs, solve the universal context-free grammar recognition problem, or solve linear systems of matrix equations.

3.3 Logics and Programming Languages for Expressing Transformer Computation

One converging theme in recent work has been attempting to propose symbolic formalisms describing computation in transformers.

[5] Here, "simulate" means that any function computed by a transformer can also be computed by such a circuit family.

[20] further refine the circuit-based upper on soft-attention transformers, showing that transformers can be simulated in log-time-uniform TC^0. This class has an equivalent characterization as the formal languages definable in first-order logic with majority quantifiers [23], or $\mathsf{FO(M)}$. This immediately implying that soft-attention transformers can be "translated" to first-order logic formulae with majority quantifiers that compute the same function.

Along similar lines, [4] propose $\mathsf{FOC}(+, \mathsf{MOD})$, or first-order logic with counting quantifiers[6], as a logical model of transformers. [4] prove that $\mathsf{FOC}(+, \mathsf{MOD})$ is an upper bound on finite-precision transformers and a lower bound on transformers with arbitrary precision, although figuring out whether there is some model of transformers for which it is a tight bound remains open.

Finally, [35] propose a programming language called RASP for expressing computation in a transformer-like way. [14] create a compiler that compiles constrained RASP programs into actual transformers. [35] shows that RASP can recognize arbitrary Dyck languages (not just 1-Dyck), and, empirically, transformers can as well.[7]

3.4 Summary and Open Questions

Counting—a key capability separating LSTMs from RNNs—also separates soft-attention transformers from hard-attention transformers. Upper bounds on the power of transformers derived via circuit complexity give us classes of problems that transformers cannot solve, but which are efficiently solvable by a recurrent model of computation like a Turing machine. The intuition behind why these problems are hard for transformers is that transformers are fundamentally constrained to parallel computation, and it is conjectured in complexity theory that certain problems are fundamentally unparallelizable.

Significant progress has been made on the analysis of transformers in recent years, and connections to deep questions in complexity theory have been revealed. Yet, there are still many things that are unclear. Is saturated attention fundamentally weaker than soft attention? Can we make upper bounds and lower bounds on transformers tighter? Can we leverage theoretical insights to extract discrete computational mechanisms from trained transformers?

4 Conclusion

There has been a wide range of work in recent years analyzing the capabilities of RNNs and transformers as formal grammars. One unifying insight that has emerged for both types of neural networks is that they can leverage counting to process structure in their input, although potentially in different ways. Transformers in particular can use counting to recognize arbitrary Dyck languages,

[6] with addition and mod but not ordering over positions.

[7] In their experiments, [35] add special regularization to the attention patterns of the transformer to get it to learn Dyck languages properly.

which are often viewed as an exemplar of hierarchical structure. Due to the transformer's lack of autoregressive structure, circuit complexity and logic have been more useful than automata theory for understanding transformers' capabilities. Hopefully, insights from these theories may continue to refine our ability to peer into the black box of transformers, and perhaps understanding transformers may even inspire new research questions or advances in these fields.

5 Resources

A key area for active research is the Formal Languages and Neural Networks (FLaNN) Discord server and talk series. For more extensive (but out of date) surveys, the reader should see [1,17]. A more up-to-date survey of the circuit complexity-based analysis of transformers is under preparation by members of FLaNN.

Acknowledgments. Thank you to Michael Hu for his feedback.

References

1. Ackerman, J., Cybenko, G.: A survey of neural networks and formal languages (2020)
2. Autebert, J.-M., Berstel, J., Boasson, L.: Context-free languages and pushdown automata. In: Rozenberg, G., Salomaa, A. (eds.) Handbook of Formal Languages, pp. 111–174. Springer, Heidelberg (1997). https://doi.org/10.1007/978-3-642-59136-5_3
3. Bhattamishra, S., Ahuja, K., Goyal, N.: On the ability and limitations of transformers to recognize formal languages. In: Proceedings of the 2020 Conference on Empirical Methods in Natural Language Processing (EMNLP), pp. 7096–7116. Association for Computational Linguistics, November 2020. https://aclanthology.org/2020.emnlp-main.576
4. Chiang, D., Cholak, P., Pillay, A.: Tighter bounds on the expressivity of transformer encoders (2023)
5. Cho, K., van Merriënboer, B., Bahdanau, D., Bengio, Y.: On the properties of neural machine translation: encoder-decoder approaches. In: Proceedings of SSST-8, Eighth Workshop on Syntax, Semantics and Structure in Statistical Translation, Doha, Qatar, pp. 103–111. Association for Computational Linguistics, October 2014. https://aclanthology.org/W14-4012
6. Deletang, G., et al.: Neural networks and the chomsky hierarchy. In: The Eleventh International Conference on Learning Representations (2023). https://openreview.net/forum?id=WbxHAzkeQcn
7. Devlin, J., Chang, M.W., Lee, K., Toutanova, K.: BERT: pre-training of deep bidirectional transformers for language understanding. In: Proceedings of the 2019 Conference of the North American Chapter of the Association for Computational Linguistics: Human Language Technologies, Volume 1 (Long and Short Papers), Minneapolis, Minnesota, pp. 4171–4186. Association for Computational Linguistics, June 2019. https://aclanthology.org/N19-1423
8. Elman, J.L.: Finding structure in time. Cogn. Sci. **14**(2), 179–211 (1990)

9. Goldberg, Y.: A primer on neural network models for natural language processing. J. Artif. Intell. Res. **57**, 345–420 (2016)
10. Hahn, M.: Theoretical limitations of self-attention in neural sequence models. Trans. Assoc. Comput. Linguist. **8**, 156–171 (2020). https://aclanthology.org/2020.tacl-1.11
11. Hao, Y., Angluin, D., Frank, R.: Formal language recognition by hard attention transformers: perspectives from circuit complexity. Trans. Assoc. Comput. Linguist. **10**, 800–810 (2022). https://aclanthology.org/2022.tacl-1.46
12. Hochreiter, S., Schmidhuber, J.: Long short-term memory. Neural Comput. **9**(8), 1735–1780 (1997)
13. Liang, P., et al.: Holistic evaluation of language models (2022)
14. Lindner, D., Kramár, J., Rahtz, M., McGrath, T., Mikulik, V.: Tracr: compiled transformers as a laboratory for interpretability (2023)
15. Mcculloch, W.S., Pitts, W.: A logical calculus of the ideas immanent in nervous activity. Bull. Math. Biophys. **5**, 115–133 (1943). https://doi.org/10.1007/BF02478259
16. Merrill, W.: Sequential neural networks as automata. In: Proceedings of the Workshop on Deep Learning and Formal Languages: Building Bridges, Florence, pp. 1–13. Association for Computational Linguistics, August 2019. https://aclanthology.org/W19-3901
17. Merrill, W.: Formal language theory meets modern NLP (2021)
18. Merrill, W., Ramanujan, V., Goldberg, Y., Schwartz, R., Smith, N.A.: Effects of parameter norm growth during transformer training: inductive bias from gradient descent. In: Proceedings of the 2021 Conference on Empirical Methods in Natural Language Processing, Punta Cana, Dominican Republic, pp. 1766–1781. Association for Computational Linguistics, November 2021. https://doi.org/10.18653/v1/2021.emnlp-main.133. https://aclanthology.org/2021.emnlp-main.133
19. Merrill, W., Sabharwal, A.: The parallelism tradeoff: limitations of log-precision transformers (2023)
20. Merrill, W., Sabharwal, A.: Transformers can be expressed in first-order logic with majority (2023)
21. Merrill, W., Sabharwal, A., Smith, N.A.: Saturated transformers are constant-depth threshold circuits. Trans. Assoc. Comput. Linguist. **10**, 843–856 (2022). https://aclanthology.org/2022.tacl-1.49
22. Merrill, W., Weiss, G., Goldberg, Y., Schwartz, R., Smith, N.A., Yahav, E.: A formal hierarchy of RNN architectures. In: Proceedings of the 58th Annual Meeting of the Association for Computational Linguistics, pp. 443–459. Association for Computational Linguistics, July 2020. https://doi.org/10.18653/v1/2020.acl-main.43. https://aclanthology.org/2020.acl-main.43
23. Mix Barrington, D.A., Immerman, N., Straubing, H.: On uniformity within NC1. J. Comput. Syst. Sci. **41**(3), 274–306 (1990). https://www.sciencedirect.com/science/article/pii/002200009090022D
24. Peters, M.E., et al.: Deep contextualized word representations. In: Proceedings of the 2018 Conference of the North American Chapter of the Association for Computational Linguistics: Human Language Technologies, Volume 1 (Long Papers), New Orleans, Louisiana, pp. 2227–2237. Association for Computational Linguistics, June 2018. https://aclanthology.org/N18-1202
25. Pérez, J., Marinković, J., Barceló, P.: On the turing completeness of modern neural network architectures. In: International Conference on Learning Representations (2019). https://openreview.net/forum?id=HyGBdo0qFm

26. Radford, A., Wu, J., Child, R., Luan, D., Amodei, D., Sutskever, I.: Language models are unsupervised multitask learners (2019)
27. Siegelmann, H., Sontag, E.: On the computational power of neural nets. J. Comput. Syst. Sci. **50**(1), 132–150 (1995). https://www.sciencedirect.com/science/article/pii/S0022000085710136
28. Srivastava, A., et al.: Beyond the imitation game: quantifying and extrapolating the capabilities of language models (2022)
29. Suzgun, M., Belinkov, Y., Shieber, S., Gehrmann, S.: LSTM networks can perform dynamic counting. In: Proceedings of the Workshop on Deep Learning and Formal Languages: Building Bridges, Florence, pp. 44–54. Association for Computational Linguistics, August 2019. https://doi.org/10.18653/v1/W19-3905. https://aclanthology.org/W19-3905
30. Tenney, I., Das, D., Pavlick, E.: BERT rediscovers the classical NLP pipeline. In: Proceedings of the 57th Annual Meeting of the Association for Computational Linguistics, Florence, Italy, pp. 4593–4601. Association for Computational Linguistics, July 2019. https://aclanthology.org/P19-1452
31. Vaswani, A., et al.: Attention is all you need. In: Advances in Neural Information Processing Systems, vol. 30 (2017)
32. Warstadt, A., Zhang, Y., Li, X., Liu, H., Bowman, S.R.: Learning which features matter: RoBERTa acquires a preference for linguistic generalizations (eventually). In: Proceedings of the 2020 Conference on Empirical Methods in Natural Language Processing (EMNLP), pp. 217–235. Association for Computational Linguistics, November 2020. https://aclanthology.org/2020.emnlp-main.16
33. Wei, J., et al.: Emergent abilities of large language models. Trans. Mach. Learn. Res. (2022). https://openreview.net/forum?id=yzkSU5zdwD. Survey Certification
34. Weiss, G., Goldberg, Y., Yahav, E.: On the practical computational power of finite precision RNNs for language recognition. In: Proceedings of the 56th Annual Meeting of the Association for Computational Linguistics (Volume 2: Short Papers), Melbourne, Australia, pp. 740–745. Association for Computational Linguistics, July 2018. https://aclanthology.org/P18-2117
35. Weiss, G., Goldberg, Y., Yahav, E.: Thinking like transformers (2021). https://openreview.net/forum?id=TmkN9JmDJx1

Jumping Automata over Infinite Words

Shaull Almagor[(✉)] and Omer Yizhaq

Technion, Haifa, Israel
shaull@technion.ac.il, omeryi@campus.technion.ac.il

Abstract. Jumping automata are finite automata that read their input in a non-consecutive manner, disregarding the order of the letters in the word. We introduce and study jumping automata over infinite words. Unlike the setting of finite words, which has been well studied, for infinite words it is not clear how words can be reordered. To this end, we consider three semantics: automata that read the infinite word in some order so that no letter is overlooked, automata that can permute the word in windows of a given size k, and automata that can permute the word in windows of an existentially-quantified bound. We study expressiveness, closure properties and algorithmic properties of these models.

Keywords: Jumping Automata · Parikh Image · Infinite Words

1 Introduction

Traditional automata read their input sequentially. Indeed, this is the case for most state-based computational models. In some settings, however, the order of the input does not matter. For example, when the input represents available resources, and we only wish to reason about their *quantity*. From a more language-theoretic perspective, this amounts to looking at the *commutative closure* of the languages, a.k.a. their *Parikh image*. To capture this notion in a computation model, *Jumping Automata* were introduced in [19]. A jumping automaton may read its input in a non-sequential manner, jumping from letter to letter, as long as every letter is read exactly once. Several works have studied the algorithmic properties and expressive power of these automata [9–11,17,23].

One of the most exciting developments in automata and language theory has been the extension to the setting of infinite words [4,15,18], which has led to powerful tools in formal methods. The infinite-word setting is far more complicated than that of finite words, involving several acceptance conditions and intricate expressiveness relationships. Most notably perhaps, nondeterministic Büchi automata cannot be determinized, but can be determinized to Rabin automata [16,18,21].

In this work, we introduce jumping automata over infinite words. The first challenge is to find meaningful definitions for this model. Indeed, the intuition

This research was supported by the ISRAEL SCIENCE FOUNDATION (grant No. 989/22).

F. Drewes and M. Volkov (Eds.): DLT 2023, LNCS 13911, pp. 9–22, 2023.
https://doi.org/10.1007/978-3-031-33264-7_2

of having the reading head "jump" to points in the word is ill-suited for infinite words, since one can construct an infinite run even without reading every letter (possibly even skipping infinitely many letters).

To this end, we propose three semantics for modes of jumping over infinite words for an automaton $\mathcal{A}$.

- In the *jumping* semantics, a word w is accepted if $\mathcal{A}$ accepts a permutation of it, i.e., a word w' that has the same number of occurrences of each letter as w for letters occurring finitely often, and the same set of letters that occur infinitely often.
- In the *k-window jumping* semantics, w is accepted if we can make $\mathcal{A}$ accept it by permuting w within contiguous windows of some fixed size k.
- In the *$\exists$-window jumping* semantics, w is accepted if there exists some k such that we can make $\mathcal{A}$ accept it by permuting w within contiguous windows of size k.

Example 1. Consider a Büchi automaton for the language $\{(ab)^\omega\}$ (see Fig. 3). Its jumping language is $\{w : w$ has infinitely many a's and b's $\}$. Its 3-window jumping language, for example, consists of words whose a's and b's can be rearranged to construct $(ab)^\omega$ in windows of size 3, such as $(aab \cdot bba)^\omega$. As for its $\exists$-window jumping language, it contains e.g., the word $(aaaabbbb)^\omega$, which is not in the 3-window language, but does not contain $aba^2b^2a^3b^3\cdots$, which is in the jumping language.

The definitions above capture different intuitive meanings for jumping: the first definition only looks at the Parikh image of the word, and corresponds to e.g., a setting where a word describes resources, with some resources being unbounded. The second (more restrictive) definition captures a setting where the word corresponds to e.g., a sequence of tasks to be handled, and we are allowed some bounded freedom in reordering them, and the third corresponds to a setting where we may introduce any finite delay to a task, but the overall delays we introduce must be bounded.

We study the expressiveness of these semantics, as well as closure properties and decision problems. Specifically, we show that languages in the jumping semantics are closed under union, intersection and complement. Surprisingly, we also show that automata admit determinization in this semantics, in contrast with standard Büchi automata. We further show that the complexity of decision problems on these automata coincide with their finite-word jumping counterparts. Technically, these results are attained by augmenting semilinear sets to accommodate ∞, and by showing that jumping languages can be described in a canonical form with such sets.

In the k-window semantics, we show a correspondence with ω-regular languages, from which we deduce closure properties and solutions for decision problems. Finally, we show that the $\exists$-window semantics is strictly more expressive than the jumping semantics.

Paper Organization. In Sect. 2 we recap some basic notions and define our jumping semantics. Section 3 is the bulk of the paper, where we study the jumping semantics. In Sect. 3.1 we introduce our variant of semilinear sets, and prove that it admits a canonical form and that it characterizes the jumping languages. Then, in Sects. 3.2 and 3.3 we study closure properties and decision problems for jumping languages. In Sects. 4 and 5 we study the k-window and $\exists$-window semantics, respectively. Finally, we conclude with future research in Sect. 6. Throughout the paper, our focus is on clean constructions and decidability. We thus defer complexity analyses to notes that follow each claim. Due to space constraints, proofs are omitted and can be found in the full version[1].

Related Work. Jumping automata were introduced in [19]. We remark that [19] contains some erroneous proofs (e.g., closure under intersection and complement, also pointed out in [11]). The works in [10,11] establish several expressiveness results on jumping automata, as well as some complexity results. In [23] many additional closure properties are established. An extension of jumping automata with a two-way tape was studied in [9].

Technically, since jumping automata correspond to the semilinear Parikh image/commutative closure [20], tools on semilinear sets are closely related. The works in [3,7,22] provide algorithmic results on semilinear sets, which we use extensively, and in [17] properties of semilinear sets are related to the state complexity of automata.

More broadly, automata and commutative closure also intersect in Parikh Automata [5,6,12,13], which read their input and accept if a certain Parikh image relating to the run belongs to a given semilinear set. In particular, [12] studies an extension of these automata to infinite words. Another related model is that of symmetric transducers – automata equipped with outputs, such that permutations in the input correspond to permutations in the output. These were studied in [2] in a jumping-flavour, and in [1] in a k-window flavour.

2 Preliminaries and Definitions

Automata. A *nondeterministic automaton* is a 5-tuple $\mathcal{A} = \langle \Sigma, Q, \delta, Q_0, \alpha \rangle$ where Σ is a finite alphabet, Q is a finite set of states, $\delta : Q \times \Sigma \to 2^Q$ is a nondeterministic transition function, $Q_0 \subseteq Q$ is a set of initial states, and $\alpha \subseteq Q$ is a set of accepting states. When $|Q_0| = 1$ and $|\delta(q,\sigma)| = 1$ for every $q \in Q$ and $\sigma \in \Sigma$, we say that $\mathcal{A}$ is *deterministic.*

We consider automata both over finite and over infinite words. We denote by Σ^* the set of finite words over Σ, and by Σ^ω the set of infinite words. For a word $w = \sigma_1, \sigma_2, \ldots$ (either finite or infinite), let $|w| \in \mathbb{N} \cup \{\infty\}$ be its length. A *run* of $\mathcal{A}$ on w is a sequence $\rho = q_0, q_1, \ldots$ such that $q_0 \in Q_0$ and for every $0 \leq i < |w|$ it holds that $q_{i+1} \in \delta(q_i, w_{i+1})$ (naturally, a run is finite if w is finite, and infinite if w is infinite). For finite words, the run ρ is *accepting* if $q_{|w|} \in \alpha$. For infinite words

[1] Full version available at https://arxiv.org/abs/2304.01278.

we use the Büchi acceptance condition, whereby the run ρ is accepting if it visits α infinitely often. Formally, define $\mathrm{Inf}(\rho) = \{q \in Q \mid \forall i \in \mathbb{N}\ \exists j > i,\ \rho_j = q\}$, then ρ is accepting if $\mathrm{Inf}(\rho) \cap \alpha \neq \emptyset$. A word w is accepted by $\mathcal{A}$ if there exists an accepting run of $\mathcal{A}$ on w.

The *language* of an automaton $\mathcal{A}$, denoted $\mathfrak{L}(\mathcal{A})$ is the set of words it accepts. We emphasize an automaton is over finite words by writing $\mathfrak{L}_{\mathrm{fin}}(\mathcal{A})$.

Remark 1. There are other acceptance conditions for automata over infinite words, e.g., parity, Rabin, Streett, co-Büchi and Muller. As we show in Proposition 3, for our purposes it is enough to consider Büchi.

Parikh Images and Permutations. Fix an alphabet Σ. We start by defining Parikh images and permutations for finite words. Consider a finite word $x \in \Sigma^*$. For each letter $\sigma \in \Sigma$ we denote by $\#_\sigma[x]$ the number of occurrences of σ in x. The *Parikh image* of x is then the vector $\Psi(x) = (\#_\sigma[x])_{\sigma\in\Sigma} \in \mathbb{N}^\Sigma$. We say that a word y is a *permutation* of x if $\Psi(y) = \Psi(x)$, in which case we write $x \sim y$ (clearly $\sim$ is an equivalence relation). We extend the Parikh image to sets of words by $\Psi(L) = \{\Psi(w) \mid w \in L\}$.

We now extend the definitions to infinite words. Let $\mathbb{N}_\infty = \mathbb{N}\cup\{\infty\}$. Consider an infinite word $w \in \Sigma^\omega$. We extend the definition of Parikh image to infinite words in the natural way: $\Psi(w) = (\#_\sigma[w])_{\sigma\in\Sigma}$ where $\#_\sigma[w]$ is the number of occurrences of σ in w if it is finite, and is ∞ otherwise. Thus, $\Psi(w) \in \mathbb{N}_\infty^\Sigma$. Moreover, since w is infinite, at least one coordinate of $\Psi(w)$ is ∞, since we often restrict ourselves to the Parikh images of infinite words, we denote by $\mathbb{M}^\Sigma = \mathbb{N}_\infty^\Sigma \setminus \mathbb{N}^\Sigma$ the set of vectors that have at least one ∞ coordinate. Note that $\boldsymbol{v} \in \mathbb{M}^\Sigma$ if and only if $\boldsymbol{v}$ is the Parikh image of some infinite word over Σ.

For words $w, w' \in \Sigma^\omega$, we abuse notation slightly and write $w \sim w'$ if $\Psi(w) = \Psi(w')$, as well as refer to w' as a *permutation* of w. We now refine the notion of permutation by restricting permutations to finite "windows": let $k \in \mathbb{N}$, we say that w is a *k-window permutation* of w', and we write $w \sim_{k\boxplus} w'$ (the symbol $\boxplus$ represents "window") if $w = x_1 \cdot x_2 \cdots$ and $w' = y_1 \cdot y_2 \cdots$ where for all $i \geq 1$ we have $|x_i| = |y_i| = k$ and $x_i \sim y_i$. That is, w' can be obtained from w by permuting contiguous disjoint windows of size k in w. Note that if $w \sim_{k\boxplus} w'$ then in particular $w \sim w'$, but the former is much more restrictive.

Semilinear Sets. Let $\mathbb{N} = \{0, 1, \ldots\}$ be the natural numbers. For dimension $d \geq 1$, we denote vectors in $\mathbb{N}^d$ in bold (e.g., $\boldsymbol{p} \in \mathbb{N}^d$). Consider a vector $\boldsymbol{b} \in \mathbb{N}^d$ and $P \subseteq \mathbb{N}^d$, the *linear set* generated by the *base* $\boldsymbol{b}$ and *periods* P is

$$\mathrm{Lin}(\boldsymbol{b}, P) = \{\boldsymbol{b} + \sum_{\boldsymbol{p}\in P} \lambda_p \boldsymbol{p} \mid \boldsymbol{b} \in \mathbb{N}^d \text{ and } \lambda_p \in \mathbb{N} \text{ for all } \boldsymbol{p} \in P\}$$

A *semilinear set* is then $\bigcup_{i\in I} \mathrm{Lin}(\boldsymbol{b}_i, P_i)$ for a finite set I and pairs $(\boldsymbol{b}_i, P_i)$.

Semilinear sets are closely related to the Parikh image of regular languages (in that the Parikh image of a regular language is semilinear) [20]. We will cite specific results relating to this connection as we use them.

Jumping Automata. Consider an automaton $\mathcal{A} = \langle \Sigma, Q, \delta, Q_0, \alpha \rangle$. Over finite words, we view $\mathcal{A}$ as a *jumping automaton* by letting it read its input in a non-sequential way, i.e. "jump" between letters as long as all letters are read exactly once. Formally, $\mathcal{A}$ accepts a word w as a jumping automaton if it accepts some permutation of w. Thus, we define the *jumping language* of $\mathcal{A}$ to be

$$\mathfrak{J}_{\text{fin}}(\mathcal{A}) = \{w \in \Sigma^* \mid \exists w' \sim w \text{ such that } w' \in \mathfrak{L}_{\text{fin}}(\mathcal{A})\}$$

We now turn to define jumping automata over infinite words. As discussed in Sect. 1, we consider three jumping semantics for $\mathcal{A}$, as follows:

Definition 1. *Consider an automaton* $\mathcal{A} = \langle \Sigma, Q, \delta, Q_0, \alpha \rangle$.

- *Its* Jumping language *is:*
  $\mathfrak{J}(\mathcal{A}) = \{w \in \Sigma^\omega \mid \exists w' \sim w \textit{ such that } w' \in \mathfrak{L}(\mathcal{A})\}.$
- *For* $k \in \mathbb{N}$, *its* k-window Jumping language *is:*
  $\mathfrak{J}_{k\boxplus}(\mathcal{A}) = \{w \in \Sigma^\omega \mid \exists w' \sim_{k\boxplus} w \textit{ such that } w' \in \mathfrak{L}(\mathcal{A})\}.$
- *Its* $\exists$-window Jumping language *is:*
  $\mathfrak{J}_{\exists\boxplus}(\mathcal{A}) = \{w \in \Sigma^\omega \mid \exists k \in \mathbb{N} \wedge \exists w' \sim_{k\boxplus} w \textit{ such that } w' \in \mathfrak{L}(\mathcal{A})\}.$

3 Jumping Languages

In this section we study the properties of $\mathfrak{J}(\mathcal{A})$ for an automaton $\mathcal{A}$. We start by characterizing $\mathfrak{J}(\mathcal{A})$ using an extension of semilinear sets.

3.1 Jumping Languages and Masked Semilinear Sets

Recall that $\mathbb{M}^\Sigma = \mathbb{N}_\infty^\Sigma \backslash \mathbb{N}^\Sigma$ and consider a vector $\boldsymbol{v} \in \mathbb{M}^\Sigma$. We separate the ∞ coordinates of $\boldsymbol{v}$ from the finite ones by, intuitively, writing $\boldsymbol{v}$ as a sum of a vector in $\mathbb{N}^\Sigma$ and a vector from $\{0, \infty\}^\Sigma \backslash \{\mathbf{0}\}$. Formally, let 🎭 $= \{0, \infty\}^\Sigma \setminus \{\mathbf{0}\}$ be the set of *masks*, namely vectors with entries 0 and ∞ that are not all 0, we refer to each $\mathfrak{m} \in$ 🎭 as a *mask*. We denote by $\mathfrak{m}|_0 = \{\sigma \in \Sigma \mid \mathfrak{m}(\sigma) = 0\}$ and $\mathfrak{m}|_\infty = \{\sigma \in \Sigma \mid \mathfrak{m}(\sigma) = \infty\}$ the 0 and ∞ coordinates of $\mathfrak{m}$, respectively.

For $\boldsymbol{x} \in \mathbb{N}^\Sigma$ and $\mathfrak{m} \in$ 🎭 let $\boldsymbol{x} \oplus \mathfrak{m} \in \mathbb{M}^\Sigma$ be the vector such that for all $\sigma \in \Sigma$ $(\boldsymbol{x} + \mathfrak{m})_\sigma = \boldsymbol{x}_\sigma$ if $\sigma \in \mathfrak{m}|_0$ and $(\boldsymbol{x} + \mathfrak{m})_\sigma = \infty$ if $\sigma \in \mathfrak{m}|_\infty$. Note that every $\boldsymbol{v} \in \mathbb{M}^\Sigma$ can be written as $\boldsymbol{x} + \mathfrak{m}$ where $\boldsymbol{x}$ is obtained from $\boldsymbol{v}$ be replacing ∞ with 0 (or indeed with any number), and having $\mathfrak{m}$ match the ∞ coordinates of $\boldsymbol{v}$.

We now augment semilinear sets with masks. A *masked semilinear set* is a union of the form $S = \bigcup_{\mathfrak{m} \in}$🎭 where $S_\mathfrak{m}$ is a semilinear set for every $\mathfrak{m}$. We also interpret S as a subset of $\mathbb{M}^\Sigma$ by interpreting addition as adding $\mathfrak{m}$ to each vector in the set $S_\mathfrak{m}$. Note that the union above is disjoint, since adding distinct masks always results in distinct vectors.

Consider a mask $\mathfrak{m} \in$ 🎭. Two vectors $\boldsymbol{u}, \boldsymbol{v} \in \mathbb{N}^\Sigma$ are called $\mathfrak{m}$-*equivalent* if $\boldsymbol{u} + \mathfrak{m} = \boldsymbol{v} + \mathfrak{m}$ (i.e., if they agree on $\mathfrak{m}|_0$). We say that a semilinear set $R \subseteq \mathbb{N}^\Sigma$ is $\mathfrak{m}$-*oblivious* if for every $\mathfrak{m}$-equivalent vectors $\boldsymbol{u}, \boldsymbol{v}$ we have $\boldsymbol{u} \in R \iff \boldsymbol{v} \in R$. Intuitively, an $\mathfrak{m}$-oblivious set does not "look" at $\mathfrak{m}|_\infty$.

We say that the masked semilinear set S above is *oblivious* if every $S_{\mathfrak{m}}$ is $\mathfrak{m}$-oblivious. Note that the property of being oblivious refers to a specific representation of S with semilinear sets $S_{\mathfrak{m}}$ for every mask $\mathfrak{m}$. In a way, it is natural for a masked semilinear set to be oblivious, since semantically, adding $\mathfrak{m}$ to a vector in $S_{\mathfrak{m}}$ already ignores the ∞ coordinates of $\mathfrak{m}$, so $S_{\mathfrak{m}}$ should not "care" about them. Moreover, if a set $S_{\mathfrak{m}}$ is $\mathfrak{m}$-oblivious, then in each of its linear components we can, intuitively, partition its period vectors to two types: those that have 0 in all the masked coordinates, and the remaining vectors that allow attaining any value in the masked coordinates. We refer to this as a *canonical* $\mathfrak{m}$-*oblivious representation*, as follows.

Definition 2. *A semilinear set* $R \subseteq \mathbb{N}^{\Sigma}$ *is in* canonical $\mathfrak{m}$-oblivious form *for a mask* $\mathfrak{m}$ *if* $R = \bigcup_{i=1}^{k} \mathrm{Lin}(\boldsymbol{b}_i, P_i)$ *such that for every* $1 \leq i \leq k$ *we have* $P_i = Q_i \cup E_{\mathfrak{m}}$ *with the following conditions:*

- $\boldsymbol{b}_i(\sigma) = 0$ *for every* $\sigma \in \mathfrak{m}|_{\infty}$.
- $\boldsymbol{p}(\sigma) = 0$ *for every* $p \in Q_i$ *and* $\sigma \in \mathfrak{m}|_{\infty}$.
- $E_{\mathfrak{m}} = \{\boldsymbol{e}_{\sigma} \mid \sigma \in \mathfrak{m}|_{\infty}\}$ *with* $\boldsymbol{e}_{\sigma}(\tau) = 1$ *if* $\sigma = \tau$ *and* 0 *otherwise.*

We now show that every masked semilinear set can be translated to an equivalent one in canonical oblivious form (see example in Fig. 1).

Lemma 1. *Consider a masked semilinear set* $S = \bigcup_{\mathfrak{m}\in \text{[illegible]}} S_{\mathfrak{m}} + \mathfrak{m}$. *Then there exists an oblivious masked semilinear set* $T = \bigcup_{\mathfrak{m}\in \text{[illegible]}} T_{\mathfrak{m}} + \mathfrak{m}$ *that represents the same subset of* $\mathbb{M}^{\Sigma}$, *and every* $T_{\mathfrak{m}}$ *is in canonical* $\mathfrak{m}$-*oblivious form.*

Proof (Sketch). Intuitively, we obtain T from S by replacing the $\mathfrak{m}|_{\infty}$ coordinates in the description of $T_{\mathfrak{m}}$ with 0, and adding the vectors $E_{\mathfrak{m}}$.

That is, for every vector $\boldsymbol{p} \in \mathbb{N}^{\Sigma}$, define $\boldsymbol{p}|_{\mathfrak{m}}$ by $\boldsymbol{p}|_{\mathfrak{m}}(\sigma) = \boldsymbol{p}(\sigma)$ if $\sigma \in \mathfrak{m}|_0$ and $\boldsymbol{p}|_{\mathfrak{m}}(\sigma) = 0$ if $\sigma \in \mathfrak{m}|_{\infty}$. That is, we replace all the ∞ coordinates of $\mathfrak{m}$ in $\boldsymbol{p}$ by 0. We lift this to sets of vectors: for a set P, let $P|_{\mathfrak{m}} = \{\boldsymbol{p}|_{\mathfrak{m}} \mid \boldsymbol{p} \in P\}$.

Now, write $S_{\mathfrak{m}} = \bigcup_{i\in I} \mathrm{Lin}(\boldsymbol{b}_i, P_i)$ and define $T_{\mathfrak{m}} = \bigcup_{i\in I} \mathrm{Lin}(\boldsymbol{b}_i|_{\mathfrak{m}}, P_i|_{\mathfrak{m}} \cup E_{\mathfrak{m}})$, we claim that $T_{\mathfrak{m}}$ is in canonical $\mathfrak{m}$-oblivious form, and is equivalent to $S_{\mathfrak{m}}$. □

Complexity Analysis 1 (of Lemma 1). The description of T involves introducing at most $|\Sigma|$ vectors to the periods of each semilinear set, and changing some entries to 0 in the existing vectors. Thus, the description of T is of polynomial size in that of S.

Our main result in this section is that the Parikh images of jumping languages coincide with masked semilinear sets. We prove this in the following lemmas.

Lemma 2. *Let* $\mathcal{A}$ *be an automaton, then* $\Psi(\mathfrak{J}(\mathcal{A}))$ *is a masked semilinear set. Moreover, we can effectively compute a representation of it from that of* $\mathcal{A}$.

$$\mathbf{Lin}\left(\begin{pmatrix}1\\0\\\mathbf{9}\\\mathbf{4}\end{pmatrix},\left\{\begin{pmatrix}1\\2\\\mathbf{5}\\\mathbf{7}\end{pmatrix},\begin{pmatrix}1\\0\\\mathbf{1}\\\mathbf{3}\end{pmatrix}\right\}\right)+\begin{pmatrix}0\\0\\\infty\\\infty\end{pmatrix} \rightsquigarrow \mathbf{Lin}\left(\begin{pmatrix}1\\0\\0\\0\end{pmatrix},\left\{\begin{pmatrix}1\\2\\0\\0\end{pmatrix},\begin{pmatrix}1\\0\\0\\0\end{pmatrix},\begin{pmatrix}0\\0\\1\\0\end{pmatrix},\begin{pmatrix}0\\0\\0\\1\end{pmatrix}\right\}\right)+\begin{pmatrix}0\\0\\\infty\\\infty\end{pmatrix}$$

Fig. 1. On the left, a masked semilinear set (actually linear), and on the right an equivalent oblivious canonical form. The bold numbers on the left get masked away, so are replaced by 0.

Proof. Consider $\mathcal{A} = \langle \Sigma, Q, \delta, Q_0, \alpha \rangle$. By Definition 1 we have that $\Psi(\mathfrak{J}(\mathcal{A})) = \Psi(\mathfrak{L}(\mathcal{A}))$. Indeed, both sets contain exactly the Parikh images of words in $\mathfrak{L}(\mathcal{A})$. Recall (or see the full version) that every ω-regular language can be (effectively) written as a union of the form $\mathfrak{L}(\mathcal{A}) = \bigcup_{i=1}^{m} S_i \cdot T_i^\omega$ where $S_i, T_i \subseteq \Sigma^*$ are regular languages over finite words.

Intuitively, we will show that $\Psi(T_i^\omega)$ can be separated to letters that are seen finitely often, and those that are seen infinitely often, where the latter will induce the mask. To this end, for every $\emptyset \neq \Gamma \subseteq \Sigma$ define $\mathfrak{m}_\Gamma \in$ [illegible] by setting $\mathfrak{m}_\Gamma(\sigma) = \infty$ if $\sigma \in \Gamma$ and $\mathfrak{m}_\Gamma(\sigma) = 0$ if $\sigma \notin \Gamma$. Now, for every regular language T_i in the union above define

$$I(T_i) = \{\mathfrak{m}_\Gamma \mid \forall \sigma \in \Gamma\ \exists w \in T_i \cap \Gamma^* \text{ s.t. } \Psi(w)(\sigma) > 0\}$$

That is, $I(T_i)$ is the set of masks $\mathfrak{m}_\Gamma$ such that every letter in Γ occurs in some word in T_i that contains only letters from Γ. Intuitively, $\mathfrak{m}_\Gamma \in I(T_i)$ can be attained by an infinite concatenation of words from T_i by covering all letters in Γ whilst not using letters outside of Γ.

In the full version we show that for every $1 \leq i \leq m$ it holds that $\Psi(S_i \cdot T_i^\omega) = \Psi(S_i \cdot T_i^*) + I(T_i)$.

It follows that $\Psi(\mathfrak{J}(\mathcal{A})) = \bigcup_{i=1}^{k} \Psi(S_i \cdot T_i^*) + I(T_i)$. We can now rearrange the sum by masks instead of by index i: for every mask $\mathfrak{m} \in$ [illegible] write $S_\mathfrak{m} = \bigcup_{i:\mathfrak{m} \in I(T_i)} \Psi(S_i \cdot T_i^*)$ (note that $S_\mathfrak{m}$ could be empty). Since $S_i \cdot T_i^*$ is a regular language, then by Parikh's Theorem [20] its Parikh image is semilinear, so $S_\mathfrak{m}$ is semilinear. Thus, $\Psi(\mathfrak{J}(\mathcal{A})) = \bigcup_{\mathfrak{m} \in} S_\mathfrak{m} + \mathfrak{m}$ is a masked semilinear set. $\square$

Complexity Analysis 2 (of Lemma 2). Writing $\mathfrak{L}(\mathcal{A})$ as a union involves only a polynomial blowup. Then, we split the resulting expression to a union over $2^{|\Sigma|}$ masks. Within the union, we convert a nondeterministic automaton for $S_i \cdot T_i^*$ to a semilinear set. By [22, Theorem 4.1] (also [14]), the resulting semilinear set has description size polynomial in the number of states n of $\mathcal{A}$ and singly-exponential in $|\Sigma|$. Moreover, the translation can be computed in time $2^{O(|\Sigma|^2 \log(n|\Sigma|))}$.

For the converse direction, we present a stronger result, namely that we can construct a *deterministic* automaton to capture a masked semilinear set.

Lemma 3. *Consider a masked semilinear set S, then there exists a deterministic automaton $\mathcal{A}$ such that $\Psi(\mathfrak{J}(\mathcal{A})) = S$.*

Proof (Sketch). By Lemma 1, we can assume $S = \bigcup_{\mathfrak{m} \in 🎭} S_{\mathfrak{m}} + \mathfrak{m}$ and that every $S_{\mathfrak{m}}$ is in canonical $\mathfrak{m}$-oblivious form. Let $\mathfrak{m} \in 🎭$ and write $S_{\mathfrak{m}} = \bigcup_{i=1}^{k} \mathrm{Lin}(\boldsymbol{b}_i, P_i \cup E_{\mathfrak{m}})$ as per Definition 2. Consider a linear set $L = \mathrm{Lin}(\boldsymbol{b}, P \cup E_{\mathfrak{m}})$ in the union above, and we omit the subscript i for brevity. We start by constructing a deterministic automaton $\mathcal{D}$ such that $\Psi(\mathfrak{J}_{\mathrm{fin}}(\mathcal{D})) = L$. To this end, let $w_{\boldsymbol{b}} \in \Sigma^*$ be a word such that $\Psi(w_{\boldsymbol{b}}) = \boldsymbol{b}$ and similarly for every $\boldsymbol{p} \in P$ let $w_{\boldsymbol{p}} \in \Sigma^*$ such that $\Psi(w_{\boldsymbol{p}}) = \boldsymbol{p}$. Note that $\Psi(\sigma) = \boldsymbol{e}_\sigma$ for every $\sigma \in \mathfrak{m}|_\infty$.

Let $\mathcal{D}'$ be a deterministic automaton for the regular expression $w_{\boldsymbol{b}} \cdot (w_{\boldsymbol{p}_1} + \ldots + w_{\boldsymbol{p}_n})^*$ where $P = \{\boldsymbol{p}_1, \ldots, \boldsymbol{p}_n\}$. Next, let $w_E = \sigma_1 \cdots \sigma_k$ be a word obtained by concatenating all the letters in $\mathfrak{m}|_\infty$ in some order. We obtain $\mathcal{D}$ from $\mathcal{D}'$ by connecting every accepting state of $\mathcal{D}'$, upon reading σ_1, to a cycle that allows reading w_E^ω. The accepting states of $\mathcal{D}$ are those on the w_E cycle. Crucially, the transition from $\mathcal{D}'$ upon reading σ_1 retains determinism, since P is in canonical form, and therefore $\boldsymbol{p}(\sigma_1) = 0$ for every $\boldsymbol{p} \in P$. That is, the letter σ_1 is not seen in any transition of $\mathcal{D}'$, allowing us to use it in the construction of $\mathcal{D}$. The construction is demonstrated in Fig. 2.

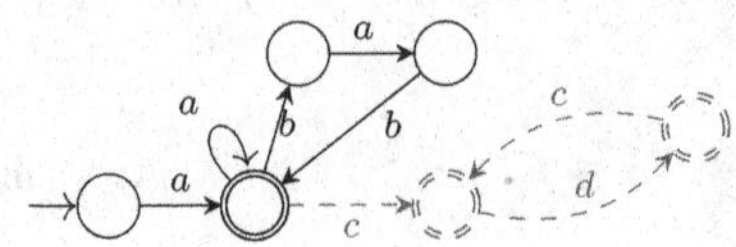

Fig. 2. The automaton $\mathcal{D}$ for the linear set from Fig. 1 (over $\Sigma = \{a, b, c, d\}$), with the representative words $w_{\boldsymbol{b}} = a$ and $\{bab, a, c, d\}$ for the period. The blue (dashed) parts are the addition of w_E and change of accepting states to obtain $\mathcal{D}$ from $\mathcal{D}'$.

In the full version we show the correctness of $\mathcal{D}$ and how a product of such automata can yield the desired automaton. □

Complexity Analysis 3 (of Lemma 3). We can naively construct an automaton for the regular expression $w_{\boldsymbol{b}} \cdot (w_{\boldsymbol{p}_1} + \ldots + w_{\boldsymbol{p}_n})^*$ of size $\|\mathrm{Lin}(\boldsymbol{b}, P)\| = \|\boldsymbol{b}\| + \sum_{i=1}^{n} \|\boldsymbol{p}_i\|$, where e.g., $\|\boldsymbol{b}\|$ is the sum of entries in $\boldsymbol{b}$ (i.e., unary representation). Then, determinization can yield a DFA of size singly exponential. Then, we take unions by the product construction, where the number of copies corresponds to the number of linear sets in S, and then to $2^{|\Sigma|} - 1$ many masks. This results in a DFA of size singly exponential in the size of S and doubly-exponential in $|\Sigma|$ (assuming S is represented in unary). Finally, adding the cycle for the mask adds at most $2^{O(|\Sigma|)}$ states (both to a deterministic or nondeterministic automaton).

Combining Lemmas 2 and 3 we have the following.

Theorem 4. *A set S is a masked semilinear set if and only if there exists an automaton $\mathcal{A}$ such that $S = \Psi(\mathfrak{J}(\mathcal{A}))$.*

3.2 Jumping Languages – Closure Properties

Using the characterizations obtained in Sect. 3.1, we can now obtain several closure properties of jumping languages. First, we remark that jumping languages are clearly closed under union, by taking the union of the automata and nondeterministically choosing which one to start at. We proceed to tackle intersection and complementation.

Note that the standard intersection construction of Büchi automata does not capture the intersection of jumping languages, as demonstrated in Fig. 3.

Fig. 3. The automata $\mathcal{A}$ and $\mathcal{B}$ satisfy $\mathfrak{L}(\mathcal{A}) = \{(ab)^\omega\}$ and $\mathfrak{L}(\mathcal{B}) = \{(ba)^\omega\}$. Thus, $\mathfrak{J}(\mathcal{A}) = \mathfrak{J}(\mathcal{B}) = \{w \in \Sigma^\omega \mid w$ has infinitely many a's and b's$\}$. However, if we start by taking the standard intersection of $\mathcal{A}$ and $\mathcal{B}$, we would end up with the empty language.

Proposition 1. *Let $\mathcal{A}$ and $\mathcal{B}$ be automata, then we can effectively construct an automaton $\mathcal{C}$ such that $\mathfrak{J}(\mathcal{C}) = \mathfrak{J}(\mathcal{A}) \cap \mathfrak{J}(\mathcal{B})$.*

Proof (Sketch). Consider a word $w \in \Sigma^\omega$, and observe that $w \in \mathfrak{J}(\mathcal{A}) \cap \mathfrak{J}(\mathcal{B})$ if and only if $\Psi(w) \in \Psi(\mathfrak{J}(\mathcal{A})) \cap \Psi(\mathfrak{J}(\mathcal{B}))$. Thus, it suffices to prove that there exists an automaton $\mathcal{C}$ such that $\Psi(\mathfrak{J}(\mathcal{C})) = \Psi(\mathfrak{J}(\mathcal{A})) \cap \Psi(\mathfrak{J}(\mathcal{B}))$. By Theorem 4 we can write $\Psi(\mathfrak{J}(\mathcal{A})) = \bigcup_{\mathfrak{m}\in \text{[illegible]}} S_\mathfrak{m} + \mathfrak{m}$ and $\Psi(\mathfrak{J}(\mathcal{B})) = \bigcup_{\mathfrak{m}\in \text{[illegible]}} T_\mathfrak{m} + \mathfrak{m}$. Furthermore, by Lemma 1 we can assume that these are oblivious masked semilinear sets.

We claim that $\Psi(\mathfrak{J}(\mathcal{A})) \cap \Psi(\mathfrak{J}(\mathcal{B})) = \bigcup_{\mathfrak{m}\in \text{[illegible]}} S_\mathfrak{m} \cap T_\mathfrak{m} + \mathfrak{m}$. That is, we can take intersection by intersecting each pair of semilinear sets, while keeping the masks. Intuitively, this holds because every word uniquely determines the mask by which it joins the union, so intersecting the unions amounts (using obliviousness) to intersecting the set of finite-valued vectors corresponding to the finite parts of the Parikh images of words in the language.

Indeed, the $\supseteq$ direction is easy, and for $\subseteq$, if $\boldsymbol{x} = \boldsymbol{u} + \mathfrak{m}$ for $\boldsymbol{u} \in S_\mathfrak{m}$ and $\boldsymbol{x} = \boldsymbol{v} + \mathfrak{m}$ for $\boldsymbol{v} \in T_\mathfrak{m}$, then $\boldsymbol{u} + \mathfrak{m} = \boldsymbol{v} + \mathfrak{m}$, and since $S_\mathfrak{m}$ and $T_\mathfrak{m}$ are $\mathfrak{m}$-oblivious, we also have that $\boldsymbol{v} \in S_\mathfrak{m}$ and $\boldsymbol{u} \in T_\mathfrak{m}$, so $\boldsymbol{x} = \boldsymbol{v} + \mathfrak{m} \in \bigcup_{\mathfrak{m}\in \text{[illegible]}} S_\mathfrak{m} \cap T_\mathfrak{m} + \mathfrak{m}$.

Finally, semilinear sets are closed under intersection [3,7]. Thus, $S_\mathfrak{m} \cap T_\mathfrak{m}$ is semilinear for every $\mathfrak{m}$, so $\bigcup_{\mathfrak{m}\in \text{[illegible]}} S_\mathfrak{m} \cap T_\mathfrak{m} + \mathfrak{m}$ is a semilinear masked set and by Lemma 3 there exists an automaton $\mathcal{C}$ such that $\mathfrak{J}(\mathcal{C}) = \mathfrak{J}(\mathcal{A}) \cap (\mathfrak{J}(\mathcal{B})$. □

Complexity Analysis 5 (of Proposition 1). The description size of the intersection of semilinear sets is polynomial in the description size of the entries of the vectors and the size of the union, and singly exponential in $|\Sigma|$ [3,7]. Since the blowup in Lemmas 1 and 2 is polynomial, then the only significant blowup is Lemma 3, due to the construction of an automaton whose size is exponential in $|\Sigma|$. However, it is not hard to see that the two exponential blowups in $|\Sigma|$ are orthogonal, and can be merged to a singly-exponential blowup.

Proposition 2. *Consider an automaton $\mathcal{A}$. There exists an automaton $\mathcal{B}$ such that $\mathfrak{J}(\mathcal{B}) = \Sigma^\omega \backslash \mathfrak{J}(\mathcal{A})$.*

Proof (Sketch). Similarly to Proposition 1, it suffices to construct $\mathcal{B}$ such that $\Psi(\mathfrak{J}(\mathcal{B})) = \mathbb{M}^{\Sigma} \setminus \Psi(\mathfrak{J}(\mathcal{A}))$. Write $\Psi(\mathfrak{J}(\mathcal{A})) = \bigcup_{\mathfrak{m} \in \text{[illegible]}} S_{\mathfrak{m}} + \mathfrak{m}$ and by Lemma 1 we assume this is an oblivious masked semilinear set. We show, using obliviousness, that $\mathbb{M}^{\Sigma} \setminus \Psi(\mathfrak{J}(\mathcal{A})) = \bigcup_{\mathfrak{m} \in \text{[illegible]}} (\mathbb{N}^{\Sigma} \setminus S_{\mathfrak{m}}) + \mathfrak{m}$ and we use this to obtain an automaton $\mathcal{B}$. □

Complexity Analysis 6 (of Proposition 2). The description size of the complement of a semilinear set is singly exponential, in both the degree and the description of the entries [3,7]. We proceed similarly to Complexity Analysis 5, ending up with a singly-exponential blowup (not just in $|\Sigma|$).

Recall that Büchi automata are generally not determinizable. That is, there exists an automaton $\mathcal{A}$ such that $\mathfrak{L}(\mathcal{A})$ is not recognizable by a deterministic Büchi automaton [16]. In stark contrast, an immediate corollary of Lemmas 2 and 3 is that jumping languages do admit determinization (also see Complexity Analysis 3).

Proposition 3. *For every automaton $\mathcal{A}$ there exists a deterministic automaton $\mathcal{D}$ such that $\mathfrak{J}(\mathcal{A}) = \mathfrak{J}(\mathcal{D})$.*

In particular, this means that additional acceptance conditions (e.g., Rabin, Streett, Muller and parity) cannot add expressiveness to the model.

Remark 2. Discussing determinization of jumping languages is slightly misleading: the definition of acceptance in a jumping language asks that *there exists* some permutation that is accepted by the automaton. This existential quantifier can be thought of as "semantic" nondeterminism. Thus, in a way, jumping languages are inherently nondeterministic, regardless of the underlying syntactic structure.

Finally, as is the case for jumping languages over finite words [10,11,19], jumping languages and ω-regular languages are incomparable. Indeed, the automata in Fig. 3 are examples of Büchi automata whose languages are not permutation-invariant, and are hence not jumping languages, and in Fig. 4 we demonstrate that the converse also does not hold.

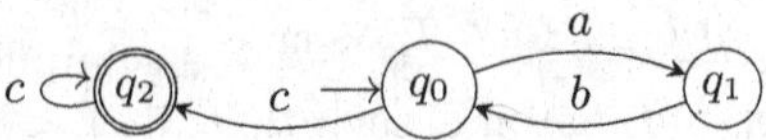

Fig. 4. An automaton $\mathcal{A}$ for which $\mathfrak{J}(\mathcal{A}) = \{u \cdot c^{\omega} \mid u \in \{a, b, c\}^* \wedge \#_a[u] = \#_b[u]\}$, which is not ω-regular by simple pumping arguments.

3.3 Jumping Languages - Algorithmic Properties

We turn our attention to algorithmic properties of jumping languages. We first notice that given an automaton $\mathcal{A}$ we have that $\mathfrak{J}(\mathcal{A}) \neq \emptyset$ if and only if $\mathfrak{L}(\mathcal{A}) \neq \emptyset$. Thus, non-emptiness of $\mathfrak{J}(\mathcal{A})$ is `NL-Complete`, as it is for Büchi automata [8,15]. Next, we consider the standard decision problems:

- **Containment**: given automata $\mathcal{A}, \mathcal{B}$, is $\mathfrak{J}(\mathcal{A}) \subseteq \mathfrak{J}(\mathcal{B})$?
- **Equivalence**: given automata $\mathcal{A}, \mathcal{B}$, is $\mathfrak{J}(\mathcal{A}) = \mathfrak{J}(\mathcal{B})$?
- **Universality**: given an automaton $\mathcal{A}$, is $\mathfrak{J}(\mathcal{A}) = \Sigma^\omega$?

In the full version we show that all of these problems reduce to their analogues in finite-word jumping automata.

Theorem 7. *Containment, Equivalence and Universality for jumping languages are* coNP*-complete (for fixed size alphabet).*

4 Fixed-Window Jumping Languages

Recall from Definition 1 that given an automaton $\mathcal{A}$ and $k \in \mathbb{N}$, the language $\mathfrak{J}_{k\boxplus}(\mathcal{A})$ consists of the words that have a k-window permutation that is accepted by $\mathcal{A}$.

Since k is fixed, we can capture k-window permutations using finite memory. Hence, as we now show, $\mathfrak{J}_{k\boxplus}(\mathcal{A})$ is ω-regular.

Theorem 8. *Consider an automaton $\mathcal{A}$, then for every $k \in \mathbb{N}$ there exists a Büchi automaton $\mathcal{B}_k$ such that $\mathfrak{J}_{k\boxplus}(\mathcal{A}) = \mathfrak{L}(\mathcal{B}_k)$.*

Proof (Sketch). Intuitively, we construct $\mathcal{B}_k$ so that it stores in a state the Parikh image of k letters, then nondeterministically simulates $\mathcal{A}$ on all words with the stored Parikh image, while noting when an accepting state may have been traversed. We illustrate the construction for $k = 2$ in Fig. 5. □

Complexity Analysis 9 (of Theorem 8). If $\mathcal{A}$ has n states, then $\mathcal{B}$ has at most $k^{|\Sigma|} \times n + n$ states. Moreover, computing $\mathcal{B}$ can be done effectively, by keeping for every Parikh image the set of states that are reachable (resp. reachable via an accepting state).

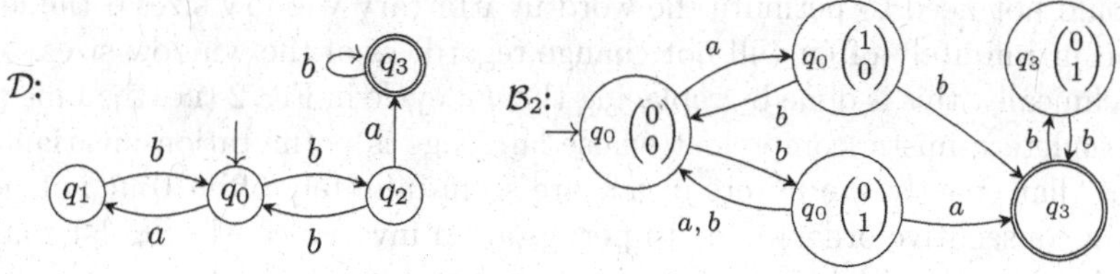

Fig. 5. An automaton $\mathcal{D}$, and the construction $\mathcal{B}_2$ for $\mathfrak{J}_{2\boxplus}(\mathcal{D})$ as per Theorem 8.

A converse of Theorem 8 also holds, in the following sense: we say that a language $L \subseteq \Sigma^\omega$ is $k\boxplus$ invariant if $w \in L \iff w' \in L$ for every $w \sim_{k\boxplus} w'$. The following result follows by definition.

Proposition 4. *Let $\mathcal{A}$ be an automaton such that $\mathfrak{L}(\mathcal{A})$ is $k\boxplus$-invariant, then $\mathfrak{L}(\mathcal{A}) = \mathfrak{J}_{k\boxplus}(\mathcal{A})$.*

Theorem 8 and Proposition 4 yield that k-window jumping languages are closed under the standard operations under which ω-regular languages are closed (union, intersection, complementation), as these properties retain $k\boxplus$ invariance. Similarly, all algorithmic problems can be reduced to the same problems on Büchi automata (by Complexity Analysis 9, assuming $|\Sigma|$ is fixed).

Notice that the construction in Theorem 8 yields a nondeterministic Büchi automaton. This is not surprising in light of Remark 2. It does raise the question of whether we can find a deterministic Büchi automaton for $\mathfrak{J}_{k\boxplus}(\mathcal{D})$ for a *deterministic* automaton $\mathcal{D}$. In the full version, we prove that this is not the case. That is, even deterministic k-window jumping has inherent nondeterminism.

Proposition 5. *There is no nondeterministic Büchi automaton whose language is $\mathfrak{J}_{k\boxplus}(\mathcal{D})$ for $k \geq 2$ and $\mathcal{D}$ as in Fig. 5.*

5 Existential-Window Jumping Languages

Recall from Definition 1 that given an automaton $\mathcal{A}$, the language $\mathfrak{J}_{\exists\boxplus}(\mathcal{A})$ contains the words that have a k-window permutation that is accepted in $\mathcal{A}$ for some k, i.e., $\mathfrak{J}_{\exists\boxplus}(\mathcal{A}) = \bigcup_{k\in\mathbb{N}} \mathfrak{J}_{k\boxplus}(\mathcal{A})$. We briefly establish some expressiveness properties of $\mathfrak{J}_{\exists\boxplus}(\mathcal{A})$.

Perhaps surprisingly, we show that $\exists\boxplus$ languages are strictly more expressive than Jumping languages.

Theorem 10. *1. Let $\mathcal{A}$ be an automaton, then there exists an automaton $\mathcal{B}$ such that $\mathfrak{J}_{\exists\boxplus}(\mathcal{B}) = \mathfrak{J}(\mathcal{A})$.*
2. There exists an automaton $\mathcal{B}$ such that $\mathfrak{J}_{\exists\boxplus}(\mathcal{B})$ is not a jumping language.

Proof (Sketch). In order to show that every jumping language is also an $\exists\boxplus$ jumping, we modify the construction of Lemma 3 to obtain an automaton for which the $\exists\boxplus$ language coincides with the given semilinear set. Intuitively, this is possible since in order to determine the set of letters that occur infinitely often, we should not need to permute the word in arbitrary window sizes – the letters that occur infinitely often will not change regardless of the window sizes.

Technically, this is done by replacing the w_E cycle in Fig. 2 (i.e. the blue part) by a nondeterministic component whose language is permutation-invariant and ensures that exactly the letters in w_E are seem infinitely often (but not necessarily in consecutive order). Due to permutation invariance, the $\exists\boxplus$ language of this component is equal to its jumping language.

Finally, it is not hard to show that the $\exists\boxplus$ language of the automaton $\mathcal{A}$ in Fig. 3 is not a jumping language, hence he strict containment. □

Finally, in the full version we show that ω-regular languages are incomparable with $\exists\boxplus$ languages.

Remark 3 (Alternative Semantics). Note that instead of the $\exists\boxplus$ semantics, we could require that there exists a partition of w into windows of size *at most k* for some $k \in \mathbb{N}$. All our proofs carry out under this definition as well.

6 Future Research

We have introduced three semantics for jumping automata over infinite words. For the existing definitions, the class of ∃⊞ semantics is yet to be explored for closure properties and decision problems.

In a broader view, it would be interesting to consider quantitative semantics for jumping – instead of the coarse separation between letters that occur finitely and infinitely often, one could envision a semantics that takes into account e.g., the frequency with which letters occur. Alternatively, one could return to the view of a "jumping reading head", and place constraints over the strategy used by the automaton to move the head (e.g., restrict to strategies implemented by one-counter machines).

References

1. Abu Nassar, A., Almagor, S.: Simulation by rounds of letter-to-letter transducers. In: 30th EACSL Annual Conference on Computer Science Logic (2022)
2. Almagor, S.: Process symmetry in probabilistic transducers. In: 40th IARCS Annual Conference on Foundations of Software Technology and Theoretical Computer Science (2020)
3. Beier, S., Holzer, M., Kutrib, M.: On the descriptional complexity of operations on semilinear sets. In: EPTCS, vol. 252, p. 41 (2017)
4. Büchi, J.R.: Symposium on decision problems: on a decision method in restricted second order arithmetic. In: Studies in Logic and the Foundations of Mathematics, vol. 44, pp. 1–11. Elsevier (1966)
5. Cadilhac, M., Finkel, A., McKenzie, P.: Affine Parikh automata. RAIRO-Theor. Inform. Appl. **46**(4), 511–545 (2012)
6. Cadilhac, M., Finkel, A., McKenzie, P.: Bounded Parikh automata. Int. J. Found. Comput. Sci. **23**(08), 1691–1709 (2012)
7. Chistikov, D., Haase, C.: The taming of the semi-linear set. In: 43rd International Colloquium on Automata, Languages, and Programming (ICALP 2016). Schloss Dagstuhl-Leibniz-Zentrum fuer Informatik (2016)
8. Emerson, E.A., Lei, C.L.: Modalities for model checking (extended abstract) branching time strikes back. In: Proceedings of the 12th ACM SIGACT-SIGPLAN Symposium on Principles of Programming Languages, pp. 84–96 (1985)
9. Fazekas, S.Z., Hoshi, K., Yamamura, A.: Two-way deterministic automata with jumping mode. Theor. Comput. Sci. **864**, 92–102 (2021)
10. Fernau, H., Paramasivan, M., Schmid, M.L.: Jumping finite automata: characterizations and complexity. In: Drewes, F. (ed.) CIAA 2015. LNCS, vol. 9223, pp. 89–101. Springer, Cham (2015). https://doi.org/10.1007/978-3-319-22360-5_8
11. Fernau, H., Paramasivan, M., Schmid, M.L., Vorel, V.: Characterization and complexity results on jumping finite automata. Theor. Comput. Sci. **679**, 31–52 (2017)
12. Guha, S., Jecker, I., Lehtinen, K., Zimmermann, M.: Parikh automata over infinite words. In: 42nd IARCS Annual Conference on Foundations of Software Technology and Theoretical Computer Science (2022)
13. Klaedtke, F., Rueß, H.: Monadic second-order logics with cardinalities. In: Baeten, J.C.M., Lenstra, J.K., Parrow, J., Woeginger, G.J. (eds.) ICALP 2003. LNCS, vol. 2719, pp. 681–696. Springer, Heidelberg (2003). https://doi.org/10.1007/3-540-45061-0_54

14. Kopczynski, E., To, A.W.: Parikh images of grammars: complexity and applications. In: 2010 25th Annual IEEE Symposium on Logic in Computer Science, pp. 80–89. IEEE (2010)
15. Kupferman, O.: Automata theory and model checking. In: Clarke, E., Henzinger, T., Veith, H., Bloem, R. (eds.) Handbook of Model Checking, pp. 107–151. Springer, Cham (2018). https://doi.org/10.1007/978-3-319-10575-8_4
16. Landweber, L.H.: Decision problems for w-automata. Technical report, University of Wisconsin-Madison, Department of Computer Sciences (1968)
17. Lavado, G.J., Pighizzini, G., Seki, S.: Operational state complexity under Parikh equivalence. In: Jürgensen, H., Karhumäki, J., Okhotin, A. (eds.) DCFS 2014. LNCS, vol. 8614, pp. 294–305. Springer, Cham (2014). https://doi.org/10.1007/978-3-319-09704-6_26
18. McNaughton, R.: Testing and generating infinite sequences by a finite automaton. Inf. Control **9**(5), 521–530 (1966)
19. Meduna, A., Zemek, P.: Jumping finite automata. Int. J. Found. Comput. Sci. **23**(07), 1555–1578 (2012)
20. Parikh, R.J.: On context-free languages. J. ACM (JACM) **13**(4), 570–581 (1966)
21. Safra, S.: On the complexity of ω-automata. In: Proceedings of the 29th IEEE Symposium on Foundations of Computer Science, pp. 319–327 (1988)
22. To, A.W.: Parikh images of regular languages: complexity and applications. arXiv preprint arXiv:1002.1464 (2010)
23. Vorel, V.: On basic properties of jumping finite automata. Int. J. Found. Comput. Sci. **29**(01), 1–15 (2018)

Isometric Words Based on Swap and Mismatch Distance

M. Anselmo[1(✉)], G. Castiglione[2(✉)], M. Flores[1,2(✉)], D. Giammarresi[3(✉)], M. Madonia[4(✉)], and S. Mantaci[2(✉)]

[1] Dipartimento di Informatica, Università di Salerno, Fisciano, Italy
{manselmo,mflores}@unisa.it

[2] Dipartimento di Matematica e Informatica, Università di Palermo, Palermo, Italy
{giuseppa.castiglione,sabrina.mantaci}@unipa.it

[3] Dipartimento di Matematica., Università Roma "Tor Vergata", Rome, Italy
giammarr@mat.uniroma2.it

[4] Dipartimento di Matematica e Informatica, Università di Catania, Catania, Italy
madonia@dmi.unict.it

Abstract. An edit distance is a metric between words that quantifies how two words differ by counting the number of edit operations needed to transform one word into the other one. A word f is said isometric with respect to an edit distance if, for any pair of f-free words u and v, there exists a transformation of minimal length from u to v via the related edit operations such that all the intermediate words are also f-free. The adjective "isometric" comes from the fact that, if the Hamming distance is considered (i.e., only mismatches), then isometric words define some isometric subgraphs of hypercubes. We consider the case of edit distance with swap and mismatch. We compare it with the case of mismatch only and prove some properties of isometric words that are related to particular features of their overlaps.

Keywords: Swap and mismatch distance · Isometric words · Overlap with errors

1 Introduction

The edit distance is a central notion in many fields of computer science. It plays a crucial role in defining combinatorial properties of families of strings (or words) as well as in designing many classical string algorithms that find applications in natural language processing, bioinformatics and, in general, in information retrieval problems. The edit distance is a string metric that quantifies how much two strings differ from each other and it is based on counting the minimum number of edit operations required to transform one string into the other one.

Partially supported by INdAM-GNCS Project 2022 and 2023, FARB Project ORSA229894 of University of Salerno, TEAMS Project and PNRR MUR Project PE0000013-FAIR University of Catania, PNRR MUR Project ITSERR CUP B53C22001770006 and FFR fund University of Palermo, MUR Excellence Department Project MatMod@TOV, CUP E83C23000330006, awarded to the Department of Mathematics, University of Rome Tor Vergata.

F. Drewes and M. Volkov (Eds.): DLT 2023, LNCS 13911, pp. 23–35, 2023.
https://doi.org/10.1007/978-3-031-33264-7_3

Different definitions of edit distance use different sets of edit operations. The operations of insertion, deletion and replacement of a character in the string characterize the Levenshtein distance which is probably the most widely known (cf. [17]). On the other hand, the most basic one is the Hamming distance which applies only to pair of strings of the same length and counts the positions where they have a mismatch; this corresponds to the restriction of using only the replacement operation. For this, the Hamming distance finds a direct application in detecting and correcting errors in strings and it is a major actor in the algorithms for string matching with mismatches (see [13]).

The notion of isometric word combines the edit distance with the property that a word does not appear as factor in other words. Note that this property is important in combinatorics as well as in the investigation on similarities or distances, on DNA sequences, where the avoided factor is referred to as an absent word [8–10].

Isometric words based on Hamming distance were first introduced in [15] as special binary strings that never appear as factors in some string transformations. A string is f-free if it does not contain f as factor. A word f is isometric if for any pair of f-free words u and v, there exists a sequence of symbol replacement operations that transform u in v where all the intermediate words are also f-free.

Isometric words are related to the definition of isometric subgraphs of the hypercubes, called generalized Fibonacci cubes. The hypercube Q_n is a graph whose vertices are the (binary) words of length n, and two vertices are adjacent when the corresponding words differ in exactly one symbol. Therefore, the distance between two vertices is the Hamming distance of the corresponding vertex-words. Let $Q_n(f)$ be the subgraph of Q_n which contains only vertices that are f-free. Then, if f is isometric, the distances of the vertices in $Q_n(f)$ are the same as calculated in the whole Q_n. Fibonacci cubes have been introduced by Hsu in [14] and correspond to the case with $f = 11$. In [15, 16, 19–21] the structure of non-isometric words for alphabets of size 2 and Hamming distance is completely characterized and related to particular properties on their overlaps. The more general case of alphabets of size greater than 2 and Lee distance is studied in [3–5]. Using these characterizations, in [7] some linear-time algorithms are given to check whether a binary word is Hamming isometric and, for quaternary words, if it is Lee isometric. These algorithms were extended to provide further information on non-isometric words, still keeping linear complexity in [4]. Binary Hamming isometric two-dimensional words have been also studied in [6].

Many challenging problems in correcting errors in strings come from computational biology. Among the chromosomal operations on DNA sequences, in gene mutations and duplication, it seems natural to consider the *swap* operation, consisting in exchanging two adjacent symbols. The Damerau-Levenshtein distance adds also the swap to all edit operations. In [18], Wagner proves that computing the edit distance with insertion, deletion, replacement, and swap, is polynomially solvable in some restriction of the problem. The swap-matching problem has been considered in [1, 12], and algorithms for computing the corresponding edit distance are given in [2, 11].

In this paper, we study the notion of binary isometric word using the edit distance based on swaps and mismatches. This distance will be referred to by using the *tilde* symbol that somehow evokes the swap operation. The tilde-distance $\text{dist}_\sim(u, v)$ of equal-length words u and v is the minimum number of replacement and swap operations to transform u into v.

It turns out that adding the swap operation to the definition makes the situation more complex, although interesting for applications. It is not a mere generalization of the Hamming case since special situations arise. A swap operation in fact is equivalent to two replacements, but it counts as one when computing the tilde-distance. Moreover, there could be different ways to transform u into v since particular triples of consecutive symbols can be managed, from left to right, either by a swap and then a replacement or by a replacement and then a swap. Note that there is also the possibility of two swaps in consecutive positions that are in fact equivalent to two replacement operations. In this paper, we do not allow transformations with such overlying swaps because they flip twice a bit in the same position, where there is not any mismatch.

The definition of *tilde-isometric* word comes in a very natural way. A word f is tilde-isometric if for any pair of equal-length words u and v that are f-free, there is a transformation from u to v that uses exactly $\text{dist}_\sim(u, v)$ replacement and swap operations and such that all the intermediate words still avoid f. We present some examples of tilde-isometric words that are not Hamming isometric and vice versa. We give necessary conditions for f to be non-isometric based on the notion of error-overlap. In order to prove that a given string f is not tilde-isometric one should exhibit a pair of f-free words $(\tilde{\alpha}, \tilde{\beta})$ such that any transformation from $\tilde{\alpha}$ to $\tilde{\beta}$ of length $\text{dist}_\sim(\tilde{\alpha}, \tilde{\beta})$ comes through words that contain f. Such a pair is called pair of *tilde-witnesses* for f. We give an explicit construction of the tilde-witnesses in many crucial cases where some constrains on the error-overlap are required.

2 Preliminaries

Let Σ be a finite alphabet. A word (or string) w of length $|w| = n$, is $w = a_1 a_2 \cdots a_n$, where $a_1, a_2, \ldots, a_n$ are symbols in Σ. The set of all words over Σ is denoted Σ^* and the set of all words over Σ of length n is denoted Σ^n. Finally, ϵ denotes the *empty word* and $\Sigma^+ = \Sigma^* - \{\epsilon\}$. For any word $w = a_1 a_2 \cdots a_n$, the *reverse* of w is the word $w^{rev} = a_n a_{n-1} \cdots a_1$. If $x \in \{0, 1\}$, we denote by $\overline{x}$ the opposite of x, i.e. $\overline{x} = 1$ if $x = 0$ and viceversa. Then we define *complement* of w the word $\overline{w} = \overline{a}_1 \overline{a}_2 \cdots \overline{a}_n$.

Let $w[i]$ denote the symbol of w in position i, i.e. $w[i] = a_i$. Then, $w[i..j] = a_i \cdots a_j$, for $1 \leq i \leq j \leq n$, is a *factor* of w. The *prefix* (resp. *suffix*) of w of length l, with $1 \leq l \leq n-1$ is $\text{pre}_l(w) = w[1..l]$ (resp. $\text{suf}_l(w) = w[n-l+1..n]$). When $\text{pre}_l(w) = \text{suf}_l(w) = u$ then u is here referred to as an *overlap* of w of length l; it is also called border, or bifix. A word w is said *f-free* if w does not contain f as a factor.

An *edit operation* is a function $O : \Sigma^* \to \Sigma^*$ that transform a word into another one. Among the most common edit operations there are the insertion, the deletion or the replacement of a character and the swap of two adjacent characters. Let OP be a *set of edit operations*. The *edit distance* of two words $u, v \in \Sigma^*$ is the minimum number of edit operations in OP needed to transform u into v.

In this paper, we consider the edit distance that only uses replacements and swaps. Note that these two operations do not change the length of the word. We give a formal definition.

Definition 1. *Let Σ be a finite alphabet and $w = a_1 a_2 \ldots a_n$ a word over Σ. The* replacement operation *(or* replacement, *for short) on w at position i with $x \in \Sigma$, $x \neq a_i$, is defined by*

$$R_{i,x}(a_1 a_2 \ldots a_{i-1} a_i a_{i+1} \ldots a_n) = a_1 a_2 \ldots a_{i-1} x a_{i+1} \ldots a_n.$$

The swap operation *(or* swap, *for short) on w at position i consists in exchanging characters at positions i and $i+1$, provided that they are different, $a_i \neq a_{i+1}$,*

$$S_i(a_1 a_2 \ldots a_i a_{i+1} \ldots a_n) = a_1 a_2 \ldots a_{i+1} a_i \ldots a_n.$$

When the alphabet $\Sigma = \{0, 1\}$ there is only a possible replacement at a given position i, so we write $R_i(w)$ instead of $R_{i,x}(w)$.

Given two equal-length words $u = a_1 \cdots a_n$ and $v = b_1 \cdots b_n$, they have a *mismatch error* (or *mismatch*) at position i if $a_i \neq b_i$ and they have a *swap error* (or *swap*) at position i if $a_i a_{i+1} = b_{i+1} b_i$, with $a_i \neq a_{i+1}$. We say that u and v have an *error* at position i if they have either a mismatch or a swap error. Note that one swap corresponds to two adjacent mismatches.

A word f is isometric if for any pair of f-free words u and v, there exists a minimal length sequence of replacement operations that transform u into v where all the intermediate words are also f-free. In this paper we refer to this definition of isometric as *Ham-isometric*. A word w has a 2-error overlap if there exists l such that $\text{pre}_l(w)$ and $\text{suf}_l(w)$ have two mismatch errors. In [20], the following characterization is given.

Proposition 2. *A word f is not Ham-isometric if and only if f has a 2-error overlap.*

3 Tilde-Distance and Tilde-Isometric Words

In this section we consider the edit distance based on swap and replacement operations that we call tilde-distance and we denote $dist_\sim$. First, we give some definitions and notations, together with some examples and the proofs of some preliminary properties.

Definition 3. *Let $u, v \in \Sigma^*$ be words of equal length. The* tilde-distance $\text{dist}_\sim(u, v)$ *between u and v is the minimum number of replacements and swaps needed to transform u into v.*

Definition 4. *Let $u, v \in \Sigma^*$ be words of equal length. A* tilde-transformation *τ of length h from u to v is a sequence of words $(w_0, w_1, \ldots, w_h)$ such that $w_0 = u$, $w_h = v$, and for any $k = 0, 1, \ldots, h-1$, $\text{dist}_\sim(w_k, w_{k+1}) = 1$. Moreover, τ is* f-free *if for any $i = 0, 1, \ldots, h$, the word w_i is f-free.*

A tilde-transformation $(w_0, w_1, \ldots, w_h)$ from u to v is associated to a sequence of h operations $(O_{i_1}, O_{i_2}, \ldots O_{i_h})$ such that, for any $k = 1, \ldots, h$, $O_{i_k} \in \{R_{i_k,x}, S_{i_k}\}$ and $w_k = O_{i_k}(w_{k-1})$; it can be represented as follows:

$$u = w_0 \xrightarrow{O_{i_1}} w_1 \xrightarrow{O_{i_2}} \cdots \xrightarrow{O_{i_h}} w_h = v.$$

With a little abuse of notation, in the sequel we will refer to a tilde-transformation both as a sequence of words and as a sequence of operations. We give some examples.

Example 5. Let $u = 1011, v = 0110$. Below we show two different tilde-transformations from u to v. Note that the length of τ_1 corresponds to $\text{dist}_\sim(u,v) = 2$.

$$\tau_1 : 1011 \xrightarrow{S_1} 0111 \xrightarrow{R_4} 0110 \qquad \tau_2 : 1011 \xrightarrow{R_1} 0011 \xrightarrow{R_2} 0111 \xrightarrow{R_4} 0110$$

Furthermore, consider the following tilde-transformations of $u' = 100$ into $v' = 001$:

$$\tau_1' : 100 \xrightarrow{S_1} 010 \xrightarrow{S_2} 001 \qquad \tau_2' : 100 \xrightarrow{R_1} 000 \xrightarrow{R_2} 001$$

Note that both τ_1' and τ_2' have the same length equal to $\text{dist}_\sim(u',v') = 2$. Interestingly, in τ_1' the bit in position 2 is flipped twice.

The next lemma shows that, in the case of a two letters alphabet, we can restrict to tilde-transformations where each character is changed at most once.

Lemma 6. *Let $u, v \in \{0,1\}^m$ with $m \geq 1$. Then, there exists a tilde-transformation of u into v of length* $\text{dist}_\sim(u,v)$ *such that for any $i = 1, 2, \ldots, m$, the character in position i is changed at most once.*

Proof. Let $u = a_1 \cdots a_m$ and $v = b_1 \cdots b_m$ and let τ be a tilde-transformation of u into v of length $d = \text{dist}_\sim(u,v)$. Suppose that, for some i, the character in position i is changed more than once by τ and let O_t and O_s be the first and the second operation, respectively, that modify the character in position i. Observe that the character in position i can be changed by the operations R_i, S_{i-1} or S_i.

Suppose that $O_t = S_i$ and $O_s = R_i$. Then, the symbol a_i is changed twice and two operations S_i and R_i could be replaced by a single R_{i+1}. This would yield a tilde-transformation of u into v of length strictly less than d; this is a contradiction to the definition of tilde-distance. Similarly for the cases where $O_t = R_i$ and $O_s = S_i$, $O_t = S_{i-1}$ and $O_s = R_i$, $O_t = R_i$ and $O_s = S_{i-1}$.

Finally, if $O_t = S_{i-1}$ and $O_s = S_i$ then the three characters in positions $i - 1$, i and $i + 1$ are changed, but the one in position i is changed twice. Hence, the two swap operations S_{i-1} and S_i can be replaced by R_{i-1} and R_{i+1} yielding a tilde-transformation of u into v of same length d which instead involves positions $i - 1$ and $i + 1$ just once (see τ_2' in Example 5). □

Remark 7. Lemma 6 only applies to a binary alphabet. Indeed, if $\Sigma = \{0,1,2\}$, and take $u = 012$ and $v = 120$, then $\text{dist}_\sim(012, 120) = 2$ because there is the tilde-transformation $012 \xrightarrow{S_1} 102 \xrightarrow{S_2} 120$. Instead, in order to change each character at most once, three replacement operations are needed.

Definition 8. *Let Σ be a finite alphabet and $u, v \in \Sigma^+$. A tilde-transformation from u to v is* minimal *if its length is equal to* $\text{dist}_\sim(u,v)$ *and characters in each position are modified at most once.*

Lemma 6 guarantees that, in the binary case, a minimal tilde-transformation always exists. In the sequel, this will be the most investigated case. Let us now define isometric words based on the swap and mismatch distance.

Definition 9. *Let $f \in \Sigma^n$, with $n \geq 1$, f is* tilde-isometric *if for any pair of f-free words u and v of length $m \geq n$, there exists a minimal tilde-transformation from u to v that is f-free. It is* tilde-non-isometric *if it is not tilde-isometric.*

In order to prove that a word is tilde-non-isometric it is sufficient to exhibit a pair (u, v) of words contradicting the Definition 9. Such a pair will be referred to as tilde-witnesses for f. Some examples follow.

Definition 10. *A pair (u, v) of words in Σ^m is a pair of* tilde-witnesses *for f if:*

1. *u and v are f-free*
2. $\text{dist}_\sim(u, v) \geq 2$
3. *there exists no minimal tilde-transformation from u to v that is f-free.*

Example 11. The word $f = 1010$ is tilde-non-isometric because $u = 11000$ and $v = 10110$ are tilde-witnesses for f. In fact, the only possible minimal tilde-transformations from u to v are $11000 \xrightarrow{S_2} 10100 \xrightarrow{R_4} 10110$ and $11000 \xrightarrow{R_4} 11010 \xrightarrow{S_2} 10110$ and in both cases 1010 appears as factor after the first step.

Remark 12. When a transformation contains a swap and a replacement that are adjacent, there could exist many distinct minimal tilde-transformations that involve different sets of operations. For instance, the pair (u, v), with $u = 010$ and $v = 101$, has the following minimal tilde-transformations:

$$010 \xrightarrow{S_1} 100 \xrightarrow{R_3} 101 \qquad 010 \xrightarrow{S_2} 001 \xrightarrow{R_1} 101$$

This fact cannot happen when only replacements are allowed. For this reason studying tilde-isometric words is more complicated than the Hamming case.

Example 11 shows a tilde-non-isometric word. Proving that a given word is tilde-isometric is much harder since it requires to give evidence that no tilde-witnesses exists. We will now prove that word 111000 is isometric with *ad-hoc* technique.

Example 13. The word $f = 111000$ is tilde-isometric. Suppose by the contrary that f is tilde-non-isometric and let (u, v) be a pair of tilde-witnesses for f of minimal tilde-distance. If u and v have only mismatch errors, this is the case of the Hamming distance and results from this theory [4, 19] show that $u = 111\mathbf{0}\mathbf{1}1000$ and $v = 111\mathbf{0}\mathbf{1}01000$; these are not tilde-witnesses since $\text{dist}_\sim(u, v) = 1$.

Therefore, u and v have a swap error in some position i; suppose $u[i..i+1] = 01$. The minimality of $\text{dist}_\sim(u, v)$ implies that $S_i(u)$ is not f-free. Then, a factor 111000 appears in $S_i(u)$ from position $i-2$, and $u[i-2..i+3] = 110100$. Since v is f-free, then there is another error in u involving some positions in $[i-2..i+3]$. It cannot be either a swap (since there are no adjacent different symbols that have not been changed yet), nor a mismatch in positions $i-1, i+2$ (since the corresponding replacement cannot let f occur). Then, it must be a mismatch either in position $i-2$ or $i+3$. Consider the case of a mismatch in position $i+3$ (the other case is analogous). Then, $u[i+3..i+8] = 011000$ and there is another error in $[i+4..i+8]$, in fact, in position $i+6$ or $i+8$. Continuing with similar reasoning, one falls back to the previous situation. This is a contradiction because the length of u is finite.

From now on, we consider only the binary alphabet $\Sigma = \{0, 1\}$ and we study isometric binary words beginning by 1, in view of the following lemma whose proof can be easily inferred by combining the definitions.

Lemma 14. *Let $f \in \{0,1\}^n$. The following statements are equivalent:*

1. *f is tilde-isometric*
2. *f^{rev} is tilde-isometric*
3. *$\overline{f}$ is tilde-isometric.*

Let us conclude the section by comparing tilde-isometric with Ham-isometric words. Although the tilde-distance is more general than the Hamming distance, they are incomparable, as stated in the following proposition.

Proposition 15. *There exists a word which is tilde-isometric but Ham-non-isometric, and a word which is tilde-non-isometric, but Ham-isometric.*

Proof. The word $f = 111000$ is tilde-isometric (see Example 13), but f is Ham-non-isometric by Proposition 2.

Conversely, $f' = 1010$ is tilde-non-isometric (see Example 11), but Ham-isometric by Proposition 2. □

4 Tilde-Isometric Words and Tilde-Error Overlaps

In this section we focus on the word property of being tilde-non-isometric and relate it to the number of errors in its overlaps. The idea reminds the characterization for Ham-isometric words of Proposition 2 but the swap operation changes all the perspectives as pointed also in Proposition 15. Recall that in this paper the alphabet is binary.

Definition 16. *Let $f \in \{0,1\}^n$. Then, f has a q-*tilde-error overlap *of length l and shift $r = n - l$, with $1 \le l \le n-1$ and $0 \le q \le l$, if $\mathrm{dist}_\sim(\mathrm{pre}_l(f), \mathrm{suf}_l(f)) = q$.*

In other words, if f has a q-tilde-error overlap of length l then there exists a minimal tilde-transformation τ from $\mathrm{pre}_l(f)$ to $\mathrm{suf}_l(f)$ of length q. In the sequel, when $q = 2$, in order to specify the kind of errors, a 2-tilde-error overlap is referred to be of type RR

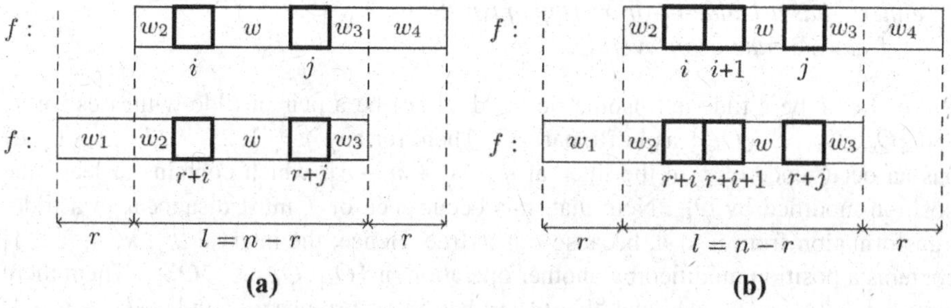

Fig. 1. A word f and its 2-tilde-error overlap of type RR (a) and SR (b)

if τ consists of two replacements, of type SS in case of two swaps, of type RS in case of replacement and swap, and of type SR in case of swap and replacement. If the two errors are in positions i and j, with $i < j$ and we say that f has a 2-tilde-error overlap in i and j or, equivalently, that i and j are the *error positions* of the 2-tilde-error overlap.

Let $f \in \{0,1\}^n$ have a 2-tilde-error overlap in positions i and j with $i < j$, of shift r and length $l = n - r$. The following situations can occur (see Fig. 1), for some $w, w_1, w_2, w_3, w_4 \in \{0,1\}^*$ and $|w_1| = |w_4| = r$.

$$
\begin{aligned}
&\text{RR: } f = w_2\boldsymbol{f}[\boldsymbol{i}]w\boldsymbol{f}[\boldsymbol{j}]w_3w_4 = w_1w_2\overline{\boldsymbol{f}[\boldsymbol{i}]}w\overline{\boldsymbol{f}[\boldsymbol{j}]}w_3\\
&\text{SR: } f = w_2\boldsymbol{f}[\boldsymbol{i}]\boldsymbol{f}[\boldsymbol{i}+\boldsymbol{1}]w\boldsymbol{f}[\boldsymbol{j}]w_3w_4 = w_1w_2\boldsymbol{f}[\boldsymbol{i}+\boldsymbol{1}]\boldsymbol{f}[\boldsymbol{i}]w\overline{\boldsymbol{f}[\boldsymbol{j}]}w_3\\
&\text{RS: } f = w_2\boldsymbol{f}[\boldsymbol{i}]w\boldsymbol{f}[\boldsymbol{j}]\boldsymbol{f}[\boldsymbol{j}+\boldsymbol{1}]w_3w_4 = w_1w_2\overline{\boldsymbol{f}[\boldsymbol{i}]}w\boldsymbol{f}[\boldsymbol{j}+\boldsymbol{1}]\boldsymbol{f}[\boldsymbol{j}]w_3\\
&\text{SS: } f = w_2\boldsymbol{f}[\boldsymbol{i}]\boldsymbol{f}[\boldsymbol{i}+\boldsymbol{1}]w\boldsymbol{f}[\boldsymbol{j}]\boldsymbol{f}[\boldsymbol{j}+\boldsymbol{1}]w_3w_4\\
&\qquad = w_1w_2\boldsymbol{f}[\boldsymbol{i}+\boldsymbol{1}]\boldsymbol{f}[\boldsymbol{i}]w\boldsymbol{f}[\boldsymbol{j}+\boldsymbol{1}]\boldsymbol{f}[\boldsymbol{j}]w_3
\end{aligned}
$$

If $w = \epsilon$ we say that the two errors are *adjacent*. In particular, in the case of a 2-tilde-error overlap in positions i and j, of type RR and RS, the two errors are adjacent if $j = i + 1$. Note that in case of adjacent errors of type RR with $f[i] \neq f[i+1]$, we have a 1-tilde-error overlap that is a swap and that we call of type S. For 2-tilde error overlap of type SR and SS in positions i and j, the two errors are adjacent if $j = i + 2$.

Remark 17. Let $f \in \{0,1\}^n$ be a tilde-non-isometric word and (u, v), with $u, v \in \Sigma^m$, be a pair of tilde-witnesses for f, with minimal $d = \text{dist}_{\sim}(u, v)$ among all pairs of tilde-witnesses of length m.

Let $\{O_{i_1}, O_{i_2}, \dots, O_{i_d}\}$ be the set of operations of a minimal tilde-transformation from u to v, $1 \leq i_1 < i_2 < \cdots < i_d \leq m$. Then, the minimality of d implies that for any $j = 1, 2, \dots, d$, $O_{i_j}(u)$ has an occurrence of f in the interval $[k_j..(k_j + n - 1)]$. Moreover such occurrence of f contains at least one position modified by O_{i_j}. In fact, if O_{i_j} is a swap operation then it changes two positions at once, positions i_j and $i_j + 1$, and the interval $[k_j..(k_j + n - 1)]$ may contain both positions or just one. Note that when only one position is contained in the interval, such position is at the boundary of the interval. This means that, although an error in a position at the boundary of a given interval may appear as caused by a replacement, this can be actually caused by a hidden swap involving positions over the boundary.

Proposition 18. *If $f \in \{0,1\}^n$ is tilde-non-isometric then*

1. *either f has a 1-tilde-error overlap of type S*
2. *or f has a 2-tilde-error overlap.*

Proof. Let f be a tilde-non-isometric word, (u, v) be a pair of tilde-witnesses for f, and $\{O_{i_1}, O_{i_2}, \dots, O_{i_d}\}$ as in Remark 17. Then, for any $j = 1, 2, \dots, d-1$, $O_{i_j}(u)$ has an occurrence of f in the interval $[k_j..k_j + n - 1]$, which contains at least one position modified by O_{i_j}. Note that, this occurrence of f must disappear in a tilde-transformation from u to v, because v is f-free. Hence, the interval $[k_j..k_j + n - 1]$ contains a position modified by another operation in $\{O_{i_1}, O_{i_2}, \dots, Oi_d\}$. Then, there exist $s, t \in \{i_1, i_2, \dots i_d\}$, such that $O_s(u)$ has an occurrence of f in $[k_s..k_s + n - 1]$ that contains at least one position modified by O_t and $O_t(u)$ has an occurrence of f

in $[k_t..k_t+n-1]$ that contains at least one position modified by O_s. Without loss of generality, suppose that $k_s < k_t$. The intersection of $[k_s..k_s+n-1]$ and $[k_t..k_t+n-1]$ contains a prefix of f in $O_t(u)$ and a suffix of f in $O_s(u)$ of some length l. Such an intersection can contain either two, or three, or four among the positions modified by O_s and O_t, of which at least one is modified by O_s and at least one by O_t.

Consider the case that the intersection of $[k_s..k_s+n-1]$ and $[k_t..k_t+n-1]$ contains two among the positions modified by O_s and O_t, and denote them i and j in f, with $1 \leq i < j \leq l$. If the positions are not adjacent, then f has a 2-tilde-error overlap (of type RR). Otherwise, if $f[i] \neq f[i+1]$ then f has a 1-tilde-error overlap of type S. If $f[i] = f[i+1]$ then f has a 2-tilde-error overlap (of type RR).

Suppose that the intersection of $[k_s..k_s+n-1]$ and $[k_t..k_t+n-1]$ contains three among the positions modified by O_s and O_t. In this case, at least one of the two operations must be a swap; suppose O_s is a swap. Then, O_t could be either a replacement on the third position, or a swap if the third position is at the boundary of $[k_t..k_t+n-1]$. In any case, f has a 2-tilde-error overlap (of type SR or SS). Suppose now that the intersection of $[k_s..k_s+n-1]$ and $[k_t..k_t+n-1]$ contains four among the positions modified by O_s and O_t. In this case, each of O_s and O_t involves two positions, and f has a 2-tilde-error overlap of type SS. □

5 Construction of Tilde-Witnesses

As already discussed in Sect. 4, in order to prove that a word is tilde-non-isometric it is sufficient to exhibit a pair of tilde-witnesses. Proposition 18 states that if a word is tilde-non-isometric then it has either a 1-tilde-error overlap of type swap or a 2-tilde-error overlap. In this section we show the construction of tilde-witnesses for a word, starting from its error overlaps. Let us start with the case of a 1-tilde-error overlap.

Proposition 19. *If f has a 1-tilde-error overlap of type S, then it is tilde-non-isometric.*

Proof. Let f have a 1-tilde-error overlap in position i of type S with shift r. The pair (u, v) with:

$$u = \text{pre}_r(f)R_i(f) \qquad v = \text{pre}_r(f)R_{i+1}(f)$$

is a pair of tilde-witnesses for f. In fact, one can prove that they satisfy the conditions in Definition 10. □

Example 20. The word $f = 101$ has a 1-tilde-error overlap of type S in position 1 therefore it is tilde-non-isometric. In fact, following the proof of previous proposition, the pair (u, v) with $u = 1001$ and $v = 1111$ is a pair of tilde-witnesses.

Let us now introduce some special words which often will serve as tilde-witnesses. Let f have a 2-tilde-error overlap of shift r in positions i and j, $i < j$, then we define

$$\tilde{\alpha}_r(f) = \text{pre}_r(f)O_i(f) \text{ and } \tilde{\beta}_r(f) = \text{pre}_r(f)O_j(f) \tag{1}$$

As an example, using the previous notations for errors of type SR we have that

$$\tilde{\alpha}_r(f) = w_1w_2f[i+1]f[i]wf[j]w_3w_4 \text{ and } \tilde{\beta}_r(f) = w_1w_2f[i]f[i+1]w\overline{f[j]}w_3w_4 \tag{2}$$

In the sequel, when there is no ambiguity, we omit f and we simply write $\tilde{\alpha}_r$ and $\tilde{\beta}_r$.

Lemma 21. *If f has a 2-tilde-error overlap of shift r, then $\tilde{\alpha}_r(f)$ is f-free.*

Proof. Suppose that $f \in \{0,1\}^n$ has a 2-tilde-error overlap. If it is of type RR then $\tilde{\alpha}_r$ is f-free by Claim 1 of Lemma 2.2 in [19], also in the case of adjacent errors. If it is of type SR, of shift r in positions i and j, with $i < j$, then, by Eq. (1), we have $\tilde{\alpha}_r = w_1 S_i(f)$ then $\tilde{\alpha}_r[r+k] = f[k]$, for any $1 \le k \le n$, with $k \neq i$ and $k \neq i+1$. If f occurs in $\tilde{\alpha}_r$ in position r_1+1 we have that $1 < r_1 < r$ (if $r_1 = 1$ then $f[i] = f[i+1]$ and there is no swap error at position i) and $\tilde{\alpha}_r[r_1+1 \ldots r_1+n] = f[1 \ldots n]$. Finally, by Eq. (2), we have that $\tilde{\alpha}_r[k] = f[k]$, for $k \neq r+j$. In conclusion, we have that $f[i] = \tilde{\alpha}_r[r_1+i] = f[r_1+i]$ (trivially, $r_1+i \neq r+j$). Furthermore $f[r_1+i] = \tilde{\alpha}_r[r+r_1+i]$ ($r_1+i \neq i$ and $r_1+i \neq i+1$ because $r_1 > 1$). But $\tilde{\alpha}_r[r+r_1+i] = f[r+i]$ then we have the contradiction that $f[i] = f[r+i]$. If the 2-tilde-error is of type RS, SS the proof is similar. For clarity, note that, also in the case of adjacent errors, supposing that f occurs in $\tilde{\alpha}_r$ leads to a contradiction in $f[i]$ that is not influenced by j. □

Note that while $\tilde{\alpha}_r$ is always f-free, $\tilde{\beta}_r$ is not. Indeed, the property $\tilde{\beta}_r$ not f-free is related to a condition on the overlap of f. We give the following definition.

Definition 22. *Let $f \in \{0,1\}^n$ and consider a 2-tilde-error overlap of f, with shift r and error positions i, j, with $1 \le i < j \le n-r$. The 2-tilde-error overlap satisfies $Condition^{\sim}$ if it is of type RR or SS and:*

$$\begin{cases} r \ is\ even \\ j - i = r/2 \\ f[i..(i+r/2-1)] = f[j..(j+r/2-1)] \end{cases} \qquad (\textbf{\textit{Condition}}^{\sim})$$

Lemma 23. *Let $f \in \{0,1\}^n$ have a 2-tilde-error overlap of shift r, then $\tilde{\beta}_r(f)$ is not f-free iff the 2-tilde-error overlap satisfies $Condition^{\sim}$.*

Proof. Suppose that $f \in \{0,1\}^n$ has a 2-tilde-error overlap that satisfies $Condition^{\sim}$. Note that a 2-tilde-error overlap of type RS or SR cannot satisfy $Condition^{\sim}$. Now, if the 2-tilde-error overlap is of type RR, then the fact that $\tilde{\beta}_r(f)$ is not f-free can be shown as in the proof of Claim 2 of Lemma 2.2 in [19]. If the 2-tilde-error overlap is of type SS, then that proof must be suitably modified. More precisely, let i, j, with $1 \le i < j \le n-r$, be the error positions of the 2-tilde-error overlap of shift r that satisfies $Condition^{\sim}$. Let $f[i] = f[j] = x$, $f[i+r] = f[j+r] = \overline{x}$, $f[i+1] = f[j+1] = \overline{x}$ and $f[i+1+r] = f[j+1+r] = x$.

It is possible to show that, for some $k_1, k_2 \ge 0$ $f = \rho(uw)^{k_1} uwuw\overline{u}w\overline{u}(w\overline{u})^{k_2}\sigma$, where $u = x\overline{x}$, $w = f[i+2..j-1]$ (w is empty, if $j = i+2$) and ρ and σ are, respectively, a suffix and a prefix of w.

Then, $\tilde{\beta}_r(f) = \rho(uw)^{k_1+1} uwuw\overline{u}w\overline{u}(w\overline{u})^{k_2+1}\sigma$ and, hence, $\tilde{\beta}_r(f)$ is not f-free.

Assume now that $\tilde{\beta}_r(f)$ is not f-free and suppose that a copy of f, say f', occurs in $\tilde{\beta}_r(f)$ at position r_1+1. A reasoning similar to the one used in the proof of Lemma 21, shows that, if i and j are the error positions, then $j - i = r_1$ and $j - i = r - r_1$. Hence $r = 2r_1$ is even and $j - i = r/2$. Therefore, $f[i+t] = f'[i+t] = f[i+t+r/2] = f[j+t]$, for $0 \le t \le r/2$, i.e. $f[i..(i+r/2-1)] = f[j..(j+r/2-1)$ and the 2-tilde-error overlap satisfies $Condition^{\sim}$. □

In the rest of the section we deal with the construction of tilde-witnesses based on 2-tilde-error overlaps. We distinguish the cases of non-adjacent and adjacent errors. Non-adjacent errors can be dealt with standard techniques, while the case of adjacent ones may show new issues. For example, when f has a 2-error-overlap of type SR with error block **101** (aligned with 010) then it can be also considered of type RS.

Moreover, note that all the adjacent pairs of errors can be listed as follows, up to complement and reverse. A 2-error overlap of type SS may have (error) block **1010** or **1001**; of type SR or RS may have block **100**, **101** or **110**; of type RR block **11** (block **10** aligned with 01 corresponds to one swap). Note that, for some error types, we need also to distinguish sub-cases related to the different characters adjacent to those error blocks. We collect all the cases in the following proposition. For lack of space, the proof is detailed only in the case 2. In the remaining cases, the proofs are sketched by exhibiting a pair of words that can be shown to be a pair of tilde-witnesses.

Theorem 24. *Let $f \in \{0,1\}^n$. Any of the following conditions, up to complement and reverse, is sufficient for f being tilde-non-isometric.*

1. *f has a 2-tilde-error overlap with not adjacent error positions*
2. *f has a 2-tilde-error overlap of type SS with adjacent error positions*
3. *f has a 2-tilde-error overlap with block* **101** *(of type SR or RS)*
4. *f has a 2-tilde-error overlap with block* **100** *(of type SR or RS) in the particular case that $f = x\mathbf{1001}z = yx\mathbf{011}$, for some $x, y, z \in \{0,1\}^*$*
5. *f has a 2-tilde-error overlap RR in the particular case that f starts with* **110** *and ends with* **100***.*

Proof. We provide, for each case in the list, a pair of tilde-witnesses for f.

Case 1. If the 2-tilde-error overlap does not satisfy *Condition*$^\sim$, following Definition 10, one can prove that the pair $(\tilde{\alpha}_r, \tilde{\beta}_r)$ as in Eq. (1) is a pair of tilde-witnesses for f. Otherwise, one can prove that $(\tilde{\eta}_r, \tilde{\gamma}_r)$ with $\tilde{\eta}_r = \text{pre}_r(f)O_i(f)\text{suf}_{r/2}(f)$ and $\tilde{\gamma}_r = \text{pre}_r(f)O_j(O_t(f))\text{suf}_{r/2}(f)$, with $t = j + r/2$, is a pair of tilde-witnesses for f.

Case 2. Proved in Lemma 25.

Case 3. We have $f = w_2\mathbf{101}w_3w_4 = w_1w_2\mathbf{010}w_3$, for some $w_1, w_2, w_3, w_4 \in \{0,1\}^*$. The pair $(\tilde{\alpha}_r, \tilde{\beta}_r)$, with $\tilde{\alpha}_r = w_1w_2\mathbf{011}w_3w_4$ and $\tilde{\beta}_r = w_1w_2\mathbf{100}w_3w_4$, is a pair of tilde-witnesses, following Definition 10.

Case 4. We have $f = w_2\mathbf{1001}w_3 = w_1w_2\mathbf{011}$, for some $w_1, w_2, w_3 \in \{0,1\}^*$. In this case we need a different technique to construct the pair of tilde-witnesses $(\tilde{\alpha}_r, \tilde{\delta}_r)$. We set $\tilde{\alpha}_r = w_1w_2\mathbf{0101}w_3$ and $\tilde{\delta}_r = w_1w_2\mathbf{1010}w_3$. Here we prove that $\tilde{\delta}_r$ is f-free. Indeed, suppose that a copy of f occurs in $\tilde{\delta}_r$ starting from position r_1. Some considerations, related to the definition of $\tilde{\delta}_r$ and to the structure of f, show that either $r_1 = 2$ or $r_1 = 3$, and one can prove that this leads to a contradiction.

Case 5. We have $f = \mathbf{110}w_1 = w_2\mathbf{100}$, for some $w_1, w_2 \in \{0,1\}^*$. By following Definition 10, one can prove that the pair $(\tilde{\alpha}_r, \tilde{\delta}_r)$, with $\tilde{\alpha}_r = w_2\mathbf{1010}w_1$ and $\tilde{\delta}_r = w_2\mathbf{0101}w_1$ is a pair of tilde-witnesses. Remark that, in such a case, the pair $(\tilde{\alpha}_r, \tilde{\beta}_r)$ of Eq. 1 is not a pair of tilde-witnesses because $\text{dist}_\sim(\tilde{\alpha}_r, \tilde{\beta}_r) = 1$. □

Lemma 25. *If f has a 2-tilde-error overlap of type SS, where the errors are adjacent, then f is tilde-non-isometric.*

Proof. Let $f \in \{0,1\}^n$ have a 2-tilde-error overlap of shift r and type SS, where the errors are adjacent. Then, two cases can occur (up to complement):

Case 1: $f = w_2\mathbf{1010}w_3w_4 = w_1w_2\mathbf{0101}w_3$, $|w_1| = r$

If the 2-tilde-error overlap does not satisfy Condition$^\sim$, then $(\tilde{\alpha}_r, \tilde{\beta}_r)$, with $\tilde{\alpha}_r = w_1w_2\mathbf{0110}w_3w_4$ and $\tilde{\beta}_r = w_1w_2\mathbf{1001}w_3w_4$, is a pair of tilde-witnesses, following Definition 10. In fact:

1. $\tilde{\alpha}_r$ is f-free thanks to Lemma 21 and $\tilde{\beta}_r$ is f-free, by Lemma 23.
2. $\text{dist}_\sim(\tilde{\alpha}_r, \tilde{\beta}_r) = 2$, straightforward.
3. a minimal tilde-transformation from $\tilde{\alpha}_r$ to $\tilde{\beta}_r$ consists of two swaps S_i and S_j with $i = |w_1w_2|+1$ and $j = i+2$. If S_i is applied to $\tilde{\alpha}_r$ as first operation, then f appears as a suffix, whereas if S_j is applied first to $\tilde{\alpha}_r$, then f appears as a prefix.

If the 2-tilde-error overlap satisfies Condition$^\sim$, then $w_3 = \mathbf{10}w_3'$ and, following Definition 10, $(\tilde{\eta}_r, \tilde{\gamma}_r)$ is a pair of tilde-witnesses, where $\tilde{\eta}_r = w_1w_2\mathbf{011010}w_3'w_4w_5$ and $\tilde{\gamma}_r = w_1w_2\mathbf{100101}w_3'w_4w_5$, with $w_5 = suf_{r/2}(f)$.

1. One can prove that $\tilde{\eta}_r$ and $\tilde{\gamma}_r$ are f-free.
2. $\text{dist}_\sim(\tilde{\eta}, \tilde{\gamma}) = 3$.
3. a minimal tilde-transformation from $\tilde{\eta}_r$ to $\tilde{\gamma}_r$ consists of three swap operations S_i, S_j and S_t with $i = |w_1w_2| + 1$, $j = i + 2$, $t = j + 2$. If S_i is applied to $\tilde{\eta}_r$ as first operation, then f occurs at position $|w_1| + 1$, if S_j is applied first then f appears as a prefix, whereas if S_t is applied first then f appears as a suffix.

Case 2: $f = w_2\mathbf{1001}w_3w_4 = w_1w_2\mathbf{0110}w_3$, $|w_1| = r$

The pair $(\tilde{\alpha}_r, \tilde{\beta}_r)$, with $\tilde{\alpha}_r = w_1w_2\mathbf{0101}w_3w_4$ and $\tilde{\beta}_r = w_1w_2\mathbf{1010}w_3w_4$, is a pair of tilde-witnesses, following Definition 10. In such a case, by Lemma 23, $\tilde{\beta}_r$ is f-free. In fact, the Condition$^\sim$ never holds, since $f[i]$ is different from $f[j]$. □

Example 26. The word $f = \mathbf{1001}0110 = 1001\mathbf{0110}$ has a 2-tilde-error overlap of type SS and shift $r = 4$ in positions $1, 3$. By Lemma 25, the pair $(\tilde{\alpha}_4, \tilde{\beta}_4)$ with $\tilde{\alpha}_4 = 1001\mathbf{0101}0110$ and $\tilde{\beta}_4 = 1001\mathbf{1010}0110$ is a pair of tilde-witnesses. Then f is tilde-non-isometric. Note that f is Ham-isometric.

Theorem 24 lists the conditions for a word f to be tilde non-isometric and the proof provides all the corresponding pairs of witnesses. The construction of $(\tilde{\alpha}_r, \tilde{\beta}_r)$ and $(\tilde{\eta}_r, \tilde{\gamma}_r)$, used so far, is inspired by an analogous construction for the Hamming distance (cf. [19]) and it is here adapted to the tilde-distance. On the contrary, in the cases 4 and 5 a new construction is needed because the usual pair of witnesses does not satisfy Definition 10 any more. The construction of $\tilde{\delta}_r$ is peculiar of the tilde-distance. It solves the situation expressed in Remark 17 when a mismatch error may appear as caused by a replacement, but it is actually caused by a hidden swap involving adjacent positions.

In conclusions, the swap and mismatch distance we adopted in this paper opens up new scenarios and presents interesting new situations that surely deserve further investigation.

References

1. Amir, A., Cole, R., Hariharan, R., Lewenstein, M., Porat, E.: Overlap matching. Inf. Comput. **181**(1), 57–74 (2003)
2. Amir, A., Eisenberg, E., Porat, E.: Swap and mismatch edit distance. Algorithmica **45**(1), 109–120 (2006)
3. Anselmo, M., Flores, M., Madonia, M.: Quaternary n-cubes and isometric words. In: Lecroq, T., Puzynina, S. (eds.) WORDS 2021. LNCS, vol. 12847, pp. 27–39. Springer, Cham (2021). https://doi.org/10.1007/978-3-030-85088-3_3
4. Anselmo, M., Flores, M., Madonia, M.: Fun slot machines and transformations of words avoiding factors. In: 11th International Conference on Fun with Algorithms. LIPIcs, vol. 226, pp. 4:1–4:15 (2022)
5. Anselmo, M., Flores, M., Madonia, M.: On k-ary n-cubes and isometric words. Theor. Comput. Sci. **938**, 50–64 (2022)
6. Anselmo, M., Giammarresi, D., Madonia, M., Selmi, C.: Bad pictures: some structural properties related to overlaps. In: Jirásková, G., Pighizzini, G. (eds.) DCFS 2020. LNCS, vol. 12442, pp. 13–25. Springer, Cham (2020). https://doi.org/10.1007/978-3-030-62536-8_2
7. Béal, M., Crochemore, M.: Checking whether a word is Hamming-isometric in linear time. Theor. Comput. Sci. **933**, 55–59 (2022)
8. Béal, M.-P., Mignosi, F., Restivo, A.: Minimal forbidden words and symbolic dynamics. In: Puech, C., Reischuk, R. (eds.) STACS 1996. LNCS, vol. 1046, pp. 555–566. Springer, Heidelberg (1996). https://doi.org/10.1007/3-540-60922-9_45
9. Castiglione, G., Mantaci, S., Restivo, A.: Some investigations on similarity measures based on absent words. Fundam. Informaticae **171**(1–4), 97–112 (2020)
10. Charalampopoulos, P., Crochemore, M., Fici, G., Mercas, R., Pissis, S.P.: Alignment-free sequence comparison using absent words. Inf. Comput. **262**, 57–68 (2018)
11. Dombb, Y., Lipsky, O., Porat, B., Porat, E., Tsur, A.: The approximate swap and mismatch edit distance. Theor. Comput. Sci. **411**(43), 3814–3822 (2010)
12. Faro, S., Pavone, A.: An efficient skip-search approach to swap matching. Comput. J. **61**(9), 1351–1360 (2018)
13. Galil, Z., Park, K.: An improved algorithm for approximate string matching. SIAM J. Comput. **19**, 989–999 (1990)
14. Hsu, W.J.: Fibonacci cubes-a new interconnection topology. IEEE Trans. Parallel Distrib. Syst. **4**(1), 3–12 (1993)
15. Ilić, A., Klavžar, S., Rho, Y.: The index of a binary word. Theor. Comput. Sci. **452**, 100–106 (2012)
16. Klavžar, S., Shpectorov, S.V.: Asymptotic number of isometric generalized Fibonacci cubes. Eur. J. Comb. **33**(2), 220–226 (2012)
17. Levenshtein, V.I.: Binary codes capable of correcting deletions, insertions, and reversals. Cybern. Control Theory **10**, 707–710 (1966)
18. Wagner, R.A.: On the complexity of the extended string-to-string correction problem. In: Rounds, W.C., Martin, N., Carlyle, J.W., Harrison, M.A. (eds.) Proceedings of the 7th Annual ACM Symposium on Theory of Computing, 5–7 May 1975, Albuquerque, New Mexico, USA, pp. 218–223 (1975)
19. Wei, J.: The structures of bad words. Eur. J. Comb. **59**, 204–214 (2017)
20. Wei, J., Yang, Y., Zhu, X.: A characterization of non-isometric binary words. Eur. J. Comb. **78**, 121–133 (2019)
21. Wei, J., Zhang, H.: Proofs of two conjectures on generalized Fibonacci cubes. Eur. J. Comb. **51**, 419–432 (2016)

Set Augmented Finite Automata over Infinite Alphabets

Ansuman Banerjee[1], Kingshuk Chatterjee[2], and Shibashis Guha[3(✉)]

[1] Indian Statistical Institute, Kolkata, India
[2] Government College of Engineering and Ceramic Technology, Kolkata, India
[3] Tata Institute of Fundamental Research, Mumbai, India
shibashis.guha@tifr.res.in

Abstract. A data language is set of finite words defined on an infinite alphabet. Data languages are used to express properties associated with data values (domain defined over a countably infinite set). In this paper, we introduce set augmented finite automata (SAFA), a new class of automata for expressing data languages. SAFA is able to recognize data languages while storing a few data values in most cases. We investigate nonemptiness, membership, closure properties, and expressiveness of SAFA.

Keywords: Automata on infinite alphabets · data languages · Register automata · Expressiveness · Closure properties

1 Introduction

A data language is a set of data words that are concatenations of data elements. A data element is a pair of an attribute and a data value. While the number of attribute items of a data language is finite, the values that these attributes hold often come from a countably infinite set (e.g. natural numbers). With large scale availability of data in recent times, there is a need for methods for modeling and analysis of data languages. Thus, there is a need for automated methods for recognizing data attribute relationships and languages defined on infinite alphabets. A data element is expressed as an attribute, data value pair (a, d) where a belongs to a finite set and d belongs to a countably infinite set. A data word is a concatenation of a finite number of data elements, i.e. a data word $w = (a_1, d_1)(a_2, d_2)...(a_{|w|}, d_{|w|})$, where $|w|$ denotes the size of w. This work introduces a new model for data languages on infinite alphabets.

The k-register automata (finite automata with k registers, each capable of storing one data value) [15] is a finite automata with the ability to handle infinite alphabets. The nonemptiness and membership problems for register automata are both NP-complete [21], however, the language recognizability is somewhat restricted, since it uses only a finite number of registers to store the data values. To increase the computational power of infinite alphabet automata, models which use hash tables to store data values, namely class memory automata

This work has been partially supported by the DST-SERB project SRG/2021/000466 *Zero-sum and Nonzero-sum Games for Controller Synthesis of Reactive Systems.*

F. Drewes and M. Volkov (Eds.): DLT 2023, LNCS 13911, pp. 36–50, 2023.
https://doi.org/10.1007/978-3-031-33264-7_4

(CMA) and class counter automata (CCA) are introduced in [2,17]. The set of languages accepted by CCA is a subset of the set of languages accepted by CMA. While both CCA, CMA can accept many important data languages, the nonemptiness problem for CCA is EXPSPACE-complete [17], while for CMA, it is non-elementary [3,8].

Our Contribution: We introduce set augmented finite automata (SAFA) that are finite automaton models equipped with a finite number of finite sets of data values. Using these sets as auxiliary storage, SAFA are able to recognize many data languages, while requiring less storage. This paper has the following contributions.

- We develop the formal definition of the SAFA model on infinite alphabets.
- We show that nonemptiness and membership are NP-complete for SAFA.
- We study the closure properties on the SAFA model. In particular, we show that SAFA are closed under union but not under complementation. We put forward the intersection problem as an open question.
- We also examine the deterministic version of our SAFA model and show that there are languages that necessarily need nondeterminism to be accepted.
- We present a study on the strict hierarchy of language acceptance ability with respect to the number of sets associated with SAFA models.
- Finally, we study the expressiveness of SAFA models. On one hand, we show that the language recognition capabilities of SAFA and register automata are incomparable. On the other hand, the set of languages recognized by SAFA is a strict subset of the set of languages recognized by CMA.

Related Work: Register automata introduced by Kaminski and Francez [15] use a finite number of registers to store data values; hence can only accept those data languages in which membership depends on remembering properties of a finite number of data values. An extension to finite register models are pushdown automata models for infinite alphabets [6]. Cheng and Kaminski [6] and Autebut et al. [1] both described context free languages for infinite alphabets. The membership problem for the grammar proposed by Cheng and Kaminski was decidable unlike Autebut's. Sakamoto and Ikeda [21] showed that the decision problems are intractable even for the simplest register finite automata model on infinite alphabets. Neven [19], Neven et al. [20] discussed the properties of register and pebble automata on infinite alphabets, in terms of their language acceptance capabilities and also established their relationship with logic. Kaminski and Tan [16] also developed a regular expression over infinite alphabets. Choffrut [7] defined finite automata on infinite alphabets which uses first order logic on transitions. This was later extended by Iosif and Xu to an alternating automata model [13] with emptiness being intractable in general but PTIME for restricted variants. Demri and Lazić [10] explored the relationship between linear temporal logic (LTL) and register automata. Grumberg et al. [12] introduced variable automata over infinite alphabets where transitions are defined over alphabets as well as over variables. Bojanczyk et al. [4] introduced data automata and the model is equivalent to class memory automata (CMA) [2] which is the first automata model over an infinite alphabet to use hashing to store data values. The set of languages recognized by CMA is a superset of the set of languages recognized by Class counter automata (CCA), another hash based finite automata on infinite alphabets [17] introduced by Manuel and Ramanujam. The paper [18] presents a detailed survey of the existing finite automata models on infinite alphabets. Tan [22] introduced a weak 2-pebble automata model whose emptiness was decidable but with significant reduction in acceptance capabilities. Bollig [5] combined CMA and register automata and showed that local existential monadic second order (MSO) logic can be converted to class register automata in time polynomial in the size of the input automaton. Figuera [11] discussed the properties of alternating register automata and also discussed restricted variants of this model where decidability is tractable. Dassow and Vaszil [9] introduced the P-automata model and established that these are equivalent to restricteds version of register automata.

The membership and nonemptiness problems are NP-complete for SAFA. On one hand, this gives us an advantage over the hash-based family of models on infinite alphabets for which nonemptiness is in EXPSPACE or higher complexity classes. On the other hand, this puts our model in the same complexity class as the k-register automata, while having the ability to accept many data languages.

2 Preliminaries

Let $\mathbb{N}$ denote the set of natural numbers, and $[k]$ the set $\{1,\ldots,k\}$ where $k > 0$. Let Σ be a finite alphabet which comprises a finite set of attributes, and D be a countably infinite set of attribute values. A *data word* $w \in (\Sigma \times D)^*$ is a concatenation of attribute, value pairs, where $*$ denotes zero or more repetitions. A data value is also known as an attribute value, that is the value associated with an attribute. An example data word is of the form $w = (a_1, d_1)\cdots(a_{|w|}, d_{|w|})$ where $a_1,\ldots,a_{|w|} \in \Sigma$, $d_1, ..., d_{|w|} \in D$, and $|w|$ denotes the length of w. An example data word w on $\Sigma = \{a, b\}$ and $D = \mathbb{N}$ is: $(a,1)(a,2)(b,1)(b,5),(a,2)(a,5)(a,7)(a,100)$ with $|w| = 8$. A *data language* $L \subseteq (\Sigma \times D)^*$ is a set of data words. Some example data languages with $\Sigma = \{a, b\}$, $D = \mathbb{N}$ are mentioned below.

- $L_{\mathsf{fd}(a)}$: language of data words, wherein the data values associated with attribute a are all distinct.
- $L_{\forall\mathsf{cnt}=2}$: language of data words wherein all data values appear exactly twice.
- $L_{\exists\mathsf{cnt}\neq 2}$: the language of data words w where there exists a data value d which does not appear twice. $L_{\exists\mathsf{cnt}\neq 2}$ is the complement of $L_{\forall\mathsf{cnt}=2}$.

For a word $w \in (\Sigma \times D)^*$, we denote by $\mathsf{proj}_\Sigma(w)$ and $\mathsf{proj}_D(w)$ the projection of w on Σ and D respectively. Let $L_{\mathsf{proj}_\Sigma(L)=regexp(r)}$ be the set of all data words w such that $\mathsf{proj}_\Sigma(w) \in L_{regexp(r)}$ where $L_{regexp(r)} \subseteq \Sigma^*$ is the set of all words over Σ generated by the regular expression r.

k-register automata [15]: A k-register automaton is a tuple $(Q, \Sigma, \delta, \tau_0, U, q_0, F)$, where Q is a finite set of states, $q_0 \in Q$ is an initial state and $F \subseteq Q$ is a set of final states, τ_0 is an initial register configuration given by $\tau_0 : [k] \to D \cup \{\perp\}$, where D is a countably infinite set, and $\perp$ denotes an uninitialized register, and U is a partial update function: $(Q \times \Sigma) \rightarrow [k]$. The transition relation is $\delta \subseteq (Q \times \Sigma \times [k] \times Q)$. The registers initially contain distinct data values other than $\perp$ which can be present in more than one register. The automaton works as follows. Consider a register automaton M in state $q \in Q$. Each of its registers r_i holds datum d_i where $0 \leq i \leq k$, $d_i \in D \cup \{\perp\}$. Let M at some instance reads the jth data element (a_j, d_j) of the input word w where $a_j \in \Sigma$, $d_j \in D$. Two cases may arise.

- Case 1: There exists an i such that $d_j = d_i$: In this case two situations may arise (i) $(q, a, i, q') \in \delta$ and (ii) $(q, a, i, q') \notin \delta$. In situation (i) the corresponding transition is executed, and in situation (ii) the automaton stops without consuming the data element.
- Case 2: There exists no register i such that $d_j = d_i$: In this case, for all i, we have $d_j \neq d_i$. We look at the partial update function U. If $U(q, a)$ is not defined, the automaton stops without consuming the data element. If $U(q, a)$ is defined, then d_j is inserted in the register $U(q, a)$ and the automaton executes the transition $(q, a, U(q, a), q')$ if $(q, a, U(q, a), q') \in \delta$, otherwise it halts if $(q, a, U(q, a), q') \notin \delta$.

The automaton M accepts an input data word w if it consumes the whole word and it ends in a final state.

Class memory automata [2]: A class memory automaton is defined as a 6-tuple $M = (Q, \Sigma, \delta, q_0, F_\ell, F_g)$ where Q is a finite set of states, $q_0 \in Q$ is an initial state and $F_g \subseteq F_\ell \subseteq Q$ are a set of global and local accepting states respectively. The transition relation is $\delta \subseteq (Q \times \Sigma \times (Q \cup \{\perp\}) \times Q)$. The automaton keeps track of the last state where a data value d is encountered. If a data value d is not yet encountered, then it is associated with $\perp$. Each transition of a CMA is dependent on the current state of the automaton and the state the automaton was in when the data value being read currently was last encountered. A data word w is accepted if the automaton reaches a state $q \in F_g$ and the last state of all the data values encountered in w are in F_ℓ.

3 Set Augmented Finite Automata

Definition 1. *A set augmented finite automaton (SAFA) is defined as a* 6*-tuple* $M = (Q, \Sigma \times D, q_0, F, H, \delta)$ *where* Q *is a finite set of states,* Σ *is a finite alphabet,* D *is a countably infinite set,* $q_0 \in Q$ *is the initial state,* $F \subseteq Q$ *is a set of final states,* H *is a finite set of finite sets of data values. The transition relation is defined as:* $\delta \subseteq Q \times \Sigma \times C \times OP \times Q$ *where* $C = \{p(h_i), !p(h_i) \mid h_i \in H\}$, h_i *denotes the* i^{th} *set in* H, *and* $OP = \{-, \ \mathsf{ins}(h_i) \mid h_i \in H\}$. □

We call a SAFA a *singleton* if $|H| = 1$. The unary Boolean predicate $p(h_i)$ evaluates to true if the data value currently being read by the automaton is in h_i. The predicate $!p(h_i)$ is true if the data value currently being read is *not* in h_i. Further, OP denotes a set of operations that a SAFA can execute on reading a symbol; the operation $\mathsf{ins}(h_i)$ inserts the data value currently being read by

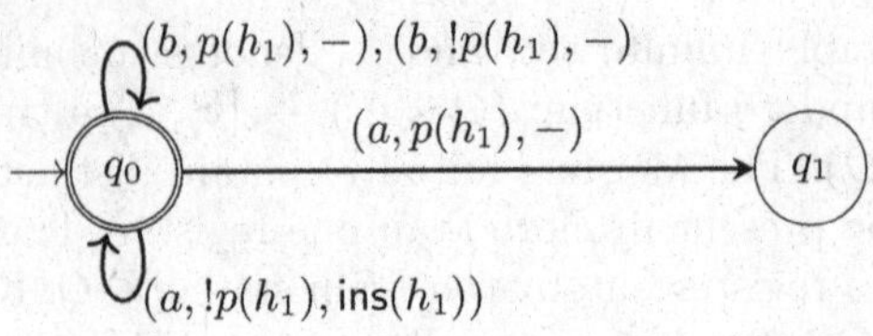

Fig. 1. SAFA for $L_{\mathsf{fd(a)}}$

the automaton into the set h_i, while $-$ denotes no such insertion is done. For any combination not in δ, we assume the transition is absent.

For a SAFA M, we define a *configuration* $(q, h) \in Q \times 2^{D^H}$ as follows: $q \in Q$ is a state of the automaton, $h = \langle h_1, ...h_{|H|}\rangle$ where each h_i for $i \in [|H|]$ is a finite subset of D, and h denotes the content of the sets in H. A *run* ρ of M on an input $w = (a_1, d_1) \cdots (a_{|w|}, d_{|w|})$ is a sequence $(q_0, h^0), \ldots, (q_{|w|}, h^{|w|})$, where $h^j = \langle h_1^j, \ldots, h_{|H|}^j\rangle$, and h_i^j for $1 \leq i \leq |H|$ is the content of the set h_i after reading the prefix $(a_1, d_1) \cdots (a_j, d_j)$ for $1 \leq j \leq |w|$. A configuration (q_{j+1}, h^{j+1}) succeeds a configuration (q_j, h^j) if there is a transition $(q_j, a_{j+1}, \alpha, \mathsf{op}, q_{j+1})$ where

- (i) for $\alpha = p(h_i)$, we have that the data value $d_{j+1} \in {h_i}^j$.
- (ii) for $\alpha = !p(h_i)$, we have that the data value $d_{j+1} \notin {h_i}^j$.

The execution of the operation $\mathsf{op} \in OP$ takes the content of the sets of data values from h^j to h^{j+1}. If op is $-$, then $h^{j+1} = h^j$. If op is $\mathsf{ins}(h_i)$, then ${h_l}^{j+1} = {h_l}^j$ for all $h_l \in H \setminus \{h_i\}$, and ${h_i}^{j+1} = {h_i}^j \cup \{d_{j+1}\}$. If the *run* consumes the whole word w, and $q_{|w|} \in F$, then the run is *accepting*, otherwise, it is rejecting. A word w is accepted by M if it has an accepting run. The language $L(M)$ accepted by M consists of all words accepted by M. We denote by $|\rho|$ the length of the run which equals the number of transitions taken. Note that for the run ρ on an input word w, we have that $|w| = |\rho|$.

Definition 2. *A SAFA* $M = (Q, \Sigma, q_0, F, H, \delta)$ *is* deterministic *(DSAFA) if for every* $q \in Q$ *and* $a \in \Sigma$*, if there is a transition* $(q, a, \alpha, \mathsf{op}, q')$*, where* $q' \in Q$*,* $\mathsf{op} \in OP$*,* $\alpha \in \{p(h_i), !p(h_i)\}$*,* $h_i \in H$*, then there cannot be any transition of the form* $(q, a, p(h_l), \mathsf{op}', q'')$*,* $(q, a, !p(h_l), \mathsf{op}', q'')$*, where* $q'' \in Q$*,* $h_l \neq h_i$*,* $h_l \in H$*,* $\mathsf{op}' \in OP$*. The only other allowed transition can be* $(q, a, \alpha', \mathsf{op}', q'')$ *for* $\alpha' \in \{p(h_i), !p(h_i)\}$*,* $\alpha' \neq \alpha$*, and* $\mathsf{op}' \in OP$*.* □

Let $\mathcal{L}_{\mathsf{SAFA}}$ and $\mathcal{L}_{\mathsf{DSAFA}}$ denote the set of all languages accepted by nondeterministic SAFA and deterministic SAFA respectively. We illustrate the SAFA model with some instances of data languages.

Example 1. The language $L_{\mathsf{fd(a)}}$ can be accepted by the DSAFA $M = (Q, \Sigma \times D, q_0, F, H, \delta)$ in Fig. 1. Here, $Q = \{q_0, q_1\}$, $\Sigma = \{a, b\}$, D is any countably infinite set, $F = \{q_0\}$, $H = \{h_1\}$, the transition relation δ consists of the transitions shown in Fig. 1. The automaton M works as follows. The set h_1 is used to store the data values encountered in the input word associated with a. At q_0, if M reads b, it remains in q_0 without modifying H. At q_0 when the automaton reads

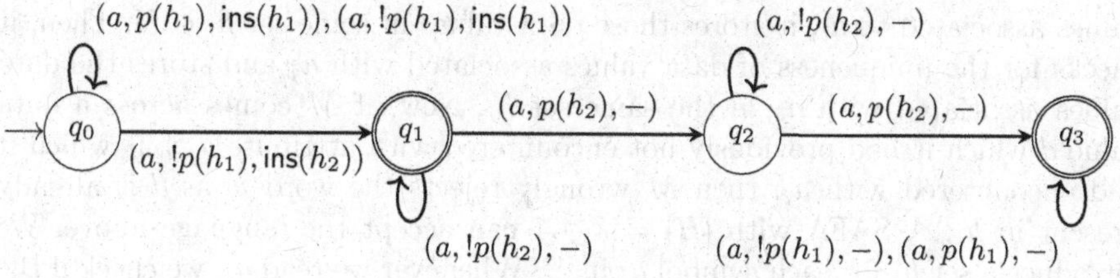

Fig. 2. SAFA for $L_{\exists \mathsf{cnt} \neq 2}$

a, it checks whether the corresponding data value is present in h_1. If present, it indicates it has already encountered this data value with attribute a before; therefore the automaton goes to q_1 which is a dead state and the input word is rejected. If the data value is not present in h_1, it implies that it has not encountered this value with a, thus it remains in q_0 and inserts the data value into h_1. Only if the automaton encounters a duplicate data value, it goes to q_1. If it does not encounter duplicate data values with respect to a in the input, the automaton remains in q_0 after consuming the entire word and it is accepted. □

Example 2. Consider the data language $L_{\exists \mathsf{cnt} \neq 2:}$ over the alphabet $\Sigma = \{a\}$: A nonempty word w is in the language if there exists a data value that appears n times in w with $n \neq 2$. This can be accepted by the nondeterministic SAFA in Fig. 2. At state q_0, the automaton nondeterministically guesses the data value that does not appear exactly twice and goes to state q_1. The automaton remains in state q_1 if the count of the data value is 1 or moves to state q_3 (via q_2) and remains there if the count of the data value is greater than 2. In both the cases, it accepts the input word if it can be consumed entirely. If the guess is incorrect, the data value appears twice, and it is in the nonaccepting state q_2 after consuming the input word. Thus, if a data word has all its data values that appear exactly twice, then all the runs end in q_2 and the input is rejected. □

The following theorem establishes a hierarchy of accepted languages by SAFA based on the size of H.

Theorem 1. *No SAFA* $M = (Q, \Sigma \times D, q_0, F, H, \delta)$ *with* $\Sigma = \{a_1, \ldots, a_{k+1}\}$, $|H| = k$ *can accept the language* $L = L_{\mathsf{fd(a_1)}} \cap \ldots L_{\mathsf{fd(a_{k+1})}} \cap L_{\mathsf{proj}_\Sigma(L) = a_1^* \cdots a_{k+1}^*}$.[1]

The language L comprises of all those data words w where the data values associated with each attribute a_i where $1 \leq i \leq k+1$ is distinct and the projection over Σ on w belongs to the language described by the regular expression $a_1^* \cdots a_{k+1}^*$. Let M be a SAFA with $|H| < |\Sigma|$ which tries to accept L. By pigeon hole principle, we can argue that when M reads a data word w, data values associated with at least two different attributes say a_i and a_j where $a_i, a_j \in \Sigma$, $i \neq j$ will be stored in the same set. When M checks for the uniqueness of data

[1] The language $L' = L_{\mathsf{fd(a_1)}} \cap \ldots L_{\mathsf{fd(a_{k+1})}}$ could have also been considered but the proof is relatively simpler if we instead consider L.

values associated to a_i, it stores those data values in some set $h_i \in H$. Then, it checks for the uniqueness of data values associated with a_j and stores the data values associated with a_j in the same set h_i. Now, if M comes across a data value d which it had previously not encountered with attribute a_j but which it had encountered with a_i, then M wrongly rejects the word w as d is already present in h_i. A SAFA with $|H| = k+1$ can accept the language above. We introduce a set h_i for each symbol a_i in Σ. Whenever we read a_i, we check if the corresponding data value is not present in h_i, and if so, we insert it in h_i and read the next data element; otherwise, we reject.

Let $\mathcal{L}_{\mathsf{SAFA}(|H|=k)}$ be the set of all languages accepted by SAFA with $|H| = k$. Since every SAFA $M = (Q, \Sigma \times D, q_0, F, H, \delta)$ with $|H| = k$ can be simulated by a SAFA $M' = (Q', \Sigma \times D, q_0, F', H', \delta')$ with $|H'| = \ell$ and $\ell > k$ by using $\ell - k$ dummy sets that are never used in an execution of M', we have the following.

Corollary 1. $\mathcal{L}_{\mathsf{SAFA}(|H|=k)} \subsetneq \mathcal{L}_{\mathsf{SAFA}(|H|=k+1)}$.

Corollary 1 shows that there is a strict hierarchy in terms of accepting capabilities of SAFA with respect to $|H|$.

4 Decision Problems and Closure Properties

We study the nonemptiness, membership problems and closure properties of SAFA.

4.1 Nonemptiness and Membership

We study the nonemptiness and the membership problems of SAFA and show that both are NP-complete. Given a SAFA M and an input word w, the *membership problem is to check if* $w \in L(M)$. Given a SAFA M, the *nonemptiness problem* is to check if $L(M) \neq \emptyset$. We start with the nonemptiness problem. To show the NP-membership, we first show that there exists a small run if the language accepted by a given SAFA is nonempty.

Lemma 1. *Every SAFA* $M = (Q, \Sigma \times D, q_0, F, H, \delta)$ *with* $L(M) \neq \emptyset$ *has a data word in* $L(M)$ *with an accepting run* ρ *such that* $|\rho| \leq |Q| \cdot (|H| + 2) - 1$.

Proof. We prove by contradiction. Assume that $L(M) \neq \emptyset$ and that for every $w \in L(M)$, for all accepting runs ρ of w, we have that $|\rho| > |Q| \cdot (|H| + 2) - 1 = |Q| \cdot (|H| + 1) + |Q| - 1$. We define an indicator function I_H which maps H to $\{0, 1\}^H$, where 0 corresponding to a set $h \in H$ denotes that h is empty while 1 denotes that h is nonempty. Since $|\rho| \geq |Q| \cdot (|H| + 2)$, the run ρ can be divided into $|H| + 2$ segments, each of length $|Q|$, that is, each segment is an infix over $|Q|$ transitions. By pigeon hole principle, in each segment, there exists a state $q \in Q$ that is visited more than once. Further, since there are $|H| + 2$ such segments, again by pigeon hole principle, there exists a segment and a state $q' \in Q$ such that q' is visited more than once in this segment and I_H does not change over the infix of the run between the two successive visits of q'. We now note that the sequence of transitions reading this infix y makes a loop over q', and thus y can be removed from w resulting into a word w' and the corresponding run is ρ' such that $|\rho'| < |\rho|$ and ρ' is accepting. Now there can be two cases if the suffix of ρ following reading y in w has a transition t with $p(h_i)$ and it reads a data value d.

- It may happen that the data value d was inserted along the infix y. Since I_H does not change while reading the infix y, it implies that the set h_i was nonempty even before the infix y was read. Let a data value d' was inserted into h_i while reading the prefix before y. Then w' may be modified to w'' so that the suffix following y in w'' reads the data value d' instead of d. Let the run corresponding to w'' be ρ'' that follows the same sequence of states as ρ'. Note that $|\rho''| = |\rho'| < |\rho|$, and that ρ'' is an accepting run.
- If while reading w, the transition t reads a data value d that was inserted while reading a prefix appearing before y, then w' does not need to be modified, and we thus have the accepting run ρ'.

Since ρ is an arbitrary accepting run of length $|Q|\cdot(|H|+2)$ or more, starting from ρ, we can remove infixes repeatedly and modify it as mentioned above if needed until we reach an accepting run of length strictly smaller than $|Q| \cdot (|H| + 2)$ without affecting acceptance, and hence the contradiction. □

Using Lemma 1 we get the following.

Lemma 2. *Nonemptiness problem for SAFA is in NP.*

Proof. Consider a SAFA M with $L(M) \neq \emptyset$. By Lemma 1, a Turing machine can nondeterministically guess a polynomial length accepting run, hence the result. □

We now show that the nonemptiness problem is NP-hard even for deterministic acyclic SAFA over an alphabet of size 3. Example 3 describes our construction.

Example 3. For a 3CNF formula $\phi = (x \vee \overline{y} \vee z) \wedge (x \vee y \vee z)$, the corresponding SAFA $M = (Q, \Sigma \times D, q_0, F, H, \delta)$ is shown in Fig. 3. We denote by $A(a,i)$ the transition $(a, !p(h_i), \mathsf{ins}(h_i))$ and by $T(a,i)$ the transition $(a, p(h_i), -)$ with $Q = \{q_0, q_x, q_y, q_z, q_{c_1}, q_{c_2}\}$, $\Sigma = \{a_1, a_2, a_3\}$, $D = \mathbb{N}$, $H = \{h_x, h_{\overline{x}}, h_y, h_{\overline{y}}, h_z, h_{\overline{z}}\}$, $F = \{q_{c_2}\}$. In particular, if there are ℓ variables in the formula, then $|H| = 2\ell$. □

Lemma 3. *The nonemptiness problem is* NP*-hard for deterministic acyclic SAFA over an alphabet of size* 3.

From Lemma 2 and Lemma 3, we have the following.

Theorem 2. *The nonemptiness problem for SAFA is* NP*-complete.*

We now show that for singleton SAFA, nonemptiness is NL-complete.

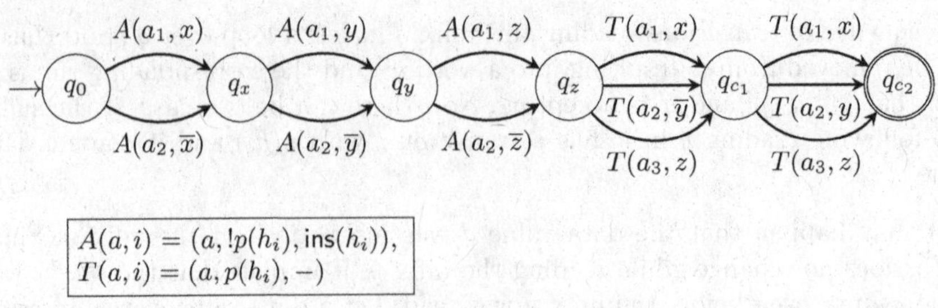

Fig. 3. The SAFA M corresponding to a 3CNF formula ϕ

Theorem 3. *The nonemptiness problem for singleton SAFA is* NL*-complete.*

Proof. We first discuss NL-membership. It can be easily shown that nonemptiness for singleton SAFA $M = (Q, \Sigma \times D, q_0, F, H, \delta)$ is in PTIME by reducing the problem to checking the nonemptiness of a nondeterministic finite automaton (NFA) with $2|Q|$ states. This NFA can be constructed on-the-fly leading to an NL-membership of the nonemptiness problem of singleton SAFA.

For NL-hardness, we show a reduction from the reachability problem on a directed graph G having vertex set $V = \{1, \ldots n\}$ which is known to be NL-complete [14]. Let G be a directed graph with $V = \{1, 2, ..., n\}$ and we are given the vertices 1 and n. We define a SAFA $M = (Q, \Sigma \times D, q_0, F, H, \delta)$ where $Q = V$, $\Sigma = \{a\}$, D is a countably infinite set, $q_0 = 1$, $F = \{n\}$, $H = \{h_1\}$ i.e. $|H| = 1$. The transitions in δ are as follows: $(i, a, !p(h_1), -, j) \in \delta$ if (i, j) is an edge in G. It is easy to see that G is reducible to M in NL and that $L(M) \neq \phi$ iff there is a path from vertex 1 to vertex n in G. Hence, the result. □

We now show that membership for SAFA is NP-complete. We first note that unlike the nonemptiness problem, for DSAFA, membership can be decided in PTIME by reading the input word and by checking if a final state is reached.

Theorem 4. *The membership problem for SAFA is* NP*-complete.*

Proof. Given a SAFA M and an input word w, if $w \in L(M)$, then a nondeterministic Turing machine can guess an accepting run in polynomial time and hence the membership problem is in NP.

For showing NP-hardness, we reduce from 3SAT for the nonemptiness problem as done for Lemma 3. Instead of a deterministic automaton that was constructed in the proof of Lemma 3, we construct a nondeterministic SAFA M with $\Sigma = \{a\}$ and all the transitions in M are labelled with the same letter $a \in \Sigma$. Everything else remains the same as in the construction in Lemma 3. Note that in the 3SAT formula ψ, if there are ℓ variables and k clauses, then there is a path of length $\ell + k$ from the initial state to the unique final state of M. We consider an input word $w = (a, d) \cdots (a, d)$, that is a word in which all the attribute, data value pairs are identical in the whole word such that $|w| = \ell + k$. It is not difficult to see that $w \in L(M)$ iff ψ is satisfiable. □

4.2 Closure Properties

We now study the closure properties of SAFA. Using a pumping argument, we show that these automata are not closed under complementation.

Lemma 4. *Let $L \in \mathcal{L}_{\mathsf{SAFA}}$. Then there exists a SAFA M with n states that accepts L such that every data word $w \in L$ of length at least n can be written as $w = xyz$ and $T_w = T_xT_yT_z$ corresponds to the sequence of transitions that M takes to accept w, where $T_x = t_{x_1} \cdots t_{x_{|x|}}$, $T_y = t_{y_1} \cdots t_{y_{|y|}}$, $T_z = t_{z_1} \cdots t_{z_{|z|}}$ is the sequence of transitions that M takes to read x, y, z respectively, and t_{u_j} denotes the jth transition of the transition sequence T_u with $u \in \{x, y, z\}$, satisfying the following:*

- $|y| \geq 1$
- $|xy| \leq n$
- *for all $\ell \geq 1$, for all words $w' = xyy'_1 \cdots y'_\ell z'$ such that $T_{w'} = T_xT_y{T_y}^\ell T_z$ is the sequence of transitions that M takes to accept w' and $\mathsf{proj}_\Sigma(y) = \mathsf{proj}_\Sigma(y'_1) = \cdots = \mathsf{proj}_\Sigma(y'_\ell)$, $\mathsf{proj}_\Sigma(z) = \mathsf{proj}_\Sigma(z')$.*
  - *if t_{y_j} has $p(h_i), h_i \in H$ then the jth datum of $\mathsf{proj}_\mathsf{D}(y'_k) \in h_i$, $1 \leq j \leq |y|$, $1 \leq k \leq \ell$.*
  - *if t_{y_j} has $!p(h_i), h_i \in H$ then the jth datum of $\mathsf{proj}_\mathsf{D}(y'_k) \notin h_i$, $1 \leq j \leq |y|$, $1 \leq k \leq \ell$.*
  - *if t_{z_j} has $p(h_i), h_i \in H$ then the jth datum of $\mathsf{proj}_\mathsf{D}(z') \in h_i$, $1 \leq j \leq |z|$.*
  - *if t_{z_j} has $!p(h_i), h_i \in H$, then the jth datum of $\mathsf{proj}_\mathsf{D}(z') \notin h_i$, $1 \leq j \leq |z|$.*
- $w' \in L$.

We now show the following.

Theorem 5. *SAFA are closed under union but not under complementation.*

Proof. SAFA are shown to be closed under union in a manner similar to NFA. To show SAFA are not closed under complementation, we first define the following functions. The function $\mathsf{cnt}(w', d)$ gives the number of times data value d is present in a data word w' and $\mathsf{uni}(w')$ gives the number of data values d with $\mathsf{cnt}(w', d) = 1$ in w'. We consider the language $L_{\exists \mathsf{cnt} \neq 2}$, which is the language of data words w where there exists a data value d such that $\mathsf{cnt}(w', d) \neq 2$. Example 2 shows a SAFA that accepts this.

Consider the complement language $L_{\forall \mathsf{cnt}=2}$ wherein all data values occur exactly twice. Using Lemma 4 we show no SAFA can accept $L_{\forall \mathsf{cnt}=2}$.

The proof is by contradiction. Suppose that there exists a SAFA M with n states accepting $L_{\forall \mathsf{cnt}=2}$. Let w be a data word such that $w \in L_{\forall \mathsf{cnt}=2}$ and $|w| = 2n$.

For every decomposition of w as $w = xyz$ and sequence $T_w = T_xT_yT_z$ of transitions that M takes to accept w with $|y| \geq 1$, we have a $w' = xyy_1y_2y_3z'$ such that $T_{w'} = T_xT_y{T_y}^3T_z$. Since $|y| \geq 1$, we have that T_y must have either a transition t with $p(h_i)$ for some $h_i \in H$ or a transition with $!p(h_i)$ for some $h_i \in H$ or both. If t has $p(h_i)$, then the first time t is executed while consuming y, assume that it consumes a data value d. It is able to consume the data value

d as it is already inserted in h_i before t is executed. Now if after consuming the word xy, T_y is executed again, then when executing the transition t it can again consume the same data value d as before. So, every time T_y is executed, the SAFA M will consume the data value d while executing the transition t. After executing T_y three times, the SAFA M executes the transition sequence T_z. All the transitions with $p(h_i)$ in T_z can be executed successfully with the same data value that they consumed when M accepted w because w' and w both have the same prefix xy. The transitions with $!p(h_i)$ in T_z consume data values that M has not encountered prior to executing these transitions. Thus, if T_y has a transition t with $p(h_i)$ for some $h_i \in H$, then M accepts the data word $w' = xyy_1 \cdots y_3 z'$ where there exists a data value d with $\mathsf{cnt}(w', d) > 3$.

If T_y has a transition, say t with $!p(h_i)$ for some $h_i \in H$, then every time T_y is executed after consuming xy, the SAFA M when executing t can always read a new data value which it has not read till executing t and that it will not read later. We can always find such data values as w' is finite whereas D is countably infinite. The sequence T_z is executed successfully due to same reasons as before. Thus, if T_y has a transition t with $!p(h_i)$, then M accepts a data word $w' = xyy_1 ... y_3 z'$ where $\mathsf{uni}(w') \geq 3$. Note that the data value consumed by M when taking the transition t while reading y may already be present in the prefix being read by the sequence of transitions prior to taking t. Therefore, $w' \notin L_{\forall \mathsf{cnt}=2}$. □

Note that from the above construction, we see that SAFA with $|H| \geq 2$ are not closed under complementation. We observe that singleton SAFA are closed under union but not under intersection. From the hierarchy theorem (Theorem 1), we see singleton SAFA cannot accept $L_{\mathsf{fd(a_1)}} \cap L_{\mathsf{fd(a_2)}} \cap L_{\mathsf{proj}_\Sigma(L)=a_1^* a_2^*}$ but singleton SAFA can accept $L_{\mathsf{fd(a_1)}} \cap L_{\mathsf{proj}_\Sigma(L)=a_1^* a_2^*}$ and $L_{\mathsf{fd(a_2)}} \cap L_{\mathsf{proj}_\Sigma(L)=a_1^* a_2^*}$. Thus, singleton SAFA are not closed under intersection, and hence also not closed under complementation. The question regarding intersection of SAFA is still open. Product automaton has most often been the tool used to establish the closure under intersection for most finite automata models. However, in case of SAFA, the product automaton construction is non-trivial as during the product construction, it is difficult to decide whether to keep the sets of both automata since SAFA does not allow update of multiple sets in a single transition. If the sets of only one automaton is kept, then we need to decide how to keep track of the data which the other automaton encounters in the product automaton simulation. However, we conjecture that SAFA are closed under intersection. Intuitively, given two SAFA M_1 and M_2 where M_1 accepts language L_1 and M_2 accepts language L_2, there exists a SAFA M accepting $L_1 \cap L_2$ that can recognise the patterns common to both L_1 and L_2.

Deterministic SAFA: Here we discuss deterministic SAFA and compare their expressiveness with SAFA. Using standard complementation construction as in deterministic finite automata (DFA), by changing non-accepting states to accepting and vice versa, we can show that DSAFA are closed under complementation. We now show that the class of languages accepted by DSAFA is strictly contained in the class of languages accepted by SAFA. Formally,

Theorem 6. $\mathcal{L}_{\mathsf{DSAFA}} \subsetneq \mathcal{L}_{\mathsf{SAFA}}$.

Proof. Recall from Example 2 that the language $L_{\exists\mathsf{cnt}\neq 2} \in \mathcal{L}_{\mathsf{SAFA}}$. On the other hand, we show in the proof of Theorem 5 that there does not exist a SAFA accepting its complement language $L_{\forall\mathsf{cnt}=2}$. This implies that the language $L_{\exists\mathsf{cnt}\neq 2}$ cannot be accepted by a DSAFA since DSAFA are closed under complementation. The result follows since every deterministic SAFA is a SAFA. □

5 Expressiveness

Let $\mathcal{L}_{\mathsf{KRFA}}$ and $\mathcal{L}_{\mathsf{CMA}}$ be the set of all languages accepted by k-register automata and CMA respectively. We compare the computational power of SAFA with k-register automata. We show that $\mathcal{L}_{\mathsf{SAFA}}$ and $\mathcal{L}_{\mathsf{KRFA}}$ are incomparable. Although $\mathcal{L}_{\mathsf{SAFA}}$ and $\mathcal{L}_{\mathsf{KRFA}}$ are incomparable, SAFA recognize many important languages which k-register automata also recognize such as L_{d_1}: wherein the first data value is repeated, $L_{a\geq 2}$: wherein attribute a is associated with more than two distinct data values. On the other hand, k-register automata fail to accept languages where we have to store more than k data values such as $L_{\mathsf{fd}(a)}$, $L_{\mathsf{even}(a)}$: wherein attribute a is associated with an even number of distinct data values [18]. SAFA can accept both these data languages. We also show below that there are languages such as L_d: the language of data words, each of which contains the data value d associated with some attribute at some position in the data word, that can be accepted by a 2-register automaton but not by SAFA.

Example 4. A 2-register automaton can accept the language L_d. Consider the 2-register automaton A with $\Sigma = \{a\}$, $Q = \{q_0, q_1\}$, $\tau_0 = \{d, \perp\}$, $F = \{q_1\}$, $U(q_0, a) = 2$, $U(q_0, b) = 2$, $U(q_1, a) = 2$, $U(q_1, b) = 2$. The transition relation δ is defined as: $\{(q_0, a, 1, q_1), (q_0, a, 2, q_0), (q_1, a, 1, q_1), (q_1, a, 2, q_1)\}$. The automaton A accepts L_d. For an input word w, the automaton checks whether the current data value of w under the head of A is equal to the content of register 1, which holds the data value d from the time of initialization of the registers. If it is equal, the automaton goes to state q_1 and consumes the word. Since q_1 is a final state, the word is accepted. If the data value d is not present in w, the automaton remains in state q_0 and rejects the input word. □

Theorem 7. *$\mathcal{L}_{\mathsf{SAFA}}$ and $\mathcal{L}_{\mathsf{KRFA}}$ are incomparable.*

Proof. We first show by contradiction that no SAFA can accept L_d. Suppose there exists a SAFA $M = (Q, \Sigma \times D, q_0, F, H, \delta)$ which accepts L_d. Now consider a word $w \in L_d$ where the data value d has occurred at the first position only. Let the sequence of transitions that M goes through to accept w be T_w. The first transition t_1 that T_w executes cannot be a transition with $p(h_i)$, $h_i \in H$ because t_1 is the first transition of T_w and there cannot be any insertion to set h_i prior to it. Therefore, t_1 is a transition with $!p(h_i)$, $h_i \in H$ and so t_1 can consume any other data value d' which is not present in w. As d does not occur anywhere else in w, it is safe to say that all other transitions in T_w with $p(h_i)$,

$h_i \in H$ have data values other than d present in their respective sets. Thus, if M accepts $w = (a, d)x$, where $x \in (\Sigma \times D)^*$ and x does not have value d in it, then M also accepts $w' = (a, d')x$, and w' does not have data value d in it, which is a contradiction. From Example 4, Example 1 and the fact that k-register automata cannot accept $L_{\mathsf{fd(a)}}$ [18] and the above, we conclude $\mathcal{L}_{\mathsf{SAFA}}$ and $\mathcal{L}_{\mathsf{KRFA}}$ are incomparable. □

Also we note that both SAFA and k-register automata have the same complexity for the nonemptiness and the membership problems.

Similar to SAFA, CCA and CMA also accept data languages such as $L_{\mathsf{fd(a)}}$, $L_{\mathsf{even}(a)}$ which k-register automata cannot, but their decision problems have higher complexity [2,17]. Further, CCA and CMA need to store a large number of values as both store every distinct data value in a given word and thus often need to store more number of data values than k-register automata or SAFA. Languages recognized by k-register automata need to store at most k values at any given instant. When SAFA recognize a language $L \in \mathcal{L}_{\mathsf{KRFA}}$ such as $L_{\exists\mathsf{cnt}\neq 2}$, they store much fewer number of data values than CCA, CMA as they selectively insert data values in the sets. For $L_{\exists\mathsf{cnt}\neq 2}$, there exists a SAFA that stores only 2 data values.

We can show that the class of languages accepted by SAFA is strictly contained in the class of languages accepted by CMA.

Theorem 8. $\mathcal{L}_{\mathsf{SAFA}} \subsetneq \mathcal{L}_{\mathsf{CMA}}$

Proof Sketch. A state of a CMA A simultaneously keeps track of the state of the SAFA M that it wants to simulate and also to which sets a particular data value has been inserted in M. Since both the number of states and the number of sets in SAFA are finite, therefore, the number of states in the CMA A is also finite. The number of states of the CMA simulating a SAFA is exponential in the size of the SAFA since a state of the CMA keeps the information of a subset of H storing a data value. Further, one can show that a CMA can accept the language $L_{\forall\mathsf{cnt}=2}$, and from Theorem 5, we know that no SAFA can accept $L_{\forall\mathsf{cnt}=2}$. □

6 Conclusion

In this paper, we introduce set augmented finite automata which use a finite set of sets of data values to accept data languages. We have shown examples of several data languages that can be accepted by our model. We compare the language acceptance capabilities of SAFA with k-register automata and CMA. The computational power and low complexity of nonemptiness and membership of SAFA makes it an useful tool for modeling and analysis of data languages. We believe our model is robust enough and can also be extended to infinite words. This model opens up some interesting avenues for future research. We would like to explore augmentations of SAFA with Boolean combinations of tests and the feature of updating multiple sets simultaneously. This may lead to having a more well-behaved model with respect to closure properties. SAFA with set initialization also form an interesting model and we would like to examine if the class of languages accepted by k-register automata is a proper subset of the class of languages accepted by such initialized SAFA. Further, we have left several open problems, for example, closure properties of deterministic SAFA, complexity of the universality problem, language containment and equivalence. We would also like to compare our model with other data automata models.

Acknowledgements. We thank Amaldev Manuel for providing useful comments on a preliminary version of this paper.

References

1. Autebert, J.M., Beauquier, J., Boasson, L.: Langages sur des alphabets infinis. Discret. Appl. Math. **2**(1), 1–20 (1980)
2. Björklund, H., Schwentick, T.: On notions of regularity for data languages. Theoret. Comput. Sci. **411**(4–5), 702–715 (2010)
3. Bojańczyk, M., Muscholl, A., Schwentick, T., Segoufin, L.: Two-variable logic on data trees and xml reasoning. J. ACM **56**(3) (2009). https://doi.org/10.1145/1516512.1516515
4. Bojanczyk, M., Muscholl, A., Schwentick, T., Segoufin, L., David, C.: Two-variable logic on words with data. In: 21st Annual IEEE Symposium on Logic in Computer Science (LICS 2006), pp. 7–16. IEEE (2006)
5. Bollig, B.: An automaton over data words that captures EMSO logic. In: Katoen, J.-P., König, B. (eds.) CONCUR 2011. LNCS, vol. 6901, pp. 171–186. Springer, Heidelberg (2011). https://doi.org/10.1007/978-3-642-23217-6_12
6. Cheng, E.Y., Kaminski, M.: Context-free languages over infinite alphabets. Acta Informatica **35**(3), 245–267 (1998)
7. Choffrut, C.: On relations of finite words over infinite alphabets. In: Dömösi, P., Iván, S. (eds.) Proceedings of 13th International Conference on Automata and Formal Languages, AFL 2011, Debrecen, Hungary, 17–22 August 2011, pp. 25–27 (2011)
8. Czerwiński, W., Lasota, S., Lazić, R., Leroux, J., Mazowiecki, F.: The reachability problem for petri nets is not elementary. In: Proceedings of the 51st Annual ACM SIGACT Symposium on Theory of Computing, STOC 2019, pp. 24–33. Association for Computing Machinery, New York (2019). https://doi.org/10.1145/3313276.3316369
9. Dassow, J., Vaszil, G.: P finite automata and regular languages over countably infinite alphabets. In: Hoogeboom, H.J., Păun, G., Rozenberg, G., Salomaa, A. (eds.) WMC 2006. LNCS, vol. 4361, pp. 367–381. Springer, Heidelberg (2006). https://doi.org/10.1007/11963516_23
10. Demri, S., Lazić, R.: LTL with the freeze quantifier and register automata. ACM Trans. Comput. Logic (TOCL) **10**(3), 1–30 (2009)
11. Figueira, D.: Alternating register automata on finite words and trees. Logical Methods Comput. Sci. **8** (2012)
12. Grumberg, O., Kupferman, O., Sheinvald, S.: Variable automata over infinite alphabets. In: Dediu, A.-H., Fernau, H., Martín-Vide, C. (eds.) LATA 2010. LNCS, vol. 6031, pp. 561–572. Springer, Heidelberg (2010). https://doi.org/10.1007/978-3-642-13089-2_47

13. Iosif, R., Xu, X.: Abstraction refinement for emptiness checking of alternating data automata. In: Beyer, D., Huisman, M. (eds.) TACAS 2018. LNCS, vol. 10806, pp. 93–111. Springer, Cham (2018). https://doi.org/10.1007/978-3-319-89963-3_6
14. Jones, N.D.: Space-bounded reducibility among combinatorial problems. J. Comput. Syst. Sci. **11**(1), 68–85 (1975)
15. Kaminski, M., Francez, N.: Finite-memory automata. Theoret. Comput. Sci. **134**(2), 329–363 (1994)
16. Kaminski, M., Tan, T.: Regular expressions for languages over infinite alphabets. Fund. Inform. **69**(3), 301–318 (2006)
17. Manuel, A., Ramanujam, R.: Class counting automata on datawords. Int. J. Found. Comput. Sci. **22**(04), 863–882 (2011)
18. Manuel, A., Ramanujam, R.: Automata over infinite alphabets. In: Modern Applications of Automata Theory, pp. 529–553. World Scientific (2012)
19. Neven, F.: Automata, logic, and XML. In: Bradfield, J. (ed.) CSL 2002. LNCS, vol. 2471, pp. 2–26. Springer, Heidelberg (2002). https://doi.org/10.1007/3-540-45793-3_2
20. Neven, F., Schwentick, T., Vianu, V.: Finite state machines for strings over infinite alphabets. ACM Trans. Comput. Logic (TOCL) **5**(3), 403–435 (2004)
21. Sakamoto, H., Ikeda, D.: Intractability of decision problems for finite-memory automata. Theoret. Comput. Sci. **231**(2), 297–308 (2000)
22. Tan, T.: On pebble automata for data languages with decidable emptiness problem. J. Comput. Syst. Sci. **76**(8), 778–791 (2010)

Fast Detection of Specific Fragments Against a Set of Sequences

Marie-Pierre Béal(✉) and Maxime Crochemore

Univ. Gustave Eiffel, CNRS, LIGM, Marne-la-Vallée 77454 Paris, France
{marie-pierre.beal,maxime.crochemore}@univ-eiffel.fr

Abstract. We design alignment-free techniques for comparing a sequence or word, called a target, against a set of words, called a reference. A target-specific factor of a target T against a reference R is a factor w of a word in T which is not a factor of a word of R and such that any proper factor of w is a factor of a word of R. We first address the computation of the set of target-specific factors of a target T against a reference R, where T and R are finite sets of sequences. The result is the construction of an automaton accepting the set of all considered target-specific factors. The construction algorithm runs in linear time according to the size of $T \cup R$. The second result consists of the design of an algorithm to compute all the occurrences in a single sequence T of its target-specific factors against a reference R. The algorithm runs in real-time on the target sequence, independently of the number of occurrences of target-specific factors.

Keywords: Specific word · Minimal forbidden word · Suffix automaton

1 Introduction

The goal of this article is to design an alignment-free technique for comparing a sequence or word, called a target, against a set of words, called a reference.

The motivation comes from the analysis of genomic sequences as done for example by Khorsand et al. in [15] in which authors introduce the notion of sample-specific strings. To avoid alignments but to extract interesting elements that differentiate the target from the reference, the chosen specific fragments are minimal forbidden factors, also called minimal absent factors. Target-specific words are factors of the target that are minimal forbidden factors of the reference. These types of factors have already been applied to compare efficiently sequences (see for example [8] and references therein), to build phylogenies of biological molecular sequences using a distance based on absent words (see [6,7],...), to discover remarkable patterns in some genomic sequences (see for example [21]) and to improve pattern matching methods (see [11]...), to quote only a few applications. In bioinformatics target-specific words act as signatures for newly sequenced biological molecules and help find their characteristics.

The notion of minimal absent factors was introduced by Mignosi et al. [18] (see also [2]) in relation to combinatorial aspects of some sequences. It has then

F. Drewes and M. Volkov (Eds.): DLT 2023, LNCS 13911, pp. 51–60, 2023.
https://doi.org/10.1007/978-3-031-33264-7_5

been extended to regular languages in [1], which obviously applies to a finite set of (finite) sequences. The first linear-time computation is described in [12] (see also [10]) and, due to the important role of the notion, the efficient computation of minimal forbidden factors has attracted quite a lot of works (see for example [20] and references therein).

In the article, we continue exploring the approach of target-specific words as done in [15] by introducing new other algorithmic techniques to detect them. See also the more general view on the usefulness of formal languages to analyze several genomes using pangenomics graphs by Bonizzoni et al. in [5].

The Results. First, we address the computation of the set of target-specific factors of a target T against a reference R, where T and R are finite sets of sequences. The result is the construction of an automaton accepting the set of all considered target-specific factors. The construction algorithm runs in linear time according to the size of $T \cup R$.

The second result consists of the design of an algorithm to compute all the occurrences in a single sequence T of its target-specific factors against a reference R. The algorithm runs in real-time on the target sequence, independently of the number of occurrences of target-specific factors, after a standard processing of the reference. This improves on the result in [15], where the running time of the main algorithm depends on the number of occurrences of sought factors.

The design of both algorithms uses the notion of suffix links that are used for building efficiently indexing data structures, like suffix trees (see [14]) and DAWGs also called suffix automata (see [3, 10]). The links can also be simulated with suffix arrays [17] and their implementations, for example, the FM-index [13]. The algorithm in [15] uses the FMD index by Li [16]. All these data structures can accommodate the sequences and their reverse complements.

Definitions. Let A be a finite alphabet and A^* be the set of the finite words drawn from the alphabet A, including the empty word ε. A *factor* of a word $u \in A^*$ is a word $v \in A^*$ that satisfies $u = wvt$ for some words $w, t \in A^*$. A *proper factor* of a word u is a factor distinct from the whole word. If P is a set of words, we denote by $\text{Fact}(P)$ the set of factors of words in P, and, if P is finite, $\text{size}(P)$ denotes the sum of lengths of the words in P.

A *minimal forbidden word* (also called a minimal absent word) for a given set of words $L \subseteq A^*$ with respect to a given alphabet B containing A is a word of B^* that does not belong to L but that all proper factors do.

Let R, T be two sets of finite words. A *T-specific word with respect to R* is a word u for which: u is a factor of a word of T, u is not a factor of a word in R and any proper factor of u is a factor of a word in R. The set R is called the *reference* and T the *target* of the problem.

Note that a word is a T-specific word with respect to R if and only if it is a minimal forbidden word of $\text{Fact}(R)$ with respect to the alphabet of letters occurring in $R \cup T$ and is also in $\text{Fact}(T)$. As a consequence, the set of T-specific words with respect to R is both prefix-free and suffix-free.

It follows from the definition that the set S of T-specific words with respect to R is:

$$A\,\mathrm{Fact}(R) \cap \mathrm{Fact}(R)A \cap (A^* - \mathrm{Fact}(R)) \cap \mathrm{Fact}(T),$$

where A is the alphabet of letters of words R and T. It is thus a regular set when R and T are regular, in particular when R and T are finite.

A *finite deterministic automaton* is denoted by $\mathcal{A} = (Q, A, i, F, \delta)$ where A is a finite alphabet, Q is a finite set of states, $i \in Q$ is the unique initial state, $F \subseteq Q$ is the set of final states and δ is the partial function from $Q \times A$ to Q representing the transitions of the automaton. The partial function δ extends to $Q \times A^*$ and a word u is accepted by $\mathcal{A}$ if and only if $\delta(i, u)$ is defined and belongs to F.

2 Background: Directed Acyclic Word Graph

In this section, we recall the definition and the construction of the directed acyclic word graph of a finite set of words. This description already appears in [1].

Let $P = \{x_1, x_2, \ldots, x_r\}$ be a finite set of words of size r. A linear-time construction of a deterministic finite state automaton recognizing $\mathrm{Fact}(P)$ has been obtained by Blumer *et al.* in [3,4], see also [19]. Their construction is an extension of the well-known incremental construction of the suffix automaton of a single word (see for instance [9,10]). The words are added one by one to the automaton. In the sequel, we call this algorithm the DAWG algorithm since it outputs a deterministic automaton called a *directed acyclic word graph.* Let us denote by $DAWG(P) = (Q, A, i, Q, \delta)$ this automaton. Let $\mathit{Suff}(v)$ denote the set of suffixes of a word v and $\mathit{Suff}(P)$ the union of all $\mathit{Suff}(v)$ for $v \in P$. The states of $DAWG(P)$ are the equivalence classes of the right invariant equivalence $\equiv_{\mathit{Suff}(P)}$ defined as follows. If $u, v \in \mathrm{Fact}(P)$,

$$u \equiv_{\mathit{Suff}(P)} v \text{ iff } \forall i 1 \le i \le r \text{ and } \quad u^{-1}\mathit{Suff}(x_i) = v^{-1}\mathit{Suff}(x_i).$$

and there is a transition labeled by a from the class of a word u to the class of ua. The automaton $DAWG(P)$ has a unique initial state, which is the class of the empty word, and all its states are final. Note that the syntactic congruence $\sim$ defining the minimal automaton of the language is

$$u \sim v \text{ iff } \bigcup_{i=1}^{r} u^{-1}\mathit{Suff}(x_i) = \bigcup_{i=1}^{r} v^{-1}\mathit{Suff}(x_i)$$

and is not the same as the above equivalence. In other words, $DAWG(P)$ is not always a minimal automaton.

The construction of $DAWG(P)$ is performed in time $O(\mathrm{size}(P) \times \log |A|)$. A time complexity of $O(\mathrm{size}(P))$ can be obtained with an implementation of automata with sparse matrices (see [10]).

Example 1. The deterministic acyclic word graph obtained with the DAWG algorithm from $P = \{abbab, abaab\}$ is displayed in Fig. 1 where dashed edges represent the suffix links. Note that this deterministic automaton is not minimal since states 3 and 7, 5 and 9, and 6 and 10 can be merged pairwise.

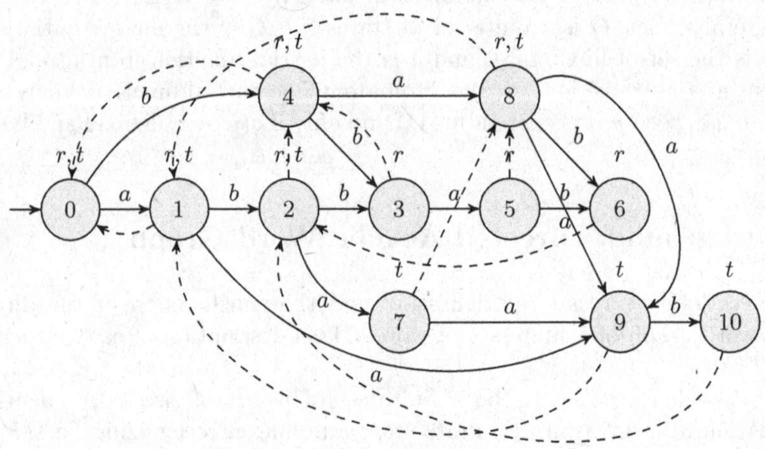

Fig. 1. Automaton $DAWG(P)$ for $P = \{abbab, abaab\}$. Marks r, t above states are defined in Sect. 3

We let s denote the suffix link function associated with $DAWG(P)$. We first define the function s' from $\text{Fact}(P) \setminus \{\varepsilon\}$ to $\text{Fact}(P)$ as follows: for $v \in \text{Fact}(P) \setminus \{\varepsilon\}$, $s'(v)$ is the longest word $u \in \text{Fact}(P)$ that is a suffix of v and for which $u \not\equiv_{Suff(P)} v$. Then, if $p = \delta(i, v)$, $s(p)$ is the state $\delta(i, s'(v))$.

3 Computing the Set of T-specific Words

In this section, we assume that the reference R and the target T are two finite sets of words and our goal is to compute the set of T-specific factors of T against R. To do so, We first compute the directed acyclic word graph $DAWG(R \cup T) = (Q, A, i, Q, \delta)$ of $R \cup T$. Further, we compute a table mark indexed by the set of states Q that satisfies: for each state p in Q, mark$[p]$ is one of the three values r, t or both r, t according to the fact that each word labeling a path from i to q is a factor of some word in R and not of a word in T, or is a factor of a word in T and not of a word in R, or is a factor of a word in R and of a word in T. This information can be obtained during the construction of the directed acyclic word graph without increasing the time and space complexity.

The following algorithm outputs a trie (digital tree) of the set of T-specific words with respect to T and R.

```
SPECIFIC-TRIE((Q, A, i, Q, δ) DAWG of (R ∪ T), s its suffix link)
  for each p ∈ Q with mark[p] = r, t in width-first search from i
      and for each a ∈ A do
      if (δ(p, a) defined and mark[δ(p, a)] = t) and ((p = i) or
          (δ(s(p), a) defined and mark[δ(s(p), a)] = r or r, t)) then
          δ′(p, a) ← new sink
      else if (δ(p, a) = q with mark[q] = r, t)
              and (q not already reached) then
          δ′(p, a) ← q
  return 𝒜, the automaton (Q, A, i, {sinks}, δ′)
```

Example 2. The automaton $DAWG(R \cup T)$ with the input $R = \{abbab\}, T = \{abaab\}$ is shown in Fig. 1. The output of algorithm SPECIFIC-TRIE on $DAWG$ $(R \cup T)$ is shown in Fig. 2 where the squares are final or sink states of the trie. The set of T-specific words with respect to R is $\{aa, aba\}$.

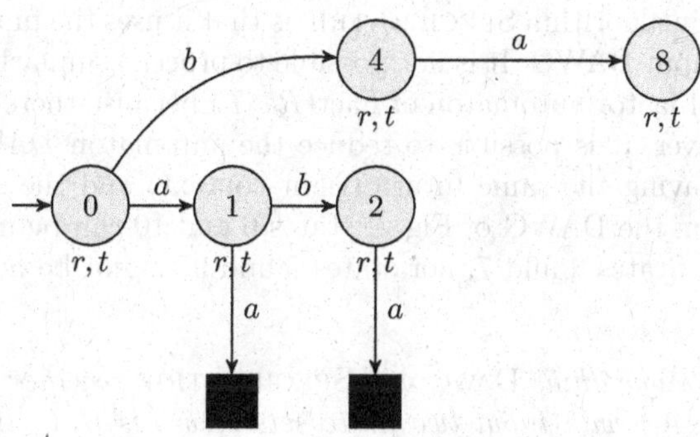

Fig. 2. The trie of T-specific words with respect to R.

Proposition 1. *Let $DAWG(R \cup T)$ be the output of algorithm* DAWG *on the finite set of words $R \cup T$, let s be its suffix function, and let* mark *be the table defined as above. Algorithm* SPECIFIC-TRIE *builds the trie recognizing the set of T-specific words with respect to R.*

Proof. Let S be the set of of T-specific words with respect to R.

Consider a word ua ($a \in A$) accepted by $\mathcal{A}$. Note that $\mathcal{A}$ accepts only nonempty words. Let $p = \delta'(i, u)$. Since the $DAWG$ automaton is processed with a width-first search, u is the shortest word for which $\delta(i, u) = p$. Therefore, if $u = bv$ with $b \in A$, we have $\delta(i, v) = s(p)$ by definition of the suffix function s. When the test "($\delta(p, a)$ defined and mark$[\delta(p, a)] = t$) and ($\delta(s(p), a)$ defined and mark$[\delta(s(p), a)] = r$ or r, t)" is satisfied, this implies that $va \in \mathrm{Fact}(R)$. Thus, $bva \notin \mathrm{Fact}(R)$, while $bv, va \in \mathrm{Fact}(R)$ and $bva \in \mathrm{Fact}(T)$. So, ua is a T-specific word with respect to R. If u is the empty word, then $p = i$. The transition from i to the sink labeled by a is created under the condition "$\delta(p, a)$ defined and mark$[\delta(p, a)] = t$", which means that $a \in \mathrm{Fact}(T)$. The word a is again a T-specific word with respect to R. Thus the words accepted by $\mathcal{A}$ are T-specific words with respect to R.

Conversely, let $ua \in S$. If u is the empty word, this means that a does not occur in $\mathrm{Fact}(R)$ and occurs in $\mathrm{Fact}(T)$ therefore there is a transition labeled by a from i in $DAWG(R \cup T)$ to a state marked t. Thus a transition from i to a sink state in $\mathcal{A}$ created Line 3 and a is accepted by $\mathcal{A}$. Now assume that $u = bv$. The word u is in $\mathrm{Fact}(R)$. So let $p = \delta(i, u)$. Note that u is the shortest word for which $p = \delta(i, u)$, because all such words are suffixes of each other in the *DAWG* automaton. The word ua is not in $\mathrm{Fact}(R)$ and is in $\mathrm{Fact}(T)$, so the condition "$\delta(p, a)$ defined and mark$[p, a] = t$" is satisfied. Let $q = s(p)$. We have $q = \delta(i, v)$ because of the minimality of the length of u and the definition of s. Since va is in $\mathrm{Fact}(R)$, the condition "$\delta(s(p), a)$ defined and mark$[\delta(s(p), a)] = r$ or r, t" at Line 2 is satisfied which yields the creation of a transition at Line 3 to make $\mathcal{A}$ accept ua as wanted.

A main point in algorithm SPECIFIC-TRIE is that it uses the function s defined on states of the input DAWG. It is not possible to proceed similarly when considering the minimal factor automaton of $\mathrm{Fact}(R \cup T)$ because there is no analogue function s. However, it is possible to reduce the automaton $DAWG(R \cup T)$ by merging states having the same future (right context) and the same image by s. For example, on the DAWG of Fig. 1, states 6 and 10 can be merged because $s(6) = s(10) = 2$. States 3 and 7, nor states 5 and 9 cannot be merged with the same argument.

Proposition 2. *Algorithms* DAWG *and* SPECIFIC-TRIE *together run in time* $O(\mathrm{size}(R \cup T) \times |A|)$ *with input two finite sets of words* R, T, *if the transition functions are implemented by transition matrices.*

If P is a set of words, we denote by A_P the set of letters occurring in P.

Proposition 3. *Let* R, T *be two finite sets of words. The number of* T*-specific words with respect to* R *is no more than* $(2\,\mathrm{size}(R) - 2)(|A_R| - 1) + |A_T \setminus A_R| - |A_R| + m$, *if* $\mathrm{size}(R) > 1$, *where* m *the number of words in* R. *The bound becomes* $|A_T \setminus A_R|$ *when* $\mathrm{size}(R) \leq 1$.

Proof. We let S denote the set of T-specific words with respect to R. Since S is included in the set of minimal forbidden words of $\mathrm{Fact}(R)$ with respect to the alphabet $A = A_R \cup A_T$, the bound comes from [1, Corollary 4.1].

4 Computing Occurrences of Target-Specific Factors: The T-specific table

In this section, we consider that R and T are just words. The goal of the section is to design an algorithm that computes all the occurrences of T-specific words in T. To do so, we define the T-specific table associated with the pair R, T of words of the problem.

A letter of T at position k is denoted by $T[k]$ and $T[i\,..\,j]$ denotes the factor $T[i]T[i+1]\cdots T[j]$ of T. Then, the T-specific table Ts is defined, for $i = 0, \ldots, |T|-1$, by

$$Ts[i] = \begin{cases} j, & \text{if } T[i\,..\,j] \text{ is } T\text{-specific}, i \leq j, \\ -1, & \text{else.} \end{cases}$$

Note 1. Since the set of T-specific factors is both prefix-free and suffix-free, for each position k on T there is at most one T-specific factor of T starting at k and for each position j on T there is at most one T-specific factor of T ending at j.

Note 2. Instead of computing the T-specific table Ts, in a straightforward way, the algorithm below can be transformed to compute the list of pairs (i, j) of positions on T for which $Ts[i] = j$ and $j \neq -1$.

To compute the table we use $\mathcal{R}$, the suffix automaton of R, with its transition function δ and equipped with both the suffix link s (used here as a failure link) and the length function ℓ defined on states by: $\ell[p] = \max\{z \in A^* \mid \delta(i, z) = p\}$. Functions s and ℓ transform the automaton into a search machine, see [10, Section 6.6].

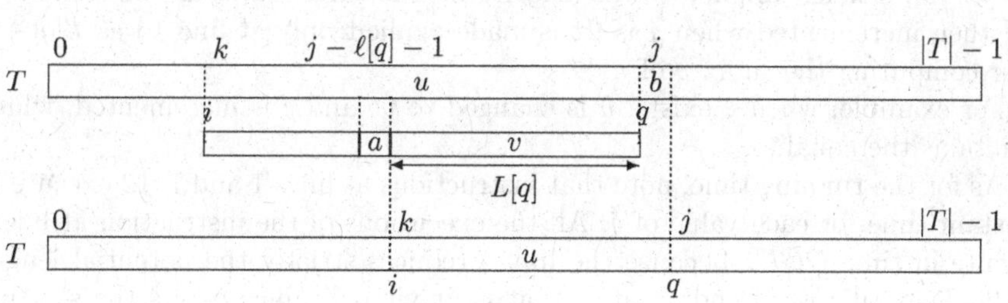

Fig. 3. A T-specific word found: when $u \in \mathrm{Fact}(R)$ and $ub \notin \mathrm{Fact}(R)$, either avb or b is a T-specific factor with respect to R (a, b are letters).

Figure 3 illustrates the principle of Algorithm TsTable. Let us assume the factor $u = T[k\,..\,j-1]$ is a factor of R but ub is not for some letter b. Then, let v be the longest suffix of u for which vb is a factor of R. If it exists, then clearly avb, with a letter preceding v, is T-specific. Indeed, $av, vb \in \mathrm{Fact}(R)$ and $avb \notin \mathrm{Fact}(R)$, which means that avb is a minimal forbidden word of R while occurring in T. Therefore, setting $q = \delta(i, u)$, $Ts[j - \ell[q] - 1] = j$ since $\ell[q] = |v|$

due to a property of the DAWG of $\mathcal{R}$. If there is no suffix of u satisfying the condition, the letter b alone is T-specific and $Ts[j] = j$.

```
TsTable(T target word, R DAWG(R), i initial(R))
 1  (q, j) ← (i, 0)
 2  while j < |T| do
 3      Ts[j] ← −1
 4      if δ(q, T[j]) undefined then
 5          while q ≠ i and δ(q, T[j]) undefined do
 6              q ← s[q]
 7          if δ(q, T[j]) undefined then  ▷ q = i
 8              Ts[j] ← j
 9              j ← j + 1
10          else Ts[j − ℓ[q] − 1] ← j
11              (q, j) ← (δ(q, T[j]), j + 1)
12      else (q, j) ← (δ(q, T[j]), j + 1)
13  return Ts
```

Theorem 1. *The DAWG of the reference set R of words being preprocessed, applied to a word T, Algorithm* TsTable *computes its T-specific table with respect to R and runs in linear time, i.e. $O(|T|)$ on a fixed-size alphabet.*

Proof. The algorithm implements the ideas detailed above. A more formal proof relies on the invariant of the while loop: $q = \delta(i, u)$, where i is the initial state of the suffix automaton of R and $u = T[k \,..\, j]$ for a position $k \leq j$. Since $k = j - |u|$, it is left implicit in the algorithm. The length $|u|$ could be computed and then incremented when j is. It is made explicit only at line 10 as $L[q] + 1$ after computing the suffix v of u.

For example, when v exists, u is changed to vb and j is incremented, which maintains the equality.

As for the running time, note that instructions at lines 1 and 7–12 execute in constant time for each value of j. All the executions of the instruction at line 6 execute in time $O(|T|)$ because the link s reduces strictly the potential length of the T-specific word ending at j, that is, it virtually increments the starting position of v in the picture.

Thus the whole execution is done in time $O(|T|)$.

Algorithm TsTable can be improved to run in real-time on a fixed-size alphabet. This is done by optimizing the suffix link s defined on the automaton $\mathcal{R}$. To do so, let us define, for each state q of $\mathcal{R}$,

$$Out(q) = \{a \mid \delta(q, a) \text{ defined for letter } a\}.$$

Then, the optimised suffix link G is defined by $G[\mathit{initial}(\mathcal{R})] = \mathbf{nil}$ and, for any other state q of $\mathcal{R}$, by

$$G[q] = \begin{cases} s[q], & \text{if } \mathit{Out}(q) \subset \mathit{Out}(s[q]), \\ G[s[q]], & \text{else.} \end{cases}$$

Note that, since we always have $\mathit{Out}(q) \subseteq \mathit{Out}(s[q])$, the definition of G can be reformulated as

$$G[q] = \begin{cases} s[q], & \text{if } \deg(q) < \deg(s[q]), \\ G[s[q]], & \text{else,} \end{cases}$$

where deg is the outgoing degree of a state. Therefore, its computation can be realized in linear time with respect to the number of states of $\mathcal{R}$. After substituting G for s in Algorithm TsTable, when the alphabet is of size α the instruction at line 6 executes no more than α times for each value of q. So the time to process a given state q is constant. This is summarized in the next corollary.

Corollary 1. *When using the optimized suffix link, Algorithm* TsTable *runs in real-time on a fixed-size alphabet.*

On a more general alphabet of size α, the processing of a given state of the automaton can be done in time $\log \alpha$.

References

1. Béal, M., Crochemore, M., Mignosi, F., Restivo, A., Sciortino, M.: Computing forbidden words of regular languages. Fundam. Informaticae **56**(1–2), 121–135 (2003)
2. Béal, M., Mignosi, F., Restivo, A., Sciortino, M.: Forbidden words in symbolic dynamics. Adv. Appl. Math. **25**(2), 163–193 (2000)
3. Blumer, A., Blumer, J., Ehrenfeucht, A., Haussler, D., McConnell, R.: Building the minimal DFA for the set of all subwords of a word on-line in linear time. In: Paredaens, J. (ed.) ICALP 1984. LNCS, vol. 172, pp. 109–118. Springer, Heidelberg (1984). https://doi.org/10.1007/3-540-13345-3_9
4. Blumer, A., Blumer, J., Haussler, D., McConnell, R., Ehrenfeucht, A.: Complete inverted files for efficient text retrieval and analysis. J. ACM **34**(3), 578–595 (1987)
5. Bonizzoni, P., Felice, C.D., Pirola, Y., Rizzi, R., Zaccagnino, R., Zizza, R.: Can formal languages help pangenomics to represent and analyze multiple genomes? In Diekert, V., Volkov, M.V. (eds.) Developments in Language Theory - 26th International Conference, DLT 2022, Tampa, FL, USA, May 9–13, 2022, Proceedings, volume 13257, LNCS, pp. 3–12. Springer, Cham (2022). https://doi.org/10.1007/978-3-031-05578-2_1
6. Castiglione, G., Gao, J., Mantaci, S., Restivo, A.: A new distance based on minimal absent words and applications to biological sequences. CoRR, abs/2105.14990 (2021)
7. Chairungsee, S., Crochemore, M.: Using minimal absent words to build phylogeny. Theor. Comput. Sci. **450**, 109–116 (2012)

8. Charalampopoulos, P., Crochemore, M., Fici, G., Mercas, R., Pissis, S.P.: Alignment-free sequence comparison using absent words. Inf. Comput. **262**, 57–68 (2018)
9. Crochemore, M.: Transducers and repetitions. Theoret. Comput. Sci. **45**(1), 63–86 (1986)
10. Crochemore, M., Hancart, C., Lecroq, T.: Algorithms on Strings. Cambridge University Press, 392p (2007)
11. Crochemore, M., et al.: Ramusat. Absent words in a sliding window with applications. Inf. Comput. **270** (2020)
12. Crochemore, M., Mignosi, F., Restivo, A.: Automata and forbidden words. Inf. Process. Lett. **67**(3), 111–117 (1998)
13. Ferragina, P., Manzini, G.: Opportunistic data structures with applications. In: 41st Annual Symposium on Foundations of Computer Science, FOCS 2000, 12–14 November 2000, Redondo Beach, California, USA, pp. 390–398. IEEE Computer Society (2000)
14. Gusfield, D.: Algorithms on Strings, Trees, and Sequences - Computer Science and Computational Biology. Cambridge University Press (1997)
15. P. Khorsand, L. Denti, H. G. S. V. Consortium, P. Bonizzoni, R. Chikhi, F. Hormozdiari: Comparative genome analysis using sample-specific string detection in accurate long reads. Bioinform. Adv. **1**(1), 05 (2021)
16. Li, H.: Exploring single-sample SNP and INDEL calling with whole-genome de novo assembly. Bioinformatics **28**(14), 1838–1844 (2012)
17. Manber, U., Myers, E.W.: Suffix arrays: A new method for on-line string searches. SIAM J. Comput. **22**(5), 935–948 (1993)
18. Mignosi, F., Restivo, A., Sciortino, M.: Forbidden factors in finite and infinite words. In: Karhumäki, J., Maurer, H.A., Paun, G., Rozenberg, G. (eds.) Jewels are Forever, Contributions on Theoretical Computer Science in Honor of Arto Salomaa, pp 339–350. Springer, Cham (1999). https://doi.org/10.1007/978-3-642-60207-8_30
19. Navarro, G., Raffinot, M.: Flexible Pattern Matching in Strings–Practical On-Line Search Algorithms for Texts and Biological Sequences. Cambridge University Press, 232p (2002)
20. Pinho, A. J., Ferreira, P. J. S. G., Garcia, S. P., Rodrigues, J. M. O. S.: On finding minimal absent words. BMC Bioinform, **10** (2009)
21. Silva, R.M., Pratas, D., Castro, L., Pinho, A.J., Ferreira, P.J.S.G.: Three minimal sequences found in ebola virus genomes and absent from human DNA. Bioinform. **31**(15), 2421–2425 (2015)

Weak Inverse Neighborhoods of Languages

Hyunjoon Cheon and Yo-Sub Han(✉)

Department of Computer Science, Yonsei University, Seoul, Republic of Korea
{hyunjooncheon,emmous}@yonsei.ac.kr

Abstract. While the edit-distance neighborhood is useful for approximate pattern matching, it is not suitable for the negative lookahead feature for the practical regex matching engines. This motivates us to introduce a new operation. We define the edit-distance interior operation on a language L to compute the largest subset $I(L)$ of L such that the edit-distance neighborhood of $I(L)$ is in L. In other words, L includes the edit-distance neighborhood of the largest edit-distance interior language. Given an edit-distance value r, we show that the radius-r edit-distance interior operation is a weak inverse of the radius-r edit-distance neighborhood operation, and vice versa. In addition, we demonstrate that regular languages are closed under the edit-distance interior operation whereas context-free languages are not. Then, we characterize the edit-distance interior languages and present a proper hierarchy with respect to the radius of operations. The family of edit-distance interior languages is closed under intersection, but not closed under union, complement and catenation.

Keywords: Edit-distance · Interior operation · Neighborhood languages · Formal languages

1 Introduction

People use regular expressions to find a specific pattern P in text T [4,6]. The practical implementations of regular expression matching engines support various additional features to concisely represent complex string patterns [10,17]. One such feature is the negative lookaheads $\neg P$ of a pattern P [11,15]; when checking a match between T and $\neg P$, if T does not match P, the negative lookahead $\neg P$ consumes nothing from T and reports a success. On the other hand, if T matches P, $\neg P$ reports a failure. For example, a pattern $\neg(ab)^*$ reports a success on a text $T_1 = abaabb$ since $(ab)^*$ does not match T_1. In contrast, on a text $T_2 = ababab$, $\neg(ab)^*$ reports a failure. Lookaheads are useful in matching text conditioned by a pattern P, and regular expressions with lookaheads are as powerful as ones without lookaheads [3]. For instance, a pattern $(\neg A)B$ recognizes exactly the strings in language $L(A)^c \cap L(B)$.

An application of regular expressions is to fix or to sanitize improper inputs given from unknown sources to have proper form [22]. Unfortunately, typography errors often occur in user inputs but patterns, like regular expressions, find an

F. Drewes and M. Volkov (Eds.): DLT 2023, LNCS 13911, pp. 61–73, 2023.
https://doi.org/10.1007/978-3-031-33264-7_6

exact match on target text. Thus, people often design a pattern tolerant on inputs with minor errors to permit such typography errors. The error measures on strings, such as the edit-distance [16] and Hamming distance [12], are one of the solutions that verify whether the input string has small deviation from the target pattern [1] by setting a distance threshold. However, it is not straightforward to construct an edit-distance neighborhood of negative patterns. Let us give an example where an input string composed of a's and b's should not have zero or an odd number of b's. With $\Sigma = \{a, b\}$, one can write a pattern $P = \neg(a^* + a^*ba^*(ba^*ba^*)^*)$ to recognize the language. Simultaneously, to relax the input condition by edit-distance of 1, the language should be the radius-1 edit-distance neighborhood $\mathsf{N}(L(P)) = L(\Sigma^*b\Sigma^*)$ of $L(P)$, whose pattern is equivalent to $P' = \neg(a^*) = \Sigma^*b\Sigma^*$. The problem is how one can find a pattern X' for $L(X') = \mathsf{N}(L(\neg X))$ without inspecting the language $L(\neg X)$ described by a pattern $\neg X$.

An idea is to find a core language $L(Y)$ of $L(X)$ that contains every string whose all possible neighbors by edit-distance of 1 are matched by X, and replace $\neg X$ by $\neg Y$. In this situation, Y should match against an input string $w \in L(X)$ as well as its all neighbors by edit-distance of 1. The operation—we call the *edit-distance interior operation*—is to compute this core language $L(Y)$ from a given original language $L(X)$.

The edit-distance was first studied in terms of self-correcting codes for errors in strings [16]. There are several researches on the edit-distance for strings [5,9] and languages [8,13,18]. Mohri [18] presented a polynomial-time algorithm to compute the edit-distance between two regular languages. Han and Ko [13] presented a polynomial time algorithm to compute the edit-distance between two visibly pushdown languages with a given upper bound k of the edit-distance [14]. Cheon and Han [8] showed that it is NEXPTIME-complete to decide if the edit-distance between a finite language and a parsing expression language is bounded above by a given positive integer. Aho and Peterson [1] proposed an error correcting parser that recognizes the edit-distance neighborhood of context-free languages. Salomaa and Schofield [21] and Ng et al. [19] investigated the edit-distance neighborhoods of regular languages and their deterministic state complexity is exponential. Okhotin and Salomaa [20] presented a construction for the edit-distance neighborhood of a visibly pushdown automata and established the state complexity.

After defining basic notions in Sect. 2, we give a formal definition of the edit-distance interior operation in Sect. 3. We study the basic Boolean operations and catenation over the edit-distance interior languages in Sect. 4. We tackle the closedness in Sect. 5, and prove that regular languages are closed whereas context-free languages are not closed under the edit-distance interior operation. We conclude the paper with a few possible research directions in Sect. 6.

2 Preliminaries

For a set A, let $|A|$ denote the cardinality of A. The empty set $\emptyset$ satisfies $|\emptyset| = 0$. For two functions $f : X \to Y$ and $g : Y \to Z$, the function $g \circ f : X \to Z$ satisfies $(g \circ f)(x) = g(f(x))$ for all $x \in X$.

An (finite) alphabet Σ is a nonempty set of symbols. A (finite) string $w = w_1 w_2 \cdots w_n$ over Σ is a finite sequence of symbols $w_i \in \Sigma$ for $1 \leq i \leq n$. The length $|w| = n$ of a string w is the number of symbols in w. For a symbol $a \in \Sigma$, $|w|_a$ is the number of a's in w. The empty string λ satisfies $|\lambda| = 0$. A language L over Σ is a set of strings over Σ. We define $L^0 = \{\lambda\}$, $L^{i+1} = L^i \cdot L$ $(i \geq 0)$ and $L^* = \bigcup_{i=0}^{\infty} L^i$. By definition, every language L over Σ is a subset of Σ^*.

For two strings x and y, the edit-distance $d(x, y)$ of x and y is the minimum number of edit operations (inserting a symbol, deleting an existing symbol and substituting a symbol with another one) to modify x into y. d is a metric; for any strings x, y and z, d satisfies the following: (1) $d(x, x) = 0$, (2) $d(x, y) = d(y, x) \geq 0$ and (3) $d(x, z) \leq d(x, y) + d(y, z)$. We extend the definition of the edit-distance to between a string and a language and between two languages as the minimum edit-distance of all possible pairs. Formally,

$$d(x, L) = \min_{y \in L} d(x, y) \text{ and } d(L_1, L_2) = \min_{x \in L_1} d(x, L_2).$$

Given a language L and a non-negative integer r, the radius-r edit-distance neighborhood $\mathsf{N}^r(L) = \{w \in \Sigma^* \mid d(w, L) \leq r\}$ is a set of strings whose edit-distance is at most r from L. We call the distance threshold r of an edit-distance neighborhood operation 'radius' since, intuitively, N^r draws a new outer boundary apart by r from the original boundary of L. It is easy to see that, for any two non-negative integers a and b, $(\mathsf{N}^a \circ \mathsf{N}^b)(L) = \mathsf{N}^{a+b}(L)$. For a string w, we use $\mathsf{N}^r(w)$ to denote $\mathsf{N}^r(\{w\})$—the edit-distance neighborhood of a singleton language $\{w\}$.

A finite-state automaton (FA) A is specified by a tuple $(Q, \Sigma, \delta, s, F)$: a finite set Q of states, a finite alphabet Σ, a (multi-valued) transition function $\delta : Q \times \Sigma \to Q$, the start state $s \in Q$ and a set $F \subseteq Q$ of final states. For an FA A, A minimal relation $\vdash_A$ over $(Q \times \Sigma^*)$ satisfies $(p, \sigma w) \vdash_A (q, w)$ if $\delta(p, \sigma) = q$ $(p, q \in Q, \sigma \in \Sigma, w \in \Sigma^*)$. $\vdash_A^*$ is the reflexive transitive closure of $\vdash_A$. We define the language $L(A) = \{w \in \Sigma^* \mid (s, w) \vdash_A^* (f, \lambda), f \in F\}$ of an FA A. For a string $w \in \Sigma^*$, we say that A accepts w if $w \in L(A)$ and A rejects w otherwise. A language L is regular if there is an FA A that satisfies $L = L(A)$. It is known that regular languages are closed under the basic Boolean operations—union, intersection and complement—as well as under catenation and N^r regardless of the radius r.

A context-free grammar (CFG) $G = (V, \Sigma, P, S)$ consists of four components: two disjoint finite alphabets V and Σ of variables (or nonterminals) and terminals, respectively, a finite set $P \subseteq V \times (V \cup \Sigma)^*$ of productions and an initial variable $S \in V$. A production $(A, \alpha) \in P$ is often denoted by $A \to \alpha$ $(A \in V$, $\alpha \in (V \cup \Sigma)^*)$. For a CFG G, a minimal relation $\vdash_G$ over $(V \cup \Sigma)^*$ satisfies $\alpha A \beta \vdash_G \alpha \gamma \beta$ if $(A, \gamma) \in P$ $(\alpha, \beta, \gamma \in (V \cup \Sigma)^*$, $A \in V)$. $\vdash_G^*$ is the reflexive transitive closure of $\vdash_G$. We define the language $L(G) = \{w \in \Sigma^* \mid S \vdash_G^* w\}$ of a CFG G. For a string $w \in \Sigma^*$, we say that G accepts w if $w \in L(G)$ and G rejects w otherwise. A language L is context-free if there is a CFG G that satisfies $L = L(G)$. Context-free languages are closed under union and catenation, but not closed under intersection and complement (and intersection and complement of context-free languages are undecidable). For more background knowledge in automata theory, the reader may refer to the textbooks [23,24].

3 Edit-Distance Neighborhoods and Edit-Distance Interiors

We propose the edit-distance interior operation, which is an inverse-like operation to the edit-distance neighborhood. For a non-negative integer r, the radius-r edit-distance interior operation is to shrink a given language L by the given radius, in contrast to the edit-distance neighborhoods, which expands L.

Question 1. For a language L, find the maximum language L_{-r} up to inclusion that satisfies $\mathsf{N}^r(L_{-r}) \subseteq L$. We call L_{-r} the radius-r edit-distance interior language (or shortly, the radius-r edit-distance interior) of L.

Note that the term 'interior' is often used in the field of mathematical topology that denotes the dual of a closure. We adopt this term despite the edit-distance interior is not an interior because the edit-distance neighborhood and the edit-distance interior expose a similar semantics to the (topological) closure and interior.

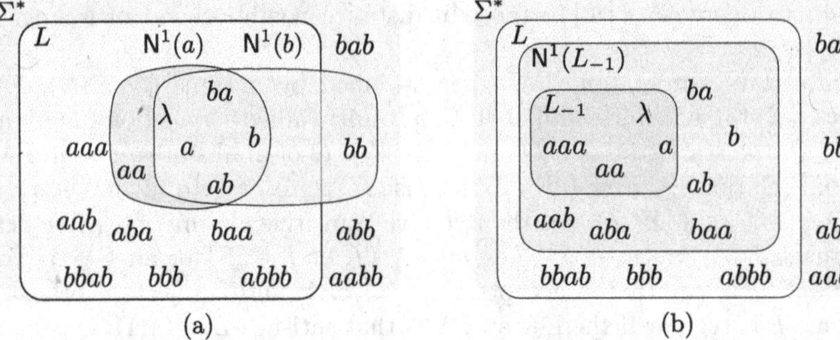

Fig. 1. Let $L = \{w \mid |w|_b = 0 \vee |w|_b \bmod 2 = 1\}$. (a) radius-1 edit-distance neighborhoods of a and b. Note that $bb \notin L$. (b) radius-1 edit-distance interior L_{-1} of L and its radius-1 edit-distance neighborhood $\mathsf{N}^1(L_{-1})$. Note that $\mathsf{N}^1(L_{-1})$ is the largest edit-distance neighborhood in L.

Figure 1(b) shows a visual representation of a language $L = \{w \mid |w|_b = 0 \vee |w|_b \bmod 2 = 1\}$, its radius-1 edit-distance interior $L_{-1} = \{w \mid |w|_b = 0\}$

and the radius-1 edit-distance neighborhood $\mathsf{N}^1(L_{-1}) = \{w \mid |w|_b \leq 1\}$ of L_{-1} over $\Sigma = \{a, b\}$. We can observe in Fig. 1(a) that $a \in L_{-1}$ because all of its edit-distance neighbors are in L. On the other hand, $b \notin L_{-1}$ since bb, one of b's edit-distance neighbors, is not in L. Before discussing the properties of edit-distance interior operation, we first show that Question 1 is sound: whether or not L_{-r} is unique.

Lemma 2. *For any language L and a non-negative integer r, the maximum language L_{-r} is unique.*

Proof. Suppose that L_{-r} is not unique. Then, there exists two distinct, maximal languages L_1 and L_2 satisfying the following conditions.

1. $\mathsf{N}^r(L_1) \subseteq L$,
2. $\mathsf{N}^r(L_2) \subseteq L$ and
3. no other languages X including L_1 or L_2 satisfy $\mathsf{N}^r(X) \subseteq L$.

These two languages L_1 and L_2 satisfy the condition in Question 1.

Since L_1 and L_2 are distinct, we can choose a string w from $L_2 \setminus L_1$ and let $X = L_1 \cup \{w\}$ be a proper superset of L_1. We demonstrate that X also satisfies the condition in Question 1. It is straightforward that $\mathsf{N}^r(w) \subseteq \mathsf{N}^r(L_2) \subseteq L$ since $w \in L_2$ and Condition 2 holds. Remark that $\mathsf{N}^r(X) = \mathsf{N}^r(L_1) \cup \mathsf{N}^r(w)$. These two results and Condition 1 give rise to $\mathsf{N}^r(X) \subseteq L$ and X satisfies the condition in Question 1. Since X is a proper superset of L_1, it contradicts to the Condition 3: maximality of L_1 and L_2. We thus conclude that there must be a unique maximum language L_{-r} satisfying Question 1. □

Due to Lemma 2, we can use a function I^r to denote the edit-distance interior operation: $\mathsf{I}^r(L) = L_{-r}$. The use of superscript is intended to denote that an edit-distance interior operation is decomposable into several edit-distance interior operations on smaller radius: $\mathsf{I}^{r_1+r_2} = \mathsf{I}^{r_1} \circ \mathsf{I}^{r_2}$—we prove this later in Lemma 9. We now give short examples for the edit-distance interiors to help to understand the operation better.

Example 3. For a language $L = \{\lambda, a\}$ over $\Sigma = \{a, b\}$, the radius-1 edit-distance neighborhood $\mathsf{N}^1(L)$ the radius-1 edit-distance interior $\mathsf{I}^1(L)$, and $(\mathsf{N}^1 \circ \mathsf{I}^1)(L)$ are $\mathsf{N}^1(L) = \{\lambda, a, b, aa, ab, ba\}$ and $\mathsf{I}^1(L) = (\mathsf{N}^1 \circ \mathsf{I}^1)(L) = \emptyset$.

Example 3 demonstrates radius-1 edit-distance neighborhood and interior operation. Note that, for any nonempty language $X \subseteq \Sigma^*$, $\mathsf{N}^1(X) \nsubseteq L = \{\lambda, a\}$—even $\mathsf{N}^1(\{\lambda\}) = \{\lambda, a, b\} \nsubseteq L$. This example also shows that L and $(\mathsf{N}^1 \circ \mathsf{I}^1)(L)$ are not always the same even the two operations are seemingly the inverse of each other. Example 4 shows that an alphabet Σ under consideration also affects the computation of the edit-distance interiors.

Example 4. For a language $L = \{\lambda, a, aa\}$, the radius-1 edit-distance interior over $\Sigma = \{a\}$ is $\mathsf{I}^1(L) = \{\lambda, a\}$ whereas, for the same language L, the radius 1 edit-distance interior over $\Sigma = \{a, b\}$ is $\mathsf{I}^1(L) = \emptyset$.

Since $\mathsf{I}^r(L)$ is unique, we now provide the language equivalent to $\mathsf{I}^r(L)$, which also matches the intuition from 1.

Definition 5. The radius-r edit-distance interior operation I^r over languages L is defined as $\mathsf{I}^r(L) = \{w \in \Sigma^* \mid \mathsf{N}^r(w) \subseteq L\}$.

As observed in Example 4, we need to specify the alphabet Σ over which the I^r or the languages are defined. However, we often do not explicitly state Σ if the exact alphabet is not relevant to the claim to write concise statements. We also consider that, if a language L is explicitly defined over an alphabet Σ, I^r receiving L is also defined over the same alphabet Σ.

We now investigate relations between the complement L^c and the complement of the edit-distance interior $\mathsf{I}^r(L)^c$ of a language L. Figure 2 depicts an example of $L = \{w \mid |w|_b = 0 \vee |w|_b \bmod 2 = 1\}$ and its radius-1 edit-distance interior $\mathsf{I}^1(L)$. We can observe that $\mathsf{I}^1(L)^c = \{w \mid |w|_b \geq 1\}$ is the radius-1 edit-distance neighborhood of $L^c = \{w \mid |w|_b > 0 \wedge |w|_b \bmod 2 = 0\}$.

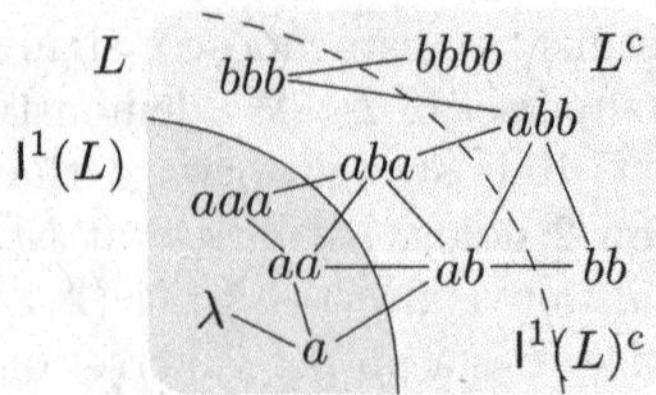

Fig. 2. An example of edit-distance interior for $L = \{w \mid |w|_b = 0 \vee |w|_b \bmod 2 = 1\}$. Note L^c and $\mathsf{I}^1(L)^c = \mathsf{N}^1(L^c)$ before and after the edit-distance interior operation. The dashed and solid curves depict the boundaries of L and $\mathsf{I}^1(L)$ in Σ^*, respectively. Lines connect two strings of edit-distance 1.

Lemma 6 shows the relation between L^c and $\mathsf{I}^r(L)^c$. We can see that the edit-distance neighborhood and the edit-distance interior are the dual of each other. Lemma 6 allows us to replace I^r by N^r by extracting a complement from the original language.

Lemma 6. *Given an arbitrary language L and a non-negative integer r, the complement of the radius-r edit-distance interior language of L is the radius-r edit-distance neighborhood language of L^c. Shortly, $\mathsf{I}^r(L)^c = \mathsf{N}^r(L^c)$.*

Proof.

$$\begin{aligned}
\mathsf{I}^r(L)^c &= \{w \in \Sigma^* \mid \mathsf{N}^r(w) \not\subseteq L\} && \text{(Definition 5)}\\
&= \{w \in \Sigma^* \mid \mathsf{N}^r(w) \cap L^c \neq \emptyset\}\\
&= \{w \in \Sigma^* \mid d(w, L^c) \leq r\} && (\exists x \in L^c : d(w,x) \leq r)\\
&= \mathsf{N}^r(L^c).
\end{aligned}$$

□

Moreover, I^r and N^r are indeed not the inverse but a weak inverse of each other. A weak inverse $g : Y \to X$ of a function $f : X \to Y$ acts like an inverse, but g (and f) may not be bijective. Formally, a function g is a weak inverse of f if and only if $g \circ f \circ g = g$. Note that g may not be unique: a constant function $g = c$ for any $c \in X$ is a weak inverse of an arbitrary function f. We first show that N^r is a weak inverse of I^r.

Lemma 7. *For any non-negative integer* r, $\mathsf{N}^r \circ \mathsf{I}^r \circ \mathsf{N}^r = \mathsf{N}^r$. *In other words,* N^r *is a weak inverse of* I^r.

Proof. For a language L over Σ, let $M = \mathsf{N}^r(L)$ and show $(\mathsf{N}^r \circ \mathsf{I}^r)(M) = M$. It is easy to verify that $(\mathsf{N}^r \circ \mathsf{I}^r)(M) \subseteq M$ since Definition 5 defines $\mathsf{I}^r(M)$ to be the maximum set satisfying the exact condition.

For the other direction $(\mathsf{N}^r \circ \mathsf{I}^r)(M) \supseteq M$, we first show that $L \subseteq \mathsf{I}^r(M)$. Because M contains all edit-distance neighbors of L by $(*)$, any string x in L satisfies $\mathsf{N}^r(x) \subseteq M$. Thus, for all x in L, $x \in \mathsf{I}^r(M)$ holds. Let w be any string in M. Since $d(w, L) \leq r$, we can always find a string $x \in L$ satisfying $d(w, x) \leq r$. Then, $w \in \mathsf{N}^r(x)$. It concludes that $w \in \mathsf{N}^r(x) \subseteq \mathsf{N}^r(L) \subseteq (\mathsf{N}^r \circ \mathsf{I}^r)(M)$ for all $w \in M$. □

It is also known that even if g is a weak inverse of f it is not necessarily true that f is a weak inverse of g: Consider the last example. If $f(x) = x$ and $g(y) = c$ for some $c \in X$, $g \circ f \circ g = c = g$ but $f \circ g \circ f = c \neq f$, thus f is not a weak inverse of g. Thus, the fact that N^r is a weak inverse of I^r is not enough to justify that I^r is a weak inverse of N^r. Instead, we apply Lemma 6 on $\mathsf{N}^r \circ \mathsf{I}^r \circ \mathsf{N}^r$ and have $(\mathsf{N}^r \circ \mathsf{I}^r \circ \mathsf{N}^r)(L)^c = (\mathsf{I}^r \circ \mathsf{N}^r \circ \mathsf{I}^r)(L^c)$. Since $\mathsf{N}^r(L)^c = \mathsf{I}^r(L^c)$, we obtain the following result by Lemma 7.

Corollary 8. *For any non-negative integer* r, I^r *is a weak inverse of* N^r.

We next study how to decompose I^r into smaller operations. Given a sequence of n edit-distance interior operations of radius $r_1, r_2, \cdots, r_n$, roughly speaking, computing $(\mathsf{I}^{r_n} \circ \cdots \circ \mathsf{I}^{r_2} \circ \mathsf{I}^{r_1})(L)$ is shrinking the language L by the sum $\sum r_i$ for all the given radius values.

Lemma 9. *For any two non-negative integers* r_1 *and* r_2, $(\mathsf{I}^{r_2} \circ \mathsf{I}^{r_1})(L) = \mathsf{I}^{r_1+r_2}(L)$.

Proof. By Lemma 6, we can move out a complement while substituting I^r by N^r.

$$(\mathsf{I}^{r_2} \circ \mathsf{I}^{r_1})(L) = \mathsf{I}^{r_2}(\mathsf{N}^{r_1}(L^c)^c) = (\mathsf{N}^{r_2} \circ \mathsf{N}^{r_1})(L^c)^c.$$

Since $(\mathsf{N}^{r_2} \circ \mathsf{N}^{r_1})(L) = \mathsf{N}^{r_1+r_2}(L)$ for a language L, we have

$$(\mathsf{N}^{r_2} \circ \mathsf{N}^{r_1})(L^c)^c = \mathsf{N}^{r_1+r_2}(L^c)^c = \mathsf{I}^{r_1+r_2}(L).$$

Thus we can decompose a radius r edit-distance interior operation I^r into two smaller edit-distance interior operations I^{r_1} and I^{r_2}, where $r = r_1 + r_2$. □

Using Lemma 9, we can rewrite $\mathsf{I}^r(L)$ into r consecutive applications of I^1 on a language L. Thus, when we say the radius-r neighborhood operations as extending the language boundary by r times, the edit-distance interior operation should shrink the language by the same amount.

4 Closure Properties on Edit-Distance Interiors

Since our main goal is to build a pattern P' that matches the edit-distance interior of the language $L(P)$ of an original pattern P, and most pattern matching expressions support basic set operations such as union, intersection and complement, we now discuss the characteristics of these basic set operations over the edit-distance interiors. Let us first define a language class $\mathcal{I}^r$ to be the image of radius-r edit-distance interior operation, namely, $\mathcal{I}^r = \{\mathsf{I}^r(L) \mid L \subseteq \Sigma^*\}$. Then, we establish the following result based on the weak inverse relations of I^r and N^r.

Proposition 10. $\mathcal{I}^r = \{L \subseteq \Sigma^* \mid (\mathsf{I}^r \circ \mathsf{N}^r)(L) = L\}$.

Proof. Let $\mathcal{X} = \{L \subseteq \Sigma^* \mid (\mathsf{I}^r \circ \mathsf{N}^r)(L) = L\}$. We can observe that every language L in $\mathcal{X}$ is an edit-distance interior $\mathsf{I}^r(M)$ of a language $M = \mathsf{N}^r(L)$. Thus, by the definition of the class $\mathcal{I}^r$, $L = \mathsf{I}^r(M) \in \mathcal{I}^r$.

For the converse, let $L \in \mathcal{I}^r$ be a language in the class. We can write L as $\mathsf{I}^r(M)$ for some language M. Then, by Corollary 8, we have $(\mathsf{I}^r \circ \mathsf{N}^r)(L) = (\mathsf{I}^r \circ \mathsf{N}^r \circ \mathsf{I}^r)(M) = \mathsf{I}^r(M) = L$. Thus $L \in \mathcal{X}$. □

Proposition 10 allows us to determine whether or not a language L is in $\mathcal{I}^r$ by comparing L to the edit-distance interior $(\mathsf{I}^r \circ \mathsf{N}^r)(L)$ of its edit-distance neighborhood. With this and Lemma 9, we can determine a strict hierarchy of the family of edit-distance interior languages.

Proposition 11. *For any non-negative integer r, $\mathcal{I}^{r+1} \subsetneq \mathcal{I}^r$.*

Proof. Since $\mathsf{I}^{r+1}(L) = (\mathsf{I}^r \circ \mathsf{I}^1)(L)$, $\mathcal{I}^{r+1}$ is a subset of $\mathcal{I}^r$ by the definition. Let $X = \{\lambda\}^c$. Then, $\mathsf{I}^r(X) = \mathsf{N}^r(\lambda)^c \in \mathcal{I}^r$. However, by Proposition 10,

$$\begin{aligned}(\mathsf{I}^{r+1} \circ \mathsf{N}^{r+1})(\mathsf{I}^r(X)) &= (\mathsf{I}^{r+1} \circ \mathsf{N}^{r+1})(\mathsf{N}^r(\lambda)^c)\\ &= \mathsf{I}^{r+1}((\mathsf{I}^{r+1} \circ \mathsf{N}^r)(\lambda)^c)\\ &= \mathsf{I}^{r+1}(\Sigma^*)\\ &= \Sigma^* \supsetneq X \supseteq \mathsf{I}^r(X).\end{aligned}$$

Thus X is not in $\mathcal{I}^{r+1}$. Note that $(\mathsf{I}^{r+1} \circ \mathsf{N}^r)(\lambda) = \emptyset$ since $\mathsf{N}^r(\lambda)$ has no strings w satisfying $\mathsf{N}^{r+1}(w) \subseteq \mathsf{N}^r(\lambda)$. □

Now we tackle the closedness of $\mathcal{I}^r$ under the basic Boolean operations. First, we can observe that, for any positive integer r, $\mathcal{I}^r$ is closed under intersection, but is not closed under union, complement and catenation. We can use Proposition 10 to show that the class $\mathcal{I}^r$ is not closed under union, complement and catenation.

Theorem 12. *For any positive integer r, $\mathcal{I}^r$ is closed under intersection but is not under union, complement, and catenation. Moreover, $\mathsf{I}^r(L_1 \cap L_2) = \mathsf{I}^r(L_1) \cap \mathsf{I}^r(L_2)$.*

Proof. We study each operation one by one. Note that, for union, complement and intersection, the results X are not equivalent to $(\mathsf{I}^r \circ \mathsf{N}^r)(X)$.
(Intersection): Since N^r is distributive over union, we have

$$\begin{aligned}\mathsf{I}^r(L_1) \cap \mathsf{I}^r(L_2) &= (\mathsf{I}^r(L_1)^c \cup \mathsf{I}^r(L_2)^c)^c \\ &= (\mathsf{N}^r(L_1^c) \cup \mathsf{N}^r(L_2^c))^c && \text{by Lemma 6} \\ &= \mathsf{N}^r(L_1^c \cup L_2^c)^c \\ &= \mathsf{I}^r(L_1 \cap L_2) && \text{by Lemma 6.}\end{aligned}$$

Therefore, $\mathcal{I}^r$ is closed under intersection.
(Union): Let L_1 and $L_2 \in \mathcal{I}^r$ be the following.

$$L_1 = \Sigma^{r+2} = \mathsf{I}^r(\bigcup_{i=2}^{2r+2} \Sigma^i) \text{ and}$$

$$L_2 = \Sigma^{3r+3} = \mathsf{I}^r(\bigcup_{i=2r+3}^{4r+3} \Sigma^i).$$

Then, $L_1 \cup L_2 \neq (\mathsf{I}^r \circ \mathsf{N}^r)(L_1 \cup L_2) = \bigcup_{i=r+2}^{3r+3} \Sigma^i$. Thus, $\mathcal{I}^r$ is not closed under union.
(Complement): For $L = \{\lambda\} \in \mathcal{I}^r$, $L^c \neq (\mathsf{I}^r \circ \mathsf{N}^r)(L^c) = \Sigma^*$. Thus $\mathcal{I}^r$ is not closed under complement.
(Catenation): Let L_1 and $L_2 \in \mathcal{I}^r$ be the following.

$$L_1 = \bigcup_{i=0}^{r+2} \Sigma^i = \mathsf{I}^r(\bigcup_{i=0}^{2r+2} \Sigma^i) \text{ and}$$

$$L_2 = \Sigma^0 \cup \Sigma^{2r+2} \setminus \{a^{2r+2}\} = \mathsf{I}^r(\bigcup_{i=0}^{r} \Sigma^i \cup \bigcup_{i=r+2}^{3r+2} \Sigma^i \setminus \{a^{3r+2}\}).$$

Then, $L_1 \cdot L_2 \neq (\mathsf{I}^r \circ \mathsf{N}^r)(L_1 \cdot L_2)$. Thus $\mathcal{I}^r$ is not closed under catenation. Note that $a^{2r+2} \in (\mathsf{I}^r \circ \mathsf{N}^r)(L_1 \cdot L_2)$ but $a^{2r+2} \notin L_1 \cdot L_2$ □

While $\mathcal{I}^r$ is not closed under union, we continue to investigate if there exists a recursive construction for the edit-distance interior for a pattern $P = A + B$ where $L(P) = L(A) \cup L(B)$. We try to find a matching pattern P' representing $\mathsf{I}^r(L(P))$ using the two expressions A' and B' each representing $\mathsf{I}^r(L(A))$ and $\mathsf{I}^r(L(B))$. Thus, we further investigate that whether $\mathsf{I}^r(A \cup B)$ is easily computable using $\mathsf{I}^r(A)$ and $\mathsf{I}^r(B)$. Unfortunately, this is not applicable in general. For example, consider two pairs of languages $(L_1, L_2) = (\Sigma^0, \Sigma^1)$ and $(L'_1, L'_2) = (\Sigma^0, \Sigma^1 \cup \Sigma^2)$ over an alphabet Σ. In both cases, the pairs radius-1 edit-distance interiors are the same: $\mathsf{I}^1(L_1) = \mathsf{I}^1(L_2) = \emptyset$ and $\mathsf{I}^1(L'_1) = \mathsf{I}^1(L'_2) = \emptyset$. However, the radius-1 edit-distance interiors of their unions are different: $\mathsf{I}^1(L_1 \cup L_2) = \Sigma^0 \neq \Sigma^0 \cup \Sigma^1 = \mathsf{I}^1(L'_1 \cup L'_2)$. This suggests that we need to know the original languages L_1 and L_2 (by symmetry) to determine the edit-distance interior of their union. Note that, by Lemma 6, there is an obvious construction for the language: $\mathsf{I}^r(A \cup B) = \mathsf{N}^r(A^c \cap B^c)^c$ —we discuss the closedness of regular languages and context-free languages in Sect. 5.

One can however determine the equivalence $\mathsf{I}^r(L_1 \cup L_2) = \mathsf{I}^r(L_1) \cup \mathsf{I}^r(L_2)$ in terms of the relative edit-distance over the languages [2] without computing $\mathsf{I}^r(L_1 \cup L_2)$ and check equivalence to the union language. The relative edit-distance $d'(A, B) = \sup_{x \in A} d(x, B)$ over two languages A and B is equivalent to $A \subseteq \mathsf{N}^{d'(A,B)}(B)$. Note that d' is not a metric: $d'(A, B) \neq d'(B, A)$.

Lemma 13. *For two languages L_1 and L_2, let $L = L_1 \cup L_2$ and $L' = \mathsf{I}^r(L_1) \cup \mathsf{I}^r(L_2)$. Then, $\mathsf{I}^r(L) = L'$ if and only if $d'(L \setminus L', L^c) \leq r$.*

Proof. It is obvious that $\mathsf{I}^r(L) \supseteq L'$. First, suppose that $X = \mathsf{I}^r(L) \setminus L'$ is empty to show the forward direction. For all $w \in L \setminus L'$ (Be aware that this set is not the same set to X.), $\mathsf{N}^r(w) \not\subseteq L$ must be true. Then, w satisfies $d(w, L^c) \leq r$, hence the statement holds.

For the opposite direction, suppose that $\mathsf{I}^r(L) \neq L'$. Then there must be a string $w \in L \setminus L'$ satisfying $\mathsf{N}^r(w) \subseteq L$. Therefore, $d(w, L^c) > r$ and $d'(L \setminus L', L^c) > r$. □

5 Edit-Distance Interiors of Regular or Context-Free Languages

Given a function f, it is an interesting question to see if $f(L)$ is in $\mathcal{X}$, where L is a language in a class $\mathcal{X}$ of languages. This question is important since we can construct a pattern P' for the language $f(L(P))$ from an original pattern P if the language class is closed under f. We first show that regular languages are closed under I^r of any radius r. Note that, since I^0 is an identity function, any class is closed under I^0.

Theorem 14. *For a regular language L and a positive integer r, $\mathsf{I}^r(L)$ is regular. In other words, regular languages are closed under I^r.*

Proof. By Lemma 6, $\mathsf{I}^r(L) = \mathsf{N}^r(L^c)^c$. Since regular languages are closed under complement and N^r, the statement holds. □

We establish the negative result for the case of context-free languages. Let $\mathbb{COPY} = \{ww \mid w \in \Sigma^*\}$ be the copy language and $\overline{\mathbb{COPY}}$ be its complement $\Sigma^* \setminus \{ww \mid w \in \Sigma^*\}$. The radius-$r$ edit-distance interior of $\overline{\mathbb{COPY}}$ is the set of strings z such that, for each possible decomposition $z = uv$, the edit-distance between u and v is larger than r. It is because, if there exists a decomposition $z = uv$ such that $d(u, v) \leq r$, then there should be a sequence of edit operations on (without loss of generality) u to become v with no more than r steps. This means that if we apply the same sequence of edit operations on uv, then we can have the string vv, which is not in $\overline{\mathbb{COPY}}$. Thus, $\mathsf{N}^r(uv) \not\subseteq \overline{\mathbb{COPY}}$ and $uv \notin \mathsf{I}^r(\overline{\mathbb{COPY}})$.

Lemma 15. *Context-free languages are not closed under* I^r *for* $r \geq 3$.

Proof. Consider the non-copy language $\overline{\mathbb{COPY}}$ over Σ, which is a well-known context-free language. By Lemma 6, we can rewrite $\mathsf{I}^r(\overline{\mathbb{COPY}})$ by $\mathsf{N}^r(\mathbb{COPY})^c = \{w \mid \exists x, y : w = xy \wedge d(x, y) \leq r\}^c = \{w \mid \forall x, y, w = xy : d(x, y) > r\}$. The resulting language is the set of all strings $w \in \Sigma^*$ such that, for any decomposition $w = xy$ into two strings x and y, x and y have edit-distance of at least $r+1$. Cheon et al. [7] showed that this language is not context-free when $r \geq 3$. Therefore, context-free languages are not closed under I^r when $r \geq 3$. □

Lemma 15 takes care of the case of $r \geq 3$. For the case of $r = 1, 2$, we rewrite I^r into a composition of r I^1's using Lemma 9. Then, the following result becomes useful to show the non-closedness of context-free languages under I^r for every positive integer r.

Lemma 16. *If a set* $\mathcal{X} \subseteq \mathcal{U}$ *is closed under a function* $f : \mathcal{U} \to \mathcal{U}$, *then* $\mathcal{X}$ *is closed under the function* f^n *for* $n \geq 1$ *as well.*

Proof. Let $f|_{\mathcal{X}} : \mathcal{X} \to \mathcal{U}$ be a restriction of f under $\mathcal{X}$: $f|_{\mathcal{X}}(x) = f(x)$ for all $x \in \mathcal{X}$. We can ensure that the codomain of $f|_{\mathcal{X}}$ is $\mathcal{X}$ since $\mathcal{X}$ is closed under f. Then, for any positive integer n, $(f|_{\mathcal{X}})^n$ is a function applying $f|_{\mathcal{X}}$ n times, whose codomain is also $\mathcal{X}$. We can easily show that $(f|_{\mathcal{X}})^n(x) = f^n(x)$ for all $x \in \mathcal{X}$. □

Theorem 17. *For any positive integer* r, *context-free languages are not closed under* I^r.

Proof. We consider two cases separately: I^r for $r \geq 3$ and I^r for $r = 1, 2$,

For $r \geq 3$, it is immediate from Lemma 15 that the statement holds.

For $r = 1$, assume that the statement is false for I^1 for the sake of contradiction. Then, because of Lemma 16, context-free languages must be closed under I^r for all $r \geq 1$. However, context-free languages are not closed under I^3 by Lemma 15, and this leads to a contradiction. A similar argument holds for the case of $r = 2$ using the fact that context-free languages are not closed under I^4 by Lemma 15. □

6 Conclusions

We have proposed the radius-r edit-distance interior operation I^r such that $\mathsf{I}^r(L) = \{w \mid \mathsf{N}^r(w) \subseteq L\}$ for a non-negative integer r. We have showed that this operation is a weak inverse of N^r and is equivalent to r applications of I^1. Then, we have demonstrated that the class of edit-distance interior languages are closed under intersection but not closed under union, complement and catenation. Moreover, the edit-distance interior of the union of two languages cannot be determined by the edit-distance interiors of each languages. We have also proved that regular languages are closed under I^r for any non-negative r but context-free languages are not closed under I^r for a positive radius r Note that I^0 is the identity function.

It is an open problem to study if the current approach of computing the complement $\mathsf{N}^r(L^c)^c$ of edit-distance neighborhood of L^c in Sect. 3 is optimal. We plan to investigate the state complexity of the edit-distance interior for regular languages.

Acknowledgments. This research was supported by the NRF grant (RS-2023-00208094). We wish to thank the anonymous reviewers for their valuable suggestions.

References

1. Aho, A.V., Peterson, T.G.: A minimum distance error-correcting parser for context-free languages. SAIM J. Comput. **1**(4), 281–353 (1972)
2. Benedikt, M., Puppis, G., Riveros, C.: Bounded repairability of word languages. J. Comput. Syst. Sci. **79**(8), 1302–1321 (2013)
3. Berglund, M., van der Merwe, B., van Litsenborgh, S.: Regular expressions with lookahead. J. Univ. Comput. Sci. **27**(4), 324–340 (2021)
4. Bispo, J., Sourdis, I., Cardoso, J.M., Vassiliadis, S.: Regular expression matching for reconfigurable packet inspection. In: Proceedings of the 2006 IEEE International Conference on Field Programmable Technology (FPT), pp. 119–126 (2006)
5. Chakraborty, D., Das, D., Goldenberg, E., Koucký, M., Saks, M.: Approximating edit distance within constant factor in truly sub-quadratic time. J. ACM **67**(6), 36:1–36:22 (2020)
6. Chapman, C., Stollee, K.T.: Exploring regular expression usage and context in Python. In: Proceedings of the 25th International Symposium on Software Testing and Analysis (ISSTA), pp. 282–293 (2016)
7. Cheon, H., Hahn, J., Han, Y.S., Ko, S.K.: Most pseudo-copy languages are not context-free. In: Proceedings of the 27th International Computing and Combinatorics Conference (COCOON), pp. 189–200 (2021)
8. Cheon, H., Han, Y.S.: Computing the shortest string and the edit-distance for parsing expression languages. In: Proceedings of the 24th International Conference on Developments in Language Theory (DLT), pp. 43–54 (2020)
9. Cormode, G., Muthukrishnan, S.: The string edit distance matching problem with moves. ACM Transactions on Algorithms 3(1), 2:1–2:19 (2007)
10. Ecma International: ECMA-262: ECMAScript(R) 2015 language specification. In: ECMA International (2015)
11. Ford, B.: Parsing expression grammars: A recognition-based syntactic foundation. In: Proceedings of the 31st ACM SIGPLAN-SIGACT Symposium on Principles of Programming Languages (POPL), pp. 111–122 (2004)
12. Hamming, R.W.: Error detecting and error correcting codes. Bell Syst. Tech. J. **29**(2), 147–160 (1950)
13. Han, Y.S., Ko, S.K.: Edit-distance between visibly pushdown languages. In: Proceedings of the 43rd International Conference on Current Trends in Theory and Practice of Computer Science (SOFSEM), pp. 196–207 (2017)
14. Han, Y.S., Ko, S.K., Salomaa, K.: The edit-distance between a regular language and a context-free language. Int. J. Found. Comput. Sci. **24**(7), 1067–1082 (2013)
15. Hazel, P.: pcre2pattern man page, https://www.pcre.org/current/doc/html/pcre2pattern.html. Accessed 23 Feb 2023

16. Levenshtein, V.I.: Binary codes capable of correcting deletions, insertions, and reversals. Soviet Phy. Doklady **10**(8), 707–710 (1966)
17. Microsoft and other contributors: Regular expression language–quick reference (2022). https://learn.microsoft.com/en-us/dotnet/standard/base-types/regular-expression-language-quick-reference. Accessed 25 Feb 2023
18. Mohri, M.: Edit-distance of weighted automata: General definitions and algorithms. Int. J. Found. Comput. Sci. **14**(6), 957–982 (2003)
19. Ng, T., Rappaport, D., Salomaa, K.: State complexity of neighborhoods and approximate patern matching. Int. J. Found. Comput. Sci. **29**(2), 315–329 (2018)
20. Okhotin, A., Salomaa, K.: Edit distance neighborhoods of input-driven pushdown automata. Theoret. Comput. Sci. **777**, 417–430 (2019)
21. Salomaa, K., Schofield, P.: State complexity of additive weighted finite automata. Int. J. Found. Comput. Sci. **18**(6), 1407–1416 (2007)
22. Shar, L.K., Tan, H.B.K.: Mining input sanitization patterns for predicting sql injection and cross site scripting vulnerabilities. In: Proceedings of the 34th International Conference on Software Engineering (ICSE), pp. 1293–1296 (2012)
23. Sipser, M.: Introduction to the Theory of Computation, 3rd edn. Cengage Learning, MA, USA (2013)
24. Wood, D.: Theory of Computation. Harper & Row, NY, USA (1987)

The Exact State Complexity for the Composition of Root and Reversal

Pascal Caron, Alexandre Durand(✉), and Bruno Patrou

LITIS, Université de Rouen, Avenue de l'Université, 76801 Saint-Étienne du Rouvray Cedex, France
{pascal.caron,bruno.patrou}@univ-rouen.fr,
alexandre.durand@etu.univ-rouen.fr

Abstract. We consider the operation obtained by composition of the two unary operations Root and reversal. We prove that its state complexity depends on the number of final states and is maximized by $n^n - (n-1)^n + 1$ when this number is one. Moreover we prove that every minimal automaton having the monster property is a witness for the reversal of Root.

1 Introduction

Mirkin in 1965 established that the reversal's state complexity is 2^n [1] and that this bound is tight by pointing out that Lupanov's ternary worst- case example is a reversal of a deterministic automaton [2]. Maslov in 1970 established that the Root's state complexity is bounded by n^n [3]. Since then, this results have been improved. In 2003, Krawetz *et al.* proved a better upper-bound of $n^n - \binom{n}{2}$ for this operation [4]. They also proved that this bound is tight. *A priori*, the worst-case state complexity of two combined operations could be as bad as the composition of the state complexity of each operation. However, in many cases it can be significantly less. A trivial example is the composition of reversal with itself. Proceeding naively would give us an upper-bound of 2^{2^n}. Yet we know that the composition of two reversals is the identity and therefore its state complexity is n. The state complexity of Kleene star and reversal combined with other operations has already been well studied. For example, union and intersection both have a state complexity of mn and Salomaa *et al.* showed in [5] that the state complexity of star of intersection is bounded by $\frac{3}{4}2^{mn}$, proven tight by Jiraskova and okhotin in [6], whereas star of union is exactly $2^{m+n-1} - 2^{m-1} - 2^{n-1} + 1$ which is significantly less. Similarly, Liu *et al.* studied in 2008 the compositions of the reversal with union, intersection, concatenation and star, finding that none of those compositions reaches its naive upper-bound [7]. Even though the operation Root has been less studied, the state complexity of Root combined with boolean operations has already been computed in [8]. In this paper we prove that the state complexity of the combination of the operations reversal and Root is $n^n - (n-1)^n + 1$ while the naive upper bound is $2^{n^n - \binom{n}{2}}$.

F. Drewes and M. Volkov (Eds.): DLT 2023, LNCS 13911, pp. 74–85, 2023.
https://doi.org/10.1007/978-3-031-33264-7_7

After brief preliminaries detailing all the needed basic notions, we focus in Sect. 3 on an algorithm producing an automaton accepting the Root reversal of any regular language. Section 4 is dedicated to the ways we can minimize such an automaton.

2 Preliminaries

For any finite set E, we denote by $\#E$ its number of elements. Let Σ be a finite alphabet. A *word* is a finite sequence of symbols of Σ. The empty word is denoted by ε. A *language* is a set of words. The concatenation of two words $u = a_0 \cdots a_i$ and $v = b_0 \cdots b_j$ is the word $u \cdot v = a_0 \cdots a_i b_0 \cdots b_j$. The concatenation of two languages L_1 and L_2 is the language $L_1 \cdot L_2 = \{w_1 \cdot w_2 \mid w_1 \in L_1, w_2 \in L_2\}$. The symbol $\cdot$ may be omitted when not necessary. Let us define $L^n = L \cdot L^{n-1}$ with $L^0 = \{\varepsilon\}$. The Kleene star of a language L is the set defined by $L^* = \bigcup_{n \in \mathbb{N}} L^n$. The set of *regular languages over* Σ, written $\mathrm{Rat}(\Sigma^*)$, is the smallest set containing $\emptyset$ and $\{a\}$ for all $a \in \Sigma$ and closed under union, concatenation and Kleene star. The *reversal* of a word w, denoted by $\overleftarrow{w}$, is defined as follows : $\overleftarrow{\varepsilon} = \varepsilon$ and for any word u of Σ^*, for any letter a of Σ we have $\overleftarrow{a \cdot u} = \overleftarrow{u} \cdot a$. We define the reversal of a language as $\overleftarrow{L} = \{\overleftarrow{w} \mid w \in L\}$. The Root of L is defined as $\mathrm{Root}(L) = \{w \mid \exists r \in \mathbb{N} \setminus \{0\} \mid w^r \in L\}$.

A *complete deterministic finite automaton* (DFA) is a quintuple $A = (\Sigma, Q, q_0, F, \delta)$ where $Q = \{q_0, q_1, \ldots, q_{\#Q-1}\}$ is a finite set of states, q_0 is the starting state, $F \subseteq Q$ is the set of final states and $\delta : Q \times \Sigma \to Q$ is the total transition function. The transition function of a letter a is denoted by δ_a. We extend δ to words by setting $\delta_\varepsilon(q) = q$ for each $q \in Q$ and $\delta_{wa} = \delta_a \circ \delta_w$ for any $w \in \Sigma^*$ and any $a \in \Sigma$. We can also denote δ_w by $[\delta_w(q_0), \delta_w(q_1), \ldots, \delta_w(q_{\#Q-1})]$. The word $w \in \Sigma^*$ is accepted by the automaton if and only if $\delta(q_0, w) \in F$. The *language of an automaton* A, denoted by $L(A)$, is the set of words accepted by A. We also define the *right language* of a state $q \in Q$ by the rule $L_{\overrightarrow{q}} = \{w \in \Sigma^* \mid \delta_w(q) \in F\}$. Two states having the same right language are (Nerode) *equivalent*. A word w *separates* two states when it is in the right language of one and not the other. By Kleene's theorem [9], the set of languages accepted by an automaton is the set of *regular languages*. For any given language L in $\mathrm{Rat}(\Sigma^*)$ there is one unique DFA (up to a renaming of states) accepting L that has a minimal number of states. This automaton is called the *minimal DFA* of L. None of its states are pairwise equivalent. The *state complexity* of L, denoted by $\mathrm{sc}(L)$, is the number of states of its minimal DFA. Let $\mathcal{L}_n \subseteq \mathrm{Rat}(\Sigma^*)$ be the set of languages of state complexity less or equal to n. The state complexity of a unary operation $\lozenge$ on regular languages denoted by $\mathrm{sc}_\lozenge(n)$ is defined by $\max(\{\mathrm{sc}(\lozenge(L)) \mid L \in \mathcal{L}_n\})$ and a language reaching this bound is called a *witness*. For example, it is well established that the state complexity of Kleene star is $\frac{3}{4}2^n$ [10]. This means that if we consider an n-state minimal DFA A, in the worst case, there exists a $\frac{3}{4}2^n$-state DFA that accepts $L(A)^*$, when $n > 1$.

3 An Automaton Accepting the Root-Reversal of a Regular Language

In this section we build an automaton accepting the composition of Root and reversal of a regular language. In the following we refer to this composition as *Root-reversal.*

Lemma 1. *Let u, v be two words of Σ^* and $L \subset \Sigma^*$ be a language.*

$$\overleftarrow{uv} = \overleftarrow{v} \cdot \overleftarrow{u} \tag{1}$$

$$\overleftarrow{u} \in L \Leftrightarrow u \in \overleftarrow{L} \tag{2}$$

Proof. Trivially shown by induction.

Lemma 2. *The Root and reversal operations commute.*

Proof. Let $L \subseteq \Sigma^*$ be a language. We have

$$w \in \mathrm{Root}\left(\overleftarrow{L}\right) \underset{\text{Def.}}{\Longleftrightarrow} \exists n \in \mathbb{N} \setminus \{0\} \mid w^n \in \overleftarrow{L} \underset{(2)}{\Longleftrightarrow} \exists n \in \mathbb{N} \setminus \{0\} \mid \overleftarrow{w^n} \in L$$
$$\underset{(1)}{\Longleftrightarrow} \exists n \in \mathbb{N} \setminus \{0\} \mid (\overleftarrow{w})^n \in L \underset{\text{Def.}}{\Longleftrightarrow} \overleftarrow{w} \in \mathrm{Root}(L) \underset{(2)}{\Longleftrightarrow} w \in \overleftarrow{\mathrm{Root}(L)}.$$

Definition 1. *Let $A = (\Sigma, Q, q_0, F, \delta)$ be a DFA. The* left transformation semi-automaton *of A (see Fig. 1), $T(A) = (\Sigma, M, \Delta)$, is defined by :*

- *its set of states M which is isomorphic to $\{\delta_w \mid w \in \Sigma^*\}$. We denote by q_w the state of M isomorphic to the function δ_w.*
- *its transition function $\Delta : M \times \Sigma \to M$, defined by $\Delta_a(q_w) = q_{aw}$.*

As shown by the two following lemmas, from the semiautomaton $T(A)$, we can compute an automaton accepting $\mathrm{Root}(\overleftarrow{L(A)})$ by choosing q_ε as initial state and $\{q_w \in M \mid \exists n > 0 \text{ such that } (\delta_w)^n(q_0) \in F\}$ as final states (See Fig. 2.) Those are the same states as for the Root operation [4].

Lemma 3. *Let $u, v \in \Sigma^*$ be two words. The equality $\Delta_{\overleftarrow{u}}(q_v) = q_{uv}$ holds.*

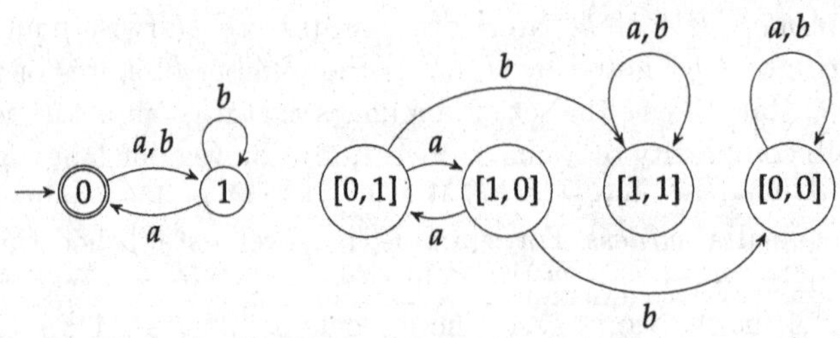

Fig. 1. An automaton B and the semiautomaton $T(B)$

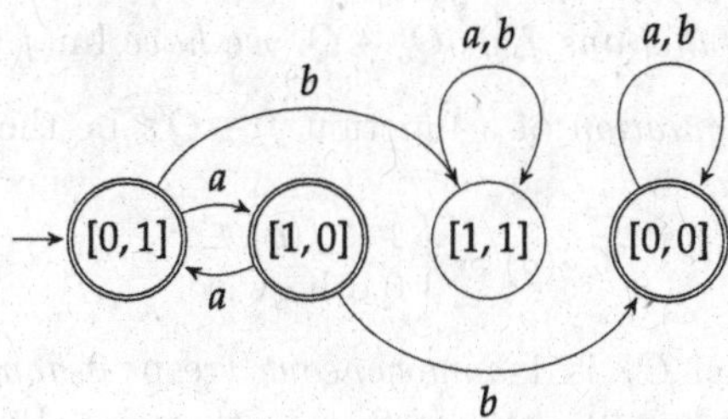

Fig. 2. An automaton that accepts the reversal of Root$(L(B))$

Proof. By induction on the size of u. If $|u| = 0$ then $u = \varepsilon$. Therefore we have $\Delta_{\overleftarrow{\varepsilon}}(q_v) = \Delta_\varepsilon(q_v) = q_v$. Let $u \in \Sigma^*$ such that $\Delta_{\overleftarrow{u}}(q_v) = q_{uv}$. We set $w = au$. Thus, we have $\Delta_{\overleftarrow{w}} = \Delta_{\overleftarrow{u}\,a} \underset{\text{Def.}}{=} \Delta_a \circ \Delta_{\overleftarrow{u}}$. Therefore $\Delta_{\overleftarrow{w}}(q_v) = (\Delta_a \circ \Delta_{\overleftarrow{u}})(q_v) \underset{\text{IH}}{=} \Delta_a(q_{uv}) \underset{\text{Def.}}{=} q_{auv} = q_{wv}$.

As a special case, we have $\Delta_{\overleftarrow{u}}(q_\varepsilon) = q_u$.

Definition 2. *Let* $A = (\Sigma, Q, q_0, F, \delta)$ *be a DFA, we denote by* $\overleftarrow{\mathcal{R}}(A) = (\Sigma, M, q_\varepsilon, F', \Delta)$ *the automaton computed from the semiautomaton* $T(A)$ *with* $F' = \{q_w \in M \mid \exists n \in \mathbb{N} \setminus \{0\},\ (\delta_w)^n(q_0) \in F\}$ *as final states and* q_ε *as initial state.*

Lemma 4. *Let* $A = (\Sigma, Q, q_0, F, \delta)$ *be a DFA. The automaton* $\overleftarrow{\mathcal{R}}(A)$ *accepts* Root$\left(\overleftarrow{L(A)}\right)$.

Proof. We have

$$\begin{aligned} w \in \text{Root}\left(\overleftarrow{L(A)}\right) &\underset{\text{Def.}}{\iff} \exists n \in \mathbb{N} \setminus \{0\} \;\; w^n \in \overleftarrow{L(A)} \\ &\iff \exists n \in \mathbb{N} \setminus \{0\} \;\; \overleftarrow{w^n} \in L(A) \\ &\underset{(2)}{\iff} \exists n \in \mathbb{N} \setminus \{0\} \;\; \overleftarrow{w}^n \in L(A) \\ &\underset{(1)}{\iff} \exists n \in \mathbb{N} \setminus \{0\} \;\; \delta_{\overleftarrow{w}^n}(q_0) \in F \\ &\underset{\text{Def.}}{\iff} \exists n \in \mathbb{N} \setminus \{0\} \;\; (\delta_{\overleftarrow{w}})^n(q_0) \in F \\ &\underset{\text{Def.}}{\iff} q_{\overleftarrow{w}} \in F'. \end{aligned}$$

By Lemma 3, we conclude

$$w \in \text{Root}\left(\overleftarrow{L(A)}\right) \iff \Delta_w(q_\varepsilon) \in F' \iff w \in L(\overleftarrow{\mathcal{R}}(A)).$$

4 State Complexity of Root-Reversal

Let $\text{Im}(f) = \{q' \in Q \mid \exists q \in Q \text{ such that } f(q) = q'\}$ be the image of a function f from Q to Q.

Lemma 5. *For any functions $f, g : Q \to Q$, we have* $\mathrm{Im}(f \circ g) \subseteq \mathrm{Im}(f)$.

Let the *final configuration* of a function $f \in Q^Q$ be the function $C_f : Q \to \{0,1\}$ defined by :

$$C_f(q) = \begin{cases} 1 \text{ if } f(q) \in F \\ 0 \text{ otherwise.} \end{cases}$$

The final configuration C_f is 1-*homogeneous* (resp. 0-*homogeneous*) if for all q in Q, $C_f(q)$ is 1 (resp. 0), *heterogeneous* otherwise. By abuse of notation, we say that the function f is 1-homogeneous, 0-homogeneous or heterogeneous. Notice that $\mathrm{Im}(f) \subseteq F$ (resp. $Q \setminus F$) is equivalent to f is 1-homogeneous (resp. 0-homogeneous).

In the next lemma we prove that all states of $\overleftarrow{\mathcal{R}}(A)$, isomorphic to a 1-homogeneous function, are equivalent.

Lemma 6. *Let $A = (\Sigma, Q, q_0, F, \delta)$ be a minimal DFA and $\overleftarrow{\mathcal{R}}(A) = (\Sigma, M, q_\varepsilon, F', \Delta)$. The right language of a state $q_w \in M$, isomorphic to a 1-homogeneous function, is Σ^*. Symmetrically the right language of a state $q_w \in M$, isomorphic to a 0-homogeneous function, is $\emptyset$. i.e.*

$$\forall w \in \Sigma^*,\ \ \mathrm{Im}(\delta_w) \subseteq F \Rightarrow L_{\overrightarrow{q_w}} = \Sigma^*. \tag{3}$$

$$\forall w \in \Sigma^*,\ \ \mathrm{Im}(\delta_w) \subseteq (Q \setminus F) \Rightarrow L_{\overrightarrow{q_w}} = \emptyset. \tag{4}$$

Proof. Consider a state $q_w \in M$ such that δ_w is 1-homogeneous (resp. 0-homogeneous). Notice that q_w is a final state (resp. a non-final state). Let $u \in \Sigma^*$ be a word. By Lemma 3, we have $\Delta_{\overline{u}}(q_w) = q_{uw}$. Lemma 5 asserts that $\mathrm{Im}(\delta_{uw}) = \mathrm{Im}(\delta_w \circ \delta_u) \subseteq \mathrm{Im}(\delta_w)$. Therefore δ_{uw} is 1-homogeneous (resp. 0-homogeneous) and q_{uw} is a final state (resp. a non-final state). Therefore $L_{\overrightarrow{q_w}} = \Sigma^*$ (resp. $\emptyset$).

Corollary 1. *In the automaton $\overleftarrow{\mathcal{R}}(A)$, all states isomorphic to a 1-homogeneous function are pairwise equivalent. Symmetrically all states isomorphic to a 0-homogeneous function, are also pairwise equivalent.*

In Theorem 1 of [11], reformulated in Theorem 1 of [12] and concurrently in [13], Caron *et al.* and Davies prove that a language, whose minimal automaton has a letter for each possible transition function, is necessarily a witness for any unary rational operation. Such automata are called monsters. Their structure induces that, for any monster A, all functions of Q^Q appear as states of $\overleftarrow{\mathcal{R}}(A)$. We extend this notion with the monster property.

Definition 3. *Let $A = (\Sigma, Q, q_0, F, \delta)$ be a DFA and $T(A) = (\Sigma, M, \Delta)$ its left transformation semiautomaton. The automaton A has the* monster property *if $M = Q^Q$.*

That is, we do not require a letter for any possible function of Q^Q, a word is enough. In the following, we focus on automata verifying the monster property in order to compute the state complexity of the Root-reversal composition. When minimizing $\overleftarrow{\mathcal{R}}(A)$, the number of pairwise non-equivalent states depends on $\#F$, the cardinality of A's final states set. We now study these parameters, considering A verifies the monster property for the rest of the paper. Notice that there exist monsters for any $n = \#Q$.

Lemma 7. *Let $A = (\Sigma, Q, q_0, F, \delta)$ be a minimal DFA having the monster property and q_{w_1}, q_{w_2} two states of $\overleftarrow{\mathcal{R}}(A)$. If $C_{\delta_{w_1}} \neq C_{\delta_{w_2}}$ then q_{w_1} and q_{w_2} are not equivalent.*

Proof. Let q_{w_1}, q_{w_2} be two states of $\overleftarrow{\mathcal{R}}(A)$ such that $C_{\delta_{w_1}} \neq C_{\delta_{w_2}}$. We then have at least one state $q_i \in Q$ for which $C_{\delta_{w_1}}(q_i) \neq C_{\delta_{w_2}}(q_i)$ (see Table 1). Without loss of generality, assume that $C_{\delta_{w_1}}(q_i) = 1$ and $C_{\delta_{w_2}}(q_i) = 0$.

Table 1. Two distinct final configurations.

	q_0	q_1	$\cdots$	q_i	$\cdots$	q_{n-1}
$C_{\delta_{w_1}}$	b_0	b_1	$\cdots$	1	$\cdots$	b_{n-1}
$C_{\delta_{w_2}}$	b'_0	b'_1	$\cdots$	0	$\cdots$	b'_{n-1}

The monster property ensures that there exists a function δ_w such that for every state $q \in Q$, we have $\delta_w(q) = q_i$ (see Table 2).

Therefore, for every state $q \in Q$, we have $C_{\delta_{w_1} \circ \delta_w}(q) = C_{\delta_{w_1}}(q_i) = 1$. That is, δ_{ww_1} is a 1-homogeneous function. From Corollary 1, we deduce $q_{ww_1} \in F'$.

We also have $C_{\delta_{w_2} \circ \delta_w}(q) = C_{\delta_{w_2}}(q_i) = 0$, that is, δ_{ww_2} is a 0-homogeneous function. From Corollary 1, we deduce $q_{ww_2} \notin F'$.

Table 2. A function δ_w separating two states with distinct final configurations.

	q_0	q_1	$\cdots$	q_i	$\cdots$	q_{n-1}
δ_w	q_i	q_i	$\cdots$	q_i	$\cdots$	q_i
$C_{\delta_{w_1} \circ \delta_w}$	1	1	$\cdots$	1	$\cdots$	1
$C_{\delta_{w_2} \circ \delta_w}$	0	0	$\cdots$	0	$\cdots$	0

Therefore

$$\overleftarrow{w} \in L_{\overrightarrow{q_{w_1}}} \setminus L_{\overrightarrow{q_{w_2}}}.$$

Thus q_{w_1} and q_{w_2} are not equivalent.

Next lemma asserts that two functions sharing the same heterogenous final configuration represent non-equivalent states of $\overleftarrow{\mathcal{R}}(A)$ if there exists a state q_i of A such that the images of q_i through both functions are different but non-final.

Lemma 8. *Let $A = (\Sigma, Q, q_0, F, \delta)$ be a minimal DFA having the monster property. Let q_{w_1}, q_{w_2} be two distinct states of $\overleftarrow{\mathcal{R}}(A)$ such that $C_{\delta_{w_1}} = C_{\delta_{w_2}}$, this common final configuration being heterogeneous. If there exists a state q_i in Q such that $\delta_{w_1}(q_i) \neq \delta_{w_2}(q_i)$ and $C_{\delta_{w_1}}(q_i) = C_{\delta_{w_2}}(q_i) = 0$ then q_{w_1} and q_{w_2} are not equivalent.*

Proof. In order to prove that q_{w_1} and q_{w_2} are not equivalent, we find a word $\overleftarrow{w}$ which is in the right language of q_{w_2} but not in the right language of q_{w_1}. Since the common final configuration is heterogeneous, there exists a state q_ℓ in Q such that $C_{\delta_{w_1}}(q_\ell) = C_{\delta_{w_2}}(q_\ell) = 1$ (see Table 3). Let us denote $\alpha = \delta_{w_1}(q_i)$ and $\beta = \delta_{w_2}(q_i)$. Notice that both α and β are non-final.

Table 3. Two functions with the same final configuration differing in q_i.

$q \in Q$	q_0	q_1	$\cdots$	q_i	$\cdots$	α	$\cdots$	β	$\cdots$	q_ℓ	$\cdots$	q_{n-1}
δ_{w_1}	q_0'	q_1'	$\cdots$	α	$\cdots$	α'	$\cdots$	β'	$\cdots$	q_ℓ'	$\cdots$	q_{n-1}'
δ_{w_2}	q_0''	q_1''	$\cdots$	β	$\cdots$	α''	$\cdots$	β''	$\cdots$	q_ℓ''	$\cdots$	q_{n-1}''
$C_{\delta_{w_1}}$	b_0	b_1	$\cdots$	0	$\cdots$	b_α	$\cdots$	b_β	$\cdots$	1	$\cdots$	b_{n-1}
$C_{\delta_{w_2}}$	b_0	b_1	$\cdots$	0	$\cdots$	b_α	$\cdots$	b_β	$\cdots$	1	$\cdots$	b_{n-1}

Consider a word w such that δ_w verifies for all q in Q :

$$\delta_w(q) = \begin{cases} q_i \text{ if } q = q_0 \text{ or } q = \alpha \\ q_\ell \text{ if } q = \beta \\ q \text{ otherwise.} \end{cases} \qquad \text{(see Table 4)}$$

Table 4. A function δ_w separating two states with the same final configuration.

$q \in Q$	q_0	q_1	$\cdots$	q_i	$\cdots$	α	$\cdots$	β	$\cdots$	q_ℓ	$\cdots$	q_{n-1}
δ_w	q_i	q_1	$\cdots$	q_i	$\cdots$	q_i	$\cdots$	q_ℓ	$\cdots$	q_ℓ	$\cdots$	q_{n-1}

The existence of w is ensured by the monster property. In the following, we compose δ_w with both functions δ_{w_1} and δ_{w_2} (see Table 5).

We have

$$\delta_{ww_1}(q_0) = \delta_{w_1}(\delta_w(q_0)) = \delta_{w_1}(q_i) = \alpha,$$

$$\delta_{ww_1}(\alpha) = \delta_{w_1}(\delta_w(\alpha)) = \delta_{w_1}(q_i) = \alpha.$$

Thus, for any positive integer k, we have $(\delta_{ww_1})^k(q_0) = \alpha$. As we know that α is a non-final state, q_{ww_1} is not a final state of $\overleftarrow{\mathcal{R}}(A)$. We have

$$\delta_{ww_2}(q_0) = \delta_{w_2}(\delta_w(q_0)) = \delta_{w_2}(q_i) = \beta.$$

Thus

$$(\delta_{ww_2})^2(q_0) = \delta_{ww_2}(\beta) = \delta_{w_2}(\delta_w(\beta)) = \delta_{w_2}(q_\ell).$$

As by assumption $\delta_{w_2}(q_\ell)$ is a final state, δ_{ww_2} is a final state of $\overleftarrow{\mathcal{R}}(A)$. As $\Delta_{\overleftarrow{w}}(q_{w_1}) = q_{ww_1}$ and $\Delta_{\overleftarrow{w}}(q_{w_2}) = q_{ww_2}$, we proved that the word $\overleftarrow{w}$ separates the states q_{w_1} and q_{w_2} (see Table 5). Therefore, they are not equivalent.

Table 5. Composition of δ_w with both functions δ_{w_1} and δ_{w_2}.

$q \in Q$	q_0	q_1	$\cdots$	q_i	$\cdots$	α	$\cdots$	β	$\cdots$	q_ℓ	$\cdots$	q_{n-1}
δ_{ww_1}	α	q_1'	$\cdots$	α	$\cdots$	α	$\cdots$	q_ℓ'	$\cdots$	q_ℓ'	$\cdots$	q_{n-1}'
δ_{ww_2}	β	q_1''	$\cdots$	β	$\cdots$	β	$\cdots$	q_ℓ''	$\cdots$	q_ℓ''	$\cdots$	q_{n-1}''
$C_{\delta_{ww_1}}$	0	b_1	$\cdots$	0	$\cdots$	0	$\cdots$	1	$\cdots$	1	$\cdots$	b_{n-1}
$C_{\delta_{ww_2}}$	0	b_1	$\cdots$	0	$\cdots$	0	$\cdots$	1	$\cdots$	1	$\cdots$	b_{n-1}

Notice that the proof remains valid if $q_\ell = \beta$. If $q_\ell = \alpha$, we need to switch the role of α and β. Last, if the only choice for q_ℓ is q_0 the proof remains valid by replacing Table 4 by Table 6.

Table 6. A function separating two states with the same final configuration when q_ℓ is q_0.

$q \in Q$	q_0	q_1	$\cdots$	q_i	$\cdots$	α	$\cdots$	β	$\cdots$	q_{n-1}
δ_w	q_i	q_1	$\cdots$	q_i	$\cdots$	q_i	$\cdots$	q_0	$\cdots$	q_{n-1}

Next lemma asserts that two functions sharing the same final configuration represent equivalent states of $\overleftarrow{\mathcal{R}}(A)$ if all states q_i of A having different images α and β through δ_{w_1} and δ_{w_2} are such that α and β are final in A.

Lemma 9. *Let $A = (\Sigma, Q, q_0, F, \delta)$ be a minimal DFA having the monster property. Let q_{w_1}, q_{w_2} be two distinct states of $\overleftarrow{\mathcal{R}}(A)$ such that $C_{\delta_{w_1}} = C_{\delta_{w_2}}$, and where this common final configuration is heterogeneous.*

If

$$\forall q \in Q, (\delta_{w_1}(q) \neq \delta_{w_2}(q) \Rightarrow (C_{\delta_{w_1}}(q) = C_{\delta_{w_2}}(q) = 1)),$$

then

$$q_{w_1} \text{ and } q_{w_2} \text{ are equivalent}.$$

Proof. Let us assume that there is a word $w \in \Sigma^*$ such that only one of $\Delta_{\overleftarrow{w}}(q_{w_1})$ and $\Delta_{\overleftarrow{w}}(q_{w_2})$ is a final state in $\overleftarrow{\mathcal{R}}(A)$.

This means that only one of q_{ww_1} and q_{ww_2} is not final in $\overleftarrow{\mathcal{R}}(A)$. Let us assume without loss of generality that q_{ww_1} is not final. By definition of F', this means that for any ℓ in $\mathbb{N} \setminus \{0\}$, we have

$$C_{(\delta_{ww_1})^\ell}(q_0) = 0.$$

As, by assumption q_{ww_2} is final, let us denote by $p > 0$ the smallest natural number such that

$$C_{(\delta_{ww_2})^p}(q_0) = 1 \quad (\neq C_{(\delta_{ww_1})^p}(q_0)). \tag{5}$$

Thus, we have

$$\forall 0 < k < p, \;\; C_{(\delta_{ww_2})^k}(q_0) = C_{(\delta_{ww_1})^k}(q_0) = 0.$$

That is

$$C_{(\delta_{w_2} \circ \delta_w)^k}(q_0) = C_{(\delta_{w_1} \circ \delta_w)^k}(q_0) = 0.$$

By contraposition of the assumption, for all $f, g \in Q^Q$ we have

$$(C_f(q_0) = C_g(q_0) = 0) \Rightarrow f(q_0) = g(q_0).$$

Then,

$$(\delta_{w_2} \circ \delta_w)^k(q_0) = (\delta_{w_1} \circ \delta_w)^k(q_0).$$

Thus, in particular

$$(\delta_{ww_2})^{p-1}(q_0) = (\delta_{ww_1})^{p-1}(q_0).$$

In the following, let us refer to $(\delta_{ww_2})^{p-1}(q_0)$ and $(\delta_{ww_1})^{p-1}(q_0)$ as q_{nf}. By the minimality of p, we have:

$$q_{\text{nf}} \notin F.$$

As $C_{\delta_{w_1}} = C_{\delta_{w_2}}$, we have

$$C_{\delta_{ww_2}}(q_{\text{nf}}) = C_{\delta_{ww_1}}(q_{\text{nf}}).$$

$$C_{\delta_{ww_2}}((\delta_{ww_2})^{p-1}(q_0)) = C_{\delta_{ww_1}}((\delta_{ww_1})^{p-1}(q_0)).$$

$$C_{(\delta_{ww_2})^p}(q_0) = C_{(\delta_{ww_1})^p}(q_0).$$

Contradiction with Eq. (5).

Let the *final rank* of C_{δ_w} be the number of states that have a final image through δ_w. Formally:

$$R(C_{\delta_w}) = \#\{q \in Q \mid C_{\delta_w}(q) = 1\}.$$

We also define the final rank of a function :

$$R(\delta_w) = \#\{q \in Q \mid \delta_w(q) \in F\} = R(C_{\delta_w}).$$

Lemma 10. *Let $A = (\Sigma, Q, q_0, F, \delta)$ be an n-state minimal DFA having the monster property. For any natural number k such that $0 < k \leq n$, there are*

$$\binom{n}{k}$$

distinct final configurations of final rank k.

Proof. This proof is straightforward. It is equivalent to the number of binary words of size n with exactly k occurrences of '1'.

Lemma 11. *Let $A = (\Sigma, Q, q_0, F, \delta)$ be an n-state minimal DFA having the monster property and C be a final configuration of rank k (where $0 < k \leq n$). The number of pairwise non-equivalent states isomorphic to functions sharing this final configuration C is*

$$(n-f)^{n-k}$$

where $f = \#F$.

Proof. Let δ_w be a function of final configuration C and final rank k. Let us compute the class of states pairwise equivalent to q_w. First, Lemma 8 ensures that any state equivalent to q_w must be isomorphic to a function that have the same non-final images as δ_w. Secondly, Lemma 9 ensures that any final image of δ_w can be replaced by any other final state while preserving the equivalence. Thus, the number of classes of states isomorphic to functions that share the final configuration C is exactly the number of mappings of zeros from C to non-final states of A. As there are $n - k$ zeros in C and $n - f$ non-final states in A the number of such mappings (thus classes of states sharing C) is $(n - f)^{n-k}$.

Notice that this property also applies for the case $k = n$ representing non-equivalent states since $(n-f)^{n-n} = 1$. This matches the result from Corollary 1.

Lemma 12. *Let $A = (\Sigma, Q, q_0, F, \delta)$ be an n-state minimal DFA having the monster property. For all natural numbers k such that $0 < k \leq n$, there are*

$$\binom{n}{k}(n-f)^{n-k}$$

non-equivalent states isomorphic to functions of final rank k, where $f = \#F$.

Proof. Lemma 7 ensures that two states cannot be equivalent if their functions don't share the same final configuration. Therefore the number of non-equivalent states, whose function is of final rank k is the straightforward product of the number of final configurations of final rank k and the number of functions sharing a specific final configuration of final rank k.

Theorem 1. *Let $A = (\Sigma, Q, q_0, F, \delta)$ be an n-state DFA having the monster property. The minimal automaton accepting* $\mathrm{Root}(\overleftarrow{L(A)})$ *has exactly*

$$(n-f+1)^n - (n-f)^n + 1$$

states, where $f = \#F$.

Proof. Lemma 12 ensures that, for any final rank $0 < k \leq n$, the number of states isomorphic to functions of final rank k is $\binom{n}{k}(n-f)^{n-k}$. Functions of final rank $k = 0$ are 0-homogeneous. Corollary 1 ensures that all the states isomorphic to those functions are equivalent. Thus, there are exactly

$$1 + \sum_{k=1}^{n} \binom{n}{k} (n-f)^{n-k}$$

pairwise non-equivalent states. Yet, we have

$$(1 + (n-f))^n = \sum_{k=0}^{n} \binom{n}{k} 1^k (n-f)^{n-k} = (n-f)^n + \sum_{k=1}^{n} \binom{n}{k} (n-f)^{n-k}.$$

Therefore

$$1 + \sum_{k=1}^{n} \binom{n}{k} (n-f)^{n-k} = (1 + n - f)^n - (n-f)^n + 1.$$

Notice that this bound is maximized for $f = 1 : n^n - (n-1)^n + 1$. Moreover this bound is minimized for $f = n - 1 : 2^n$.

Conclusion

In this paper we have studied the composition of reversal and Root. We have proven that $(n - f + 1)^n - (n - f)^n + 1$ is the state complexity of this composition (f being the number of final states). This is exponentially less than the naive composition of their respective state complexity. Furthermore we have established that every minimal DFA having the monster property reaches this bound. As three generators are necessary and enough to generate all functions over a finite set [14], thus three letters are necessary and enough to produce witnesses for the reversal Root composition. Notice that the state complexity of Root is bigger than the state complexity of reversal Root. As it can be seen in [4] witnesses for Root are also witnesses for reversal Root. Therefore, we can deduce that in the worst cases, it is more efficient to study the reversal of Root than the Root of a regular language.

References

1. Mirkin, B.G.: On dual automata. Kibernetica **2**(1), 7–10 (1966). http://www.academia.edu/download/46550561/bf0107224720160616-14181-oqti80.pdf
2. Lupanov, O.B.: A comparison of two types of finite automata. Probl. Kibernetiki **9**, 321–326 (1963)
3. Maslov, A.N.: Estimates of the number of states of finite automata. Soviet Math. Dokl. **11**, 1373–1375 (1970)

4. Krawetz, B., Lawrence, J., Shallit, J.: State complexity and the monoid of transformations of a finite set. Int. J. Found. Comput. Sci. **16**(3), 547–563 (2005). https://doi.org/10.1142/S0129054105003157
5. Salomaa, A., Salomaa, K., Yu, S.: State complexity of combined operations. Theor. Comput. Sci. **383**(2–3), 140–152 (2007). https://doi.org/10.1016/j.tcs.2007.04.015
6. Jirásková, G., Okhotin, A.: On the state complexity of star of union and star of intersection. Fundam. Inform. **109**(2), 161–178 (2011). https://doi.org/10.3233/FI-2011-502
7. Liu, G., Martín-Vide, C., Salomaa, A., Yu, S.: State complexity of basic language operations combined with reversal. Inf. Comput. **206**(9–10), 1178–1186 (2008). https://doi.org/10.1016/j.ic.2008.03.018
8. Caron, P., Hamel-De le Court, E., Luque, J.-G.: Combination of roots and Boolean operations: an application to state complexity. Inf. Comput. **289**(Part), 104961 (2022). https://doi.org/10.1016/j.ic.2022.104961
9. Kleene, S.C.: Representation of events in nerve nets and finite automata. Automata Stud. Ann. Math. Stud. **34**, 3–41 (1956)
10. Yu, S., Zhuang, Q., Salomaa, K.: The state complexities of some basic operations on regular languages. Theor. Comput. Sci. **125**(2), 315–328 (1994). https://doi.org/10.1016/0304-3975(92)00011-F
11. Caron, P., Hamel-De le Court, E., Luque, J.-G., Patrou, B.: New tools for state complexity, Discret. Math. Theor. Comput. Sci. **22**(1) (2020). http://dmtcs.episciences.org/6179
12. Caron, P., Hamel-De le Court, E., Luque, J.-G.: State complexity of the star of a Boolean operation. CoRR abs/2206.05100 (2022). https://doi.org/10.48550/arXiv.2206.05100
13. Davies, S.: A general approach to state complexity of operations: formalization and limitations. In: Hoshi, M., Seki, S. (eds.) DLT 2018. LNCS, vol. 11088, pp. 256–268. Springer, Cham (2018). https://doi.org/10.1007/978-3-319-98654-8_21
14. Ganyushkin, O., Mazorchuk, V.: Classical Finite Transformation Semigroups: An Introduction. Algebra and Applications, Springer, Dordrecht (2009). https://doi.org/10.1007/978-1-84800-281-4

Bit Catastrophes for the Burrows-Wheeler Transform

Sara Giuliani[1], Shunsuke Inenaga[2], Zsuzsanna Lipták[1], Giuseppe Romana[3], Marinella Sciortino[3](✉), and Cristian Urbina[4]

[1] University of Verona, Verona, Italy
{sara.giuliani_01,zsuzsanna.liptak}@univr.it
[2] Kyushu University, Fukuoka, Japan
inenaga.shunsuke.380@m.kyushu-u.ac.jp
[3] University of Palermo, Palermo, Italy
{giuseppe.romana01,marinella.sciortino}@unipa.it
[4] University of Chile, Santiago, Chile
crurbina@dcc.uchile.cl

Abstract. A bit catastrophe, loosely defined, is when a change in just one character of a string causes a significant change in the size of the compressed string. We study this phenomenon for the Burrows-Wheeler Transform (BWT), a string transform at the heart of several of the most popular compressors and aligners today. The parameter determining the size of the compressed data is the number of equal-letter runs of the BWT, commonly denoted r.

We exhibit infinite families of strings in which insertion, deletion, resp. substitution of one character increases r from constant to $\Theta(\log n)$, where n is the length of the string. These strings can be interpreted both as examples for an increase by a multiplicative or an additive $\Theta(\log n)$-factor. As regards multiplicative factor, they attain the upper bound given by Akagi, Funakoshi, and Inenaga [Inf & Comput. 2023] of $\mathcal{O}(\log n \log r)$, since here $r = \mathcal{O}(1)$.

We then give examples of strings in which insertion, deletion, resp. substitution of a character increases r by a $\Theta(\sqrt{n})$ additive factor. These strings significantly improve the best known lower bound for an additive factor of $\Omega(\log n)$ [Giuliani et al., SOFSEM 2021].

Keywords: Burrows-Wheeler transform · Equal-letter run · Repetitiveness measure · Sensitivity

1 Introduction

The Burrows-Wheeler Transform (BWT) [5] is a reversible transformation of a string, consisting of a permutation of its characters. It can be obtained by sorting all of its rotations and then concatenating their last characters. The BWT is present in several compressors, such as *bzip* [27]. It also lies at the heart of some of the most powerful compressed indexes in terms of query time and functionality,

F. Drewes and M. Volkov (Eds.): DLT 2023, LNCS 13911, pp. 86–99, 2023.
https://doi.org/10.1007/978-3-031-33264-7_8

such as the well-known *FM-index* [9], and the more recent *RLBWT* [22] and *r-index* [2,11,12]. Some of the most commonly used bioinformatics tools such as `bwa` [20], `bowtie` [19], and SOAP2 [18] also use the BWT at their core.

Given a string w, the measure $r = r(w)$ is defined as $r(w) = runs(\text{BWT}(w))$, where $runs(v)$ denotes the number of equal-letter runs of a string v. It is well known that r tends to be small on repetitive inputs. This is because, on texts with many repeated substrings, the BWT tends to create long runs of equal characters (so-called *clustering effect*). Due to the widespread use of runlength-compressed BWT-based data structures, the BWT can thus be viewed as a *compressor*, with the number of runs r the compression size. On the other hand, r is also increasingly being used as a *repetitiveness measure*, i.e. as a parameter of the input text itself. In [25], the relationships between many repetitiveness measures, among these r, is explored in detail. In particular, all repetitiveness measures considered are lower bounded by δ [16], a measure closely related to the *factor complexity* function [21]. It was further shown in [14] that r is within a polylog(n) factor of z, the number of phrases of LZ77-compressor.

In [13], the ratio $r(w)/r(\text{rev}(w))$ was studied, and infinite families of strings given for which this ratio is $\Theta(\log n)$, where n is the length of the string. This family can also serve as examples of strings in which appending one character can cause r to increase from $\mathcal{O}(1)$ to $\Theta(\log n)$. In this paper we further explore this effect, extending it to the other edit operations of deletion and substitution, for which we also give examples of a change from $\mathcal{O}(1)$ to $\mathcal{O}(\log n)$. Note that this attains the known upper bound of $\mathcal{O}(\log r \log n)$ [1].

Akagi et al. [1] refer to the question of how changes of just one character affect the compression ratio of a compressor as *sensitivity*. In particular, the sensitivity of a compressor is the maximum *multiplicative factor* by which a single-character edit operation can change the size of the output of the compressor. In addition, they also study the maximum *additive factor* an edit operation may cause in the output. Our second family of strings falls in this category: these are strings with r in $\Theta(\sqrt{n})$ on which an edit operation (insertion, deletion, or substitution) can cause r to increase by a further additive factor of $\Theta(\sqrt{n})$. This is a significant improvement over the previous lower bound of $\Omega(\log n)$ [13].

Lagarde and Perifel in [17] show that Lempel-Ziv 78 (LZ78) compression suffers from the so-called "one-bit catastrophe": they give an infinite family of strings for which prepending a character causes a previously compressible string to become incompressible. They also show that this "catastrophe" is not a "tragedy": they prove that this can only happen when the original string was already poorly compressible.

Here we are using the term "one-bit catastrophe" in a looser meaning, namely simply to denote the effect that an edit operation may change the compression size significantly, i.e. increase it such that $r(w'_n) = \omega(r(w_n))$, for an infinite family $(w_n)_{n>0}$, where w'_n is the word resulting from applying a single edit operation to w_n. For a stricter terminology we would need to decide for one of the different definitions of BWT-compressibility currently in use. In particular, a string may be called compressible if r is in $\mathcal{O}(n/\text{polylog}(n))$ [14], or if $r(w)/runs(w) \to 0$ [10], or even as soon as $runs(w) > r(w)$ [23].

Some of our bit catastrophes can also be thought of as "tragedies", since the example families of the first group are precisely those with the best possible compression: their BWT has 2 runs. In this sense our result on the BWT is even more surprising than that of [17] on LZ78.

Note that, in contrast to Lempel-Ziv compression, for the BWT, appending, prepending, and inserting are equivalent operations, since the BWT is invariant w.r.t. conjugacy. This means that, if there exists a word w and a character c s.t. appending c to w causes a certain change in r, then this immediately implies the existence of equivalent examples for prepending and inserting character c. This is because $r(wc)/r(w) = x$ (appending) implies that $r(cw)/r(w) = x$ (prepending), as well as $r(ucv)/r(uv) = x$, for every conjugate uv of $w = vu$ (insertion).

Finally, the BWT comes in two variants: in one, the transform is applied directly on the input string: this is the preferred variant in literature on combinatorics on words. In the other, the input string is assumed to have an end-of-string marker, usually denoted \$: this variant is common in the string algorithms and data structures literature. We show that there can be a multiplicative $\Theta(\log n)$, or an additive $\Theta(\sqrt{n})$ factor difference between the two transforms. It is interesting to note that the previous remark about the equivalence of insertion in different places in the text does not extend to the variant with the final dollar. We show, however, that our results regarding the $\Theta(\sqrt{n})$ additive factor extend also to this variant, for all three edit operations. This is in stark contrast with the known fact that *prepending* a character only changes the number of runs of the \$-variant by at most 2.

Due to space restrictions, some proofs are deferred to the full version of this work.

2 Preliminaries

Basics on words. Given a finite and ordered *alphabet* Σ, let σ be the size of the alphabet, i.e. the number of symbols $\{c_0, c_1, \ldots, c_{\sigma-1}\}$ in Σ, called *characters*. A finite sequence of characters $v = v[0]v[1]\cdots v[n-1]$ (or just $v = v[0..n-1]$) of length n from Σ is a *word* (or *string*) over Σ. We denote by $|v|$ the length of the word v. The set of words over Σ is denoted Σ^*. We denote by ϵ the empty string of length 0. We call the *reverse* of a word, the word read right-to-left, and we denote it by $\text{rev}(v) = v[n-1..0]$. If $n > 0$, we denote by $\widehat{v}$ the word $v[0..n-2]$, i.e. v without its last character. Given a word $v = pts$, p is a *prefix*, s a *suffix* and t a *factor* of v. Let $v' = tsp$, then v and v' are said to be *conjugates* (or *rotations* one of the other), and $v' = \text{conj}_{|p|}(v)$. In other words, $v = v[0..n-1]$ is a conjugate of v' if and only if there exists an integer $i \leq n-1$ such that $v' = v[i..n-1]v[0..i-1]$. A word u is a *circular factor* of v if it is a prefix of some conjugate of v. A power of a word v, denoted by v^k for some integer $k > 0$, is the word obtained by concatenating v with itself k times. Moreover, we denote by $\Pi_{i=1}^{k} v_i$ the concatenation of the words $v_1, v_2, \ldots, v_k$.

An *equal-letter run* (or simply *run*) is a maximal substring consisting of the same character, and *runs*(v) is the number of equal-letter runs in the word v. For example, the word `catastrophic` of length 12 has 12 runs, having all runs of length 1, while the word `mississippi` has 8 runs.

Given two words $v = uw$, $v' = uw'$ such that $v[|u|] \neq v'[|u|]$ and for $0 \leq i < |u|$, $v[i] = v'[i]$, we call u the *longest common prefix* of v, v', and $u = \text{lcp}(v, v')$. The *lexicographic order* on Σ^* is defined by: $v < v'$ if either v is a proper prefix of v', or ua is a prefix of v and ub is a prefix of w, where $u = \text{lcp}(v, v')$ and $a, b \in \Sigma$ with $a < b$.

Burrows-Wheeler Transform. We call *conjugate array* the permutation CA of the positions of v such that $i < j$ if and only if $\text{conj}_{\text{CA}[i]}(v) < \text{conj}_{\text{CA}[j]}(v)$, or $\text{conj}_{\text{CA}[i]}(v) = \text{conj}_{\text{CA}[j]}(v)$ and $\text{CA}[i] < \text{CA}[j]$. In other words, CA contains the starting position of all conjugates of v in lexicographic order. Going back to our example, the positions of the word `catastrophic` sorted w.r.t. the lexicographic order of its rotation are $[3, 1, 0, 11, 9, 10, 7, 8, 6, 4, 2, 5]$. This means that the rotation starting in position 0 (the word itself) is the third smallest rotation, the rotation starting in position 3 is the smallest one, and so on.

Let $w = \text{BWT}(v)$ be the *Burrows-Wheeler Transform* of the word v defined as follows: $w[i] = v[\text{CA}[i] - 1]$ when $\text{CA}[i] > 0$, $w[i] = v[n-1]$ when $\text{CA}[i] = 0$. The BWT of the word `catastrophic` is `tcciphrotaas`. Clearly, the rotations of every word in the same conjugacy class will have the same lexicographic ordering, therefore the order of the positions in CA will be different for each rotation, but the order of the characters in the BWT will be the same. Let us consider the rotation of our example starting in position 7. The word is `ophiccatastr`, and sorting its rotations, we have the following sorted positions $[8, 6, 5, 4, 2, 3, 0, 1, 11, 9, 7, 10]$, but the BWT is again `tcciphrotaas`.

We denote with $r(v) = runs(\text{BWT}(v))$ the number of runs in the BWT of the word v. For example, $r(\texttt{catastrophic}) = runs(\texttt{tcciphrotaas}) = 10$.

In the context of string algorithms and data structures, it is usually assumed that each string terminates with an end-of-string symbol (denoted by \$) that is smaller than all other symbols in the alphabet. In fact, with this assumption, sorting the conjugates of $w\$$ can be reduced to lexicographically sorting its suffixes. Note that appending the character \$ to the word w changes the output of BWT. We denote by $r_\$(v) = runs(\text{BWT}(v\$))$. For example, $\text{BWT}(\texttt{catastrophic\$}) = \texttt{ctci\$phrotaas}$ and $r_\$(\texttt{catastrophic}) = 12$.

Standard Words. Given an infinite sequence of integers $(d_0, d_1, d_2, \ldots)$, with $d_0 \geq 0, d_i > 0$ for all $i > 0$, called a *directive sequence*, define a sequence of words s_i with $i \geq 0$ of increasing length as follows: $s_0 = \texttt{b}$, $s_1 = \texttt{a}$, $s_{i+1} = s_i^{d_{i-1}} s_{i-1}$, for $i \geq 1$. The words s_i are called *standard words*. The index i is referred to as the *order* of s_i. Without loss of generality, here we can assume that $d_0 > 0$. Each standard word s_i, with $i \geq 2$, can be written as $s_i = x_i\texttt{ab}$ if i is even, $s_i = x_i\texttt{ba}$ if i is odd. The factor x_i is a palindrome [7].

Standard words represent a very well known family of binary words that have several combinatorial properties and characterizations and appear as extreme case in many contexts [6,15,24,26]. In particular, in [24], it has been proved that there exist binary words whose BWT consists of exactly two runs, meaning that their characters are as clustered as possible. They are precisely all the conjugates of powers of standard words.

Fibonacci words are a particular case of standard words, having the directive sequence consisting of only ones. More precisely, Fibonacci words can be defined as follows: $s_0 = \mathtt{b}$, $s_1 = \mathtt{a}$, $s_{i+1} = s_i s_{i-1}$, for $i \geq 1$. Denoted by F_k the length of the Fibonacci word s_k, F_k is a Fibonacci number. By using the properties of the Fibonacci numbers, $k = \Theta(\log n)$, where $n = |s_k|$. Moreover, for all $k \geq 1$, $s_{2k} = x_{2k}\mathtt{ab}$ and $s_{2k+1} = x_{2k+1}\mathtt{ba}$, where x_{2k} and x_{2k+1} are palindromes ($x_2 = \epsilon$). The recursive structure of the words x_{2k} and x_{2k+1} is also known [8]. In fact, $x_{2k} = x_{2k-1}\mathtt{ba}x_{2k-2} = x_{2k-2}\mathtt{ab}x_{2k-1}$ and $x_{2k+1} = x_{2k}\mathtt{ab}x_{2k-1} = x_{2k-1}\mathtt{ba}x_{2k}$.

3 Increasing r by a $\Theta(\log n)$-Factor

In this section we show some infinite families of words such that a single edit operation, such as insertion, deletion or substitution of a character, causes an increment of r from constant to $\Theta(\log n)$, where n is the length of the word. The impact on the BWT of the three edit operations is shown in Fig. 1.

3.1 Inserting a Character

Recall that appending a character to the reverse of a Fibonacci word can increase the number of runs by a logarithmic factor [13].

Proposition 1 ([13], Prop. 3). *Let v be the reverse of a Fibonacci word. If $rev(v)$ is of even order $2k$, then $r(v\mathtt{b}) = 2k$. If $rev(v)$ is of odd order $2k+1$, then $r(v\mathtt{a}) = 2k$.*

The following proposition can be deduced from the previous result and shows that there exist at least two positions in a Fibonacci word of even order, where adding a character causes a logarithmic increment of r.

Proposition 2. *Let s be the Fibonacci word of even order $2k$. Adding a* b *at position $F_{2k-1} - 2$ (or an* a *at position $F_{2k} - 2$), then the BWT of the resulting word has $\Theta(\log n)$ runs, where $n = |s|$.*

Proof. It is known that each standard word and its reverse are conjugate. Let us consider $s = x_{2k}\mathtt{ab}$, where x_{2k} is palindrome. Moreover, from known properties on Fibonacci words, $x_{2k} = x_{2k-1}\mathtt{ba}x_{2k-2} = x_{2k-2}\mathtt{ab}x_{2k-1}$ with x_{2k-1} and x_{2k-2} palindrome, as well. Let $v = \mathtt{ba}x_{2k}$ be the reverse of s. From Proposition 1, we know that $r(v\mathtt{b}) = 2k$. One can verify that appending a b to v is equivalent to inserting b at position $F_{2k-1} - 2$ of s. On the other hand $x_{2k}\mathtt{ba}$ is a conjugate of both v and s and it is a standard word of odd order $2k - 1$. Let us denote it t. It is easy to verify that $\mathrm{rev}(t) = \mathrm{conj}_{F_{2k}-2}(s)$. By [13, Proposition 8], $r(\mathrm{rev}(t)\mathtt{a}) = 2k - 2 = \Theta(\log n)$, and for similar considerations as above, appending an a to $\mathrm{rev}(t)$ is equivalent to inserting a at position $F_{2k} - 2$ in s. □

An analogous result can be proved for Fibonacci words of odd order.

Note that, by using similar arguments as in Proposition 1, it is possible to prove that adding a further b at the end of the reverse of a Fibonacci word of even order has the same effect as adding a character c greater than b and not present in the word. This is because in both cases a new factor is introduced in the word, which are bb respectively c. Both these factors are greater than all the other factors of the word, and they are the only change in the word. The analogous happens adding a further a to the reverse of a Fibonacci word of odd order or a character smaller than a, say $. We formalize this in the following:

Proposition 3. *Let v be the reverse of the Fibonacci word of even order $2k$ and $c > \mathtt{b}$, then $r(vc) = \Theta(k)$.*

Proposition 4. *Let v be the reverse of the Fibonacci word of odd order $2k+1$ and $c < \mathtt{a}$, then $r(vc) = \Theta(k)$.*

3.2 Deleting a Character

We next show that deleting a character can result in a logarithmic increment in r. In particular, we consider a Fibonacci word of even order and compute the form of its BWT, as shown in the following proposition.

Proposition 5. *Let s be the Fibonacci word of length n and order $2k > 4$, and $v = s[0..n-2] = \widehat{s}$. Then $BWT(v)$ has the following form: $BWT(v) = \mathtt{b}^{k-1}\mathtt{ab}^{F_{2k-3}-k+1}\mathtt{ab}^{F_{2k-5}} \ldots \mathtt{b}^{F_5}\mathtt{ab}^{F_3}\mathtt{aba}^{F_{2k-1}-k+1}$. In particular, it has $2k$ runs.*

To give the proof, we divide the BWT matrix of the word v in three parts: *top*, *middle* and *bottom part*, showing that the top part consists in 3 runs of the form $\mathrm{BWT}_{\mathrm{top}}(v) = \mathtt{b}^{k-1}\mathtt{ab}^{F_{2k-3}-k+1}$, the middle part in $2(k-2)$ runs of the form $\mathrm{BWT}_{\mathrm{mid}}(v) = \mathtt{ab}^{F_{2k-5}}\mathtt{ab}^{F_{2k-7}} \cdots \mathtt{ab}$, and the bottom part $\mathrm{BWT}_{\mathrm{bot}}(v)$ is just one long run of a's. Altogether, we then have $3 + 2(k-2) + 1 = 2k$ runs.

We identify 3 conjugates of the word v of length $n-1$ that delimit the 3 parts of the BWT matrix of the word: $\mathrm{conj}_{n-3}(v) = \mathtt{aa}x_{2k-2}\mathtt{ab}x_{2k-3}\mathtt{ba}\cdots\mathtt{ab}$, $\mathrm{conj}_{n-5}(v) = \mathtt{abaaab}\cdots x_3\mathtt{ba}$ and $\mathrm{conj}_0(v) = x_{2k}\mathtt{a}$, such that $\mathrm{conj}_{n-3}(v) < \mathrm{conj}_{n-5}(v) < \mathrm{conj}_0(v)$.

The first mentioned rotation is the smallest rotation in the matrix due to the unique aaa prefix. The second rotation indicates the beginning of the middle part, and it is the smallest rotation starting with ab. Finally, the word itself $v = x_{2k}\mathtt{a}$ determines the beginning of the bottom part, namely the last long run of a's in the BWT.

The top part of the matrix consists of all rotations of the word starting with aa. We are going to show that only one of these rotations ends with an a, and we show where the a in the BWT of the top part breaks the run of b's.

Lemma 1 (Top part). *Given v the Fibonacci word of order $2k$ without the last character, then the first k rotations in the BWT matrix are $\mathtt{aa}\cdots\mathtt{b} < \mathtt{a}x_4\mathtt{aa}\cdots\mathtt{b} < \mathtt{a}x_6\mathtt{aa}\cdots\mathtt{b} < \ldots < \mathtt{a}x_{2k-2}\mathtt{aa}\cdots\mathtt{b} < \mathtt{a}x_{2k}$. All other $F_{2k-3}-k+1$ rotations starting with* aa *end with a* b.

Proof. There are $F_{2k-3}+1$ occurrences of aa in v. This is because there are F_{2k-1} a's, of which $F_{2k-2}-1$ are followed by a b, therefore there are $F_{2k-1}-F_{2k-2}+1 = F_{2k-3}+1$ occurrences of a followed by an a.

Among all rotations starting by aa the k smallest ones are those starting with the rightmost occurrence of ax_h for each even h in increasing order of h. This is because of the single occurrence of aaa, consisting in the rightmost occurrence of x_3 followed by aa. Finally, only the largest ax_h is preceded by an a, therefore the k smallest rotations of v are all preceded by a b except for the largest of them which is preceded by an a. This shows that the k smallest rotations in the BWT matrix form two runs: $\mathtt{b}^{k-1}\mathtt{a}$.

There are no other occurrence of aaa, therefore all the remaining $F_{2k-3}-k+1$ rotations starting with aa are preceded by b by construction. Therefore, we have $\mathtt{b}^{k-1}\mathtt{ab}^{F_{2k-3}-k+1}$ in the top part of the BWT matrix of v. □

In order to prove the number of runs in the middle part of the matrix, we first draw the attention to the fact that, for each x_h, all rotations starting with x_h appear together in the BWT matrix. In particular, they occur in blocks such that rotations starting with x_ha, h odd, are immediately preceded by the unique rotation starting with x_{h-1}aa, and immediately followed by that starting with x_{h+1}aa. This is because x_ha, h odd, is prefixed by x_{h-1}ab.

Note that every two lexicographically consecutive rotations of a Fibonacci word s differ by just two characters [3]. The word v separates the middle and the bottom part. Therefore, for all $\mathrm{conj}_i(v)$ such that $\mathrm{lcp}(v, \mathrm{conj}_i(v)) = x_h$ for some h, where $\mathrm{conj}_i(v)$ is prefixed by x_ha while v is prefixed by x_hb, then $\mathrm{conj}_i(v)$ is in the middle part (i.e. smaller than v).

Lemma 2. *There are $F_{2h}-1$ occurrences of* b$x_{2(k-h)}$a *and F_{2h+1} occurrences of* b$x_{2(k-h)-1}$a *as circular factors in $v = x_{2k}$*a

Lemma 3 (Middle part). *The middle part contributes to $r(v)$ with $2(k-2)$ runs in the following form:* $\mathtt{ab}^{F_{2k-5}}\mathtt{ab}^{F_{2k-7}}\cdots\mathtt{ab}^{F_3}\mathtt{ab}$

Proof. One can prove that, among rotations starting with the same x_h, $h \geq 4$ even, the smallest one is preceded by a. All the following F_{2k-h-1} rotations starting with x_{h+1}a are preceded by b (Lemma 2). The fact that there exist exactly $k-2$ such x_h proves the claim. □

The rotations that divide the middle part from the bottom part are the two rotations prefixed by the two occurrences of x_{2k-1}. By properties of Fibonacci words, one rotation is prefixed by x_{2k-1}a (end middle part) and the other by x_{2k-1}b (beginning bottom part). The latter follows the first in lexicographic order. Note that the rotation starting with x_{2k-1}b is v, namely x_{2k-1}bax_{2k-2}a.

Lemma 4 (Bottom part). *All rotations greater than $v = x_{2k-1}$bax_{2k-2}a end with* a.

Proof. From Lemma 1 we have that $k-1+F_{2k-3}-k+1$ rotations ending with b have already appeared in the matrix, and from Lemma 3 $F_{2k-5}+\ldots+F_3+F_1$

rotations ending with b have already appeared in the matrix. Summing the number of b's we have $k-1+F_{2k-3}-k+1+F_{2k-5}+\ldots+F_3+F_1 = F_{2k-3}+F_{2k-5}+\ldots+F_3+F_1$. We can decompose each odd Fibonacci number F_{2x+1} in the sum $F_{2x}+F_{2x-1}$. Therefore, the previous sum becomes $F_{2k-4}+F_{2k-5}+F_{2k-6}+F_{2k-7}\cdots+F_2+F_1+F_1$. For every Fibonacci number F_x, it holds that $F_x = F_{x-2}+F_{x-3}+F_{x-4}+\ldots+F_2+F_1+2$. Therefore, $F_{2k-4}+F_{2k-5}+F_{2k-6}+F_{2k-7}\cdots+F_2+F_1+F_1 = F_{2k-2}-1$, which is exactly the number of b's in v. Therefore all the remaining rotations must end with a. □

We have thus shown that appending or deleting a single character can substantially increase the parameter r. This implies the following known result:

Corollary 1. *The measure r is not monotone.*

3.3 Substituting a Character

In this subsection, we show how to increment r by a logarithmic factor by substituting a character. Consider a modification of the reverse of Fibonacci words of even order in which we replace the last a by a b. More formally, let $s = x_{2k}\mathtt{ab}$ be the Fibonacci word of order $2k$, and consider $v = \mathtt{ba}\widehat{x_{2k}}\mathtt{b}$. Recall that replacing the last a in rev(s) with a b is equivalent to replace in the Fibonacci word s the character a at position $F_{2k-1}-3$ by b. Then BWT(v) has $\Theta(\log n)$ runs, where n is the length of the word, as shown in the following proposition.

Proposition 6. *Let s be the Fibonacci word of order $2k$ with a* b *instead of the first* a*, and v = rev(s) = $\mathtt{ba}\widehat{x}\mathtt{b}$. Then BWT($v$) has the form BWT($v$) =* $\mathtt{b}^{F_{2k-2}-k+1}\mathtt{a}^{F_0}\mathtt{ba}^{F_2}\mathtt{b}\cdots\mathtt{a}^{F_{2k-4}}\mathtt{ba}^{F_{2k-2}-2}\mathtt{ba}$. *In particular, BWT($v$) has $2k$ runs.*

To prove the proposition we can use a similar strategy as previously, dividing the BWT matrix of v in three parts: *top, middle* and *bottom* part.

4 Additive $\Omega(\sqrt{n})$ Factor

We know that there exist infinite families of words such that $r = \Theta(\log n)$. For instance, we can consider the Thue-Morse words [4]. Moreover, there exist words such that r is maximal. For instance, if w = aaaabbababbbbaabab, then $r(w) = 18 = |w|$ [23]. Here, we show that there is no gap between these two scenarios, i.e., it is possible to construct infinite families of words w such that $r(w) = \Theta(n^{1/k})$, for any $k > 1$.

Proposition 7. *Let k be a positive integer. There exists an infinite family T_k of binary words such that $r(w) = \Theta(n^{1/k})$, for any $w \in T_k$.*

Proof. We can define the set $T_k = \{w_{i,k} = \prod_{j=1}^{i} \mathtt{ab}^{j^k} \mid i \geq 1\}$. We can state that $|w_{i,k}| = \Theta(i^{k+1})$. Moreover, BWT($w_{i,k}$) = $\Theta(i)$. In fact, it is possible to prove that, when $k=1$, BWT($w_{i,1}$) = $\mathtt{bb}^i\mathtt{ab}^{i-1}\mathtt{a}\cdots\mathtt{ab}^3\mathtt{ab}^2\mathtt{aa}$. Hence, $r(w_{i,1}) = 2i-2 = \Theta(i)$. If $k > 1$, it is possible to verify that two consecutive a's never appear in BWT($w_{i,k}$), therefore $r(w_{i,k}) = 2i = \Theta(i) = \Theta(n^{\frac{1}{k+1}})$, where $n = |w_{i,k}|$. □

	CA	rotations of abaababaabaab	
0	7	aabaababaabab	b
1	2	aababaabaabab	b
2	10	aababaababaab	b
3	5	abaabaababaab	b
4	0	abaababaabaab	b
5	8	abaababaababa	a
6	3	ababaabaababa	a
7	11	ababaababaaba	a
8	6	baabaababaaba	a
9	1	baababaabaaba	a
10	9	baababaababaa	a
11	4	babaabaababaa	a
12	12	babaababaabaa	a

(a) Fibonacci word of order 6

	CA	rotations of abaababbaabaab	
0	8	aabaababaababb	b
1	11	aababaababbaab	b
2	2	aababbaabaabab	b
3	9	abaababaababba	a
4	0	abaababbaabaab	b
5	12	ababaababbaaba	a
6	3	ababbaabaababa	a
7	5	abbaabaababaab	b
8	7	baabaababaabab	b
9	10	baababaababbaa	a
10	1	baababbaabaaba	a
11	13	babaababbaabaa	a
12	4	babbaabaababaa	a
13	6	bbaabaababaaba	a

(b) Insertion

	CA	rotations of abaababaabaa	
0	10	aaabaababaab	b
1	7	aabaaabaabab	b
2	11	aabaababaaba	a
3	2	aababaabaaab	b
4	8	abaaabaababa	a
5	5	abaabaaabaab	b
6	0	abaababaabaa	a
7	3	ababaabaaaba	a
8	9	baaabaababaa	a
9	6	baabaaabaaba	a
10	1	baababaabaaa	a
11	4	babaabaaabaa	a

(c) Deletion

	CA	rotations of abaabbbaabaab	
0	7	aabaababaabbb	b
1	10	aababaabbbaab	b
2	2	aabbbaabaabab	b
3	8	abaababaabbba	a
4	0	abaabbbaabaab	b
5	11	ababaabbbaaba	a
6	3	abbbaabaababa	a
7	6	baabaababaabb	b
8	9	baababaabbbaa	a
9	1	baabbbaabaaba	a
10	12	babaabbbaabaa	a
11	5	bbaabaababaab	b
12	4	bbbaabaababaa	a

(d) Substitution

Fig. 1. The BWT matrix of the Fibonacci word of order 6 (a), and that of the result for 3 bit-catastrophes: (b) inserting a character in position $6 = F_{6-1} - 2$, (c) deleting the last character (d) substituting the character in position $5 = F_{6-1} - 3$.

Furthermore, whereas in the previous section we proved that a single edit operation can cause a multiplicative increase by a logarithmic factor in the number of runs, here we exhibit an infinite family of words in which a single edit operation causes an additive increment of r by $\Theta(\sqrt{n})$.

Given an integer k, let $s_i = \mathtt{ab}^i\mathtt{aa}$, $e_i = \mathtt{ab}^i\mathtt{aba}^{i-2}$ for all $2 \leq i \leq k-1$, and $q_k = \mathtt{ab}^k\mathtt{a}$. The words for which the gap announced holds is for $w_k = (\prod_{i=2}^{k-1} s_i e_i) q_k$, for all $k > 5$.

Lemma 5. *Let $n = |w_k|$ for some $k > 2$. It holds that $n = \sum_{i=2}^{k-1}(3i+4) + (k+2) = (3k^2 + 7k - 18)/2$. Moreover, it holds that $k = \Theta(\sqrt{n})$.*

Table 1. Scheme of the BWT-matrix of a word w_k with $k > 5$. The *block prefix* column shows the common prefix shared by all the rotations in a block. The *ordering factor* column shows the factor following the block prefix of a rotation, which decides its relative order inside its block. The BWT column shows the last character of each rotation. The dashed lines divide sub-ranges of rotations for which the BWT follows distinct patterns.

Block prefix	Ordering factor	BWT
$\mathtt{a}^{k-2}\mathtt{b}$	$\mathtt{b}^{k-1}\mathtt{a}$	b
$\mathtt{a}^{k-3}\mathtt{b}$	$\mathtt{b}^{k-2}\mathtt{aa}$	b
	$\mathtt{b}^{k-1}\mathtt{a}$	a
⋮	⋮	⋮
$\mathtt{a}^{4}\mathtt{b}$	$\mathtt{b}^{5}\mathtt{aa}$	b
	$\mathtt{b}^{6}\mathtt{aa}$	a
	⋮	⋮
	$\mathtt{b}^{k-1}\mathtt{a}$	a
aaab	bab	b
	bbaba	b
	bbbabaa	b
	bbbbaa	b
	bbbbabaaa	b
	bbbbbaa	a
	bbbbbabaaaa	b
	⋮	⋮
	$\mathtt{b}^{k-2}\mathtt{aa}$	a
	$\mathtt{b}^{k-2}\mathtt{aba}^{k-3}$	b
	$\mathtt{b}^{k-1}\mathtt{a}$	a
aab	baa	b
	bab	a
	bbaba	a
	bbbaa	b
	bbbabaa	b
	bbbbaa	a
	bbbbabaaa	a
	⋮	⋮
	$\mathtt{b}^{k-2}\mathtt{aa}$	a
	$\mathtt{b}^{k-2}\mathtt{aba}^{k-3}$	a
	$\mathtt{b}^{k-1}\mathtt{a}$	a
ab	$\mathtt{a}^{k-3}q_k$	b
	$\mathtt{a}^{k-4}s_{k-1}$	b
	⋮	⋮
	s_3	b
	baa	a
	bab	b
	bbaa	b
	bbaba	a
	bbbaa	a
	bbbabaa	a
	⋮	⋮
	$\mathtt{b}^{k-1}\mathtt{a}$	a
ba	$\mathtt{a}^{k-4}q_k$	a
	$\mathtt{a}^{k-5}s_{k-1}$	a
	⋮	⋮
	$\mathtt{a}^{2}s_6$	a
	$\mathtt{a}e_2$	b
	$\mathtt{a}e_3$	b
	$\mathtt{a}e_4$	a
	$\mathtt{a}s_5$	a
	$\mathtt{a}e_5$	b
	$\mathtt{a}e_6$	b
	⋮	⋮
	$\mathtt{a}e_{k-1}$	b
	s_2	b
	s_4	a
	$\mathtt{ba}^{k-3}q_k$	b
	$\mathtt{ba}^{k-4}s_{k-1}$	b
	⋮	⋮
	$\mathtt{b}s_3$	b
	bbbaa	a
bba	$\mathtt{a}e_2$	a
	$\mathtt{a}e_3$	b
	⋮	⋮
	$\mathtt{a}e_{k-1}$	b
	s_2	b
	$\mathtt{ba}^{k-3}q_k$	b
	$\mathtt{ba}^{k-4}s_{k-1}$	b
	⋮	⋮
	$\mathtt{ba}s_4$	b
	$\mathtt{ba}s_3$	a
⋮	⋮	⋮
$\mathtt{b}^{k-1}\mathtt{a}$	$\mathtt{a}e_{k-1}$	a
	s_2	b
	$\mathtt{ba}^{k-3}q_k$	a
$\mathtt{b}^{k}\mathtt{a}$	s_2	a

The following lemma will be used to show how the rotations of w_k can be sorted according to the factorization $s_2e_2 \cdots s_{k-1}e_{k-1}q_k$.

Lemma 6. *Let $k > 2$ be an integer. Then, $s_2 < e_2 < s_3 < e_3 < \ldots < s_{k-1} < e_{k-1} < q_k$. Moreover the set $\bigcup_{i=2}^{k-1}\{s_i, e_i\} \cup \{q_k\}$ is prefix-free.*

In order to compute the BWT for the word $w_k = (\prod_{i=2}^{k-1} s_i e_i) q_k = (\prod_{i=2}^{k-1} \mathtt{ab}^i\mathtt{aa} \cdot \mathtt{ab}^i\mathtt{aba}^{i-2}) \cdot \mathtt{ab}^k\mathtt{a}$, we divide BWT into disjoint ranges of consecutive rotations sharing the same prefix. Given $x, w \in \Sigma^*$, we denote by $\beta(x, w)$ the block of BWT(w) corresponding to the rotations prefixed by x. We can omit the second parameter if it is clear from the context.

The structure of the whole BWT matrix is summarized in Table 1. The following proposition puts together the BWT computations carried out for all blocks of consecutive rows, highlighting which prefixes are shared.

Proposition 8. *Given an integer $k > 5$, let $w_k = (\prod_{i=2}^{k-1} s_i e_i) q_k$. Then, $\beta(\mathtt{a}^i\mathtt{b}) = \mathtt{ba}^{k-i-2}$ for all $4 \leq i \leq k-2$, $\beta(\mathtt{a}^3\mathtt{b}) = \mathtt{b}^5(\mathtt{ab})^{k-6}\mathtt{a}$, $\beta(\mathtt{a}^2\mathtt{b}) = \mathtt{baaba}^{2k-8}$, $\beta(\mathtt{ab}) = \mathtt{b}^{k-2}\mathtt{aaba}^{2k-6}$, $\beta(\mathtt{ba}) = \mathtt{a}^{k-5}\mathtt{bbbab}^{k-4}\mathtt{ab}^{k-2}\mathtt{a}$, $\beta(\mathtt{b}^j\mathtt{a}) =$*

Table 2. BWTs of the word w_k and its variants after different edit operations. The word in the intersection of the column $\beta(x)$ with the row w is the range of BWT(w) corresponding to all the rotations that have x as a prefix. The columns $\beta(\mathtt{a}^i\mathtt{b})$ and $\beta(\mathtt{b}^j\mathtt{a})$ represent ranges of columns from $i \in [k-2, 4]$ (in that order) and $j \in [2, k-1]$, respectively. Note that the prefixes in the columns are disjoint, and cover all the possible ranges for the set of words considered. The BWT of each word is the concatenation of all the words in its row from left to right. In the last column appears the number of BWT runs of each of these words.

Word	$\beta(\$)$	$\beta(\mathtt{a}\$)$	$\beta(\mathtt{aa}\$)$	$\beta(\mathtt{a}^i\mathtt{b})$	$\beta(\mathtt{a}^3\mathtt{b})$	$\beta(\mathtt{a}^2\mathtt{b})$	$\beta(\mathtt{ab})$
w_k	ϵ	ϵ	ϵ	$\mathtt{ba}^{k-i-2}$	$\mathtt{b}^5(\mathtt{ab})^{k-6}\mathtt{a}$	$\mathtt{baaba}^{2k-8}$	$\mathtt{b}^{k-2}\mathtt{aaba}^{2k-6}$
$w_k\mathtt{a}$	ϵ	ϵ	ϵ	$\mathtt{ba}^{k-i-2}$	$\mathtt{bb}^5(\mathtt{ab})^{k-6}\mathtt{a}$	$\mathtt{aaaba}^{2k-8}$	$\mathtt{b}^{k-2}\mathtt{aaba}^{2k-6}$
$\widehat{w_k}$	ϵ	ϵ	ϵ	$\mathtt{ba}^{k-i-2}$	$\mathtt{b}^5(\mathtt{ab})^{k-6}\mathtt{a}$	$\mathtt{aaba}^{2k-8}$	$\mathtt{b}^{k-2}\mathtt{baba}^{2k-6}$
$\widehat{w_k}\mathtt{b}$	ϵ	ϵ	ϵ	$\mathtt{ba}^{k-i-2}$	$\mathtt{b}^5(\mathtt{ab})^{k-6}\mathtt{a}$	$\mathtt{aaba}^{2k-8}$	$\mathtt{b}^{k-2}\mathtt{baba}^{2k-6}$
$w_k\$$	a	b	ϵ	$\mathtt{ba}^{k-i-2}$	$\mathtt{b}^5(\mathtt{ab})^{k-6}\mathtt{a}$	$\mathtt{aaba}^{2k-8}$	$\mathtt{b}^{k-2}\$\mathtt{aba}^{2k-6}$
$w_k\mathtt{b}\$$	b	ϵ	ϵ	$\mathtt{ba}^{k-i-2}$	$\mathtt{b}^5(\mathtt{ab})^{k-6}\mathtt{a}$	$\mathtt{aaba}^{2k-8}$	$\mathtt{bb}^{k-2}\$\mathtt{aba}^{2k-6}$
$w_k\mathtt{bb}\$$	b	ϵ	ϵ	$\mathtt{ba}^{k-i-2}$	$\mathtt{b}^5(\mathtt{ab})^{k-6}\mathtt{a}$	$\mathtt{aaba}^{2k-8}$	$\mathtt{bb}^{k-2}\$\mathtt{aba}^{2k-6}$
$w_k\mathtt{a}\$$	a	a	b	$\mathtt{ba}^{k-i-2}$	$\mathtt{b}^5(\mathtt{ab})^{k-6}\mathtt{a}$	$\mathtt{aaba}^{2k-8}$	$\mathtt{b}^{k-2}\$\mathtt{aba}^{2k-6}$

Word	$\beta(\mathtt{b}\$)$	$\beta(\mathtt{ba})$	$\beta(\mathtt{bb}\$)$	$\beta(\mathtt{b}^j\mathtt{a})$	$\beta(\mathtt{b}^k\mathtt{a})$	$\beta(\mathtt{b}^{k+1})$	$r(\cdot)$
w_k	ϵ	$\mathtt{a}^{k-5}\mathtt{bbbab}^{k-4}\mathtt{ab}^{k-2}\mathtt{a}$	ϵ	$\mathtt{ab}^{2k-2j-1}\mathtt{a}$	a	ϵ	$6k-12$
$w_k\mathtt{a}$	ϵ	$\mathtt{a}^{k-5}\mathtt{bbbbab}^{k-5}\mathtt{ab}^{k-2}\mathtt{a}$	ϵ	$\mathtt{bab}^{2k-2j-2}\mathtt{a}$	a	ϵ	$8k-20$
$\widehat{w_k}$	ϵ	$\mathtt{a}^{k-5}\mathtt{bbbab}^{k-5}\mathtt{ab}^{k-2}\mathtt{ba}$	ϵ	$\mathtt{ab}^{2k-2j-2}\mathtt{ab}$	a	ϵ	$8k-20$
$\widehat{w_k}\mathtt{b}$	ϵ	$\mathtt{a}^{k-5}\mathtt{bbbab}^{k-5}\mathtt{ab}^{k-2}\mathtt{ba}$	ϵ	$\mathtt{ab}^{2k-2j-2}\mathtt{ab}$	b	a	$8k-20$
$w_k\$$	ϵ	$\mathtt{ba}^{k-5}\mathtt{bbbab}^{k-5}\mathtt{ab}^{k-2}\mathtt{a}$	ϵ	$\mathtt{bab}^{2k-2j-2}\mathtt{a}$	a	ϵ	$8k-16$
$w_k\mathtt{b}\$$	a	$\mathtt{a}^{k-5}\mathtt{bbbab}^{k-5}\mathtt{abb}^{k-2}\mathtt{a}$	ϵ	$\mathtt{ab}^{2k-2j-1}\mathtt{a}$	a	ϵ	$6k-13$
$w_k\mathtt{bb}\$$	b	$\mathtt{a}^{k-5}\mathtt{bbbab}^{k-5}\mathtt{abb}^{k-2}\mathtt{a}$	a	$\mathtt{ab}^{2k-2j-2}\mathtt{ab}$	a	ϵ	$8k-17$
$w_k\mathtt{a}\$$	ϵ	$\mathtt{ba}^{k-5}\mathtt{bbbab}^{k-5}\mathtt{ab}^{k-2}\mathtt{a}$	ϵ	$\mathtt{bab}^{2k-2j-2}\mathtt{a}$	a	ϵ	$8k-16$

$\mathtt{ab}^{2k-2j-1}\mathtt{a}$ *for all* $2 \le j \le k-1$, *and* $\beta(\mathtt{b}^k\mathtt{a}) = \mathtt{a}$, *and the BWT of the* w_k *is* $BWT(w_k) = \prod_{i=2}^{k-1} \beta(\mathtt{a}^{k-i}\mathtt{b}) \cdot \prod_{i=1}^{k} \beta(\mathtt{b}^i\mathtt{a})$. *Moreover,* $r(w_k) = 6k - 12$.

For a given word $w \neq \epsilon$, let w^{ins}, w^{del}, and w^{sub} be the words obtained by applying on w an insertion, a deletion, and a substitution of a character respectively. The structure of the BWT of w_k and other words obtained by applying one or more edit operations on w_k are summed up in Table 2.

Proposition 9. *There exists an infinite family of words* w *such that: (i)* $r(w^{ins}) - r(w) = \Theta(\sqrt{n})$*; (ii)* $r(w^{del}) - r(w) = \Theta(\sqrt{n})$*; (iii)* $r(w^{sub}) - r(w) = \Theta(\sqrt{n})$.

Proof. The family is composed of the words w_k with $k > 5$. Let $n = |w_k|$. If $w_k^{ins} = w_k\mathtt{a}$, $w_k^{del} = \widehat{w_k}$, and $w_k^{sub} = \widehat{w_k}\mathtt{b}$, from Table 2 we have $r(w_k\mathtt{a}) = r(\widehat{w_k}) = r(\widehat{w_k}\mathtt{b}) = r(w_k) + (2k-8)$. From Lemma 5, that $2k - 8 = \Theta(\sqrt{n})$. □

5 Bit Catastrophes for $r_\$$

In this section we discuss bit catastrophes when the parameter $r_\$$ is considered. In fact, not only are the measures r and $r_\$$ not equal over the same input, but they may differ by a $\Theta(\log n)$ multiplicative, or by a $\Theta(\sqrt{n})$ additive factor.

Proposition 10. *There exists an infinite family of words w such that $r_\$(w)/r(w) = \Theta(\log n)$, where $n = |w|$. Moreover, there exists another infinite family of words w' such that $r_\$(w') - r(w') = \Theta(\sqrt{n'})$, where $n' = |w'|$.*

Proof. The first family consists of the Fibonacci words of odd order. The proof can be derived from Proposition 4. The second family consists of the words w_k for all $k > 5$ (Sect. 4); proof can be derived from Table 2 and Lemma 5. □

First let us consider the case where a symbol $c \in \Sigma$ is prepended to a word v. As recently noted in [1], it is well known that in this case the value $r_\$$ can only vary by a constant value. For the sake of completeness, we include a proof.

Proposition 11. *For any* $\mathtt{c} \in \Sigma$, *we have* $r_\$(v) - 1 \leq r_\$(\mathtt{c}v) \leq r_\$(v) + 2$.

Proof. Let us consider the list of lexicographically sorted cyclic rotations or, equivalently, the list of lexicographically sorted suffixes of $cv\$$. Such a list is obtained from the list of suffixes of $v\$$, to which the suffix $cv\$$ is added. Note that for all suffixes other than $cv\$$, their relative order will remain the same. Moreover, the corresponding symbol in the BWT also remains the same, except that the character c takes the place of \$. This could decrease the number of BWT-runs by at most 2 or no change could occur. The symbol corresponding to the new suffix $cv\$$ (which produces the insertion of \$ in the corresponding position in the BWT) could either increase the number of BWT-runs by 1 or break a BWT-run of $v\$$, thus increasing $r_\$$ by 2. □

The following proposition shows that there are some cases in which $r_\$$ is not affected by any bit catastrophe.

Proposition 12. *Let* $\mathtt{c}$ *be smaller than or equal to the smallest character in a word* v, *then* $r_\$(v) \leq r_\$(v\mathtt{c}) \leq r_\$(v) + 1$.

Proof. The lexicographic order of the rotations (or equivalently, the suffixes) of $v\mathtt{c}\$$ can be obtained by that of $v\$$. Since the new character c is smaller or equal than the smallest character of v, but still greater than \$, then the relative order of the rotations of v does not change after $\mathtt{c}$ is appended. For the same reason, then the new rotation $\mathtt{c}\$v$ must be the second smallest rotation of $v\mathtt{c}\$$, therefore BWT($v\$$) is equal to the $(n-1)$-length suffix of BWT($v\mathtt{c}\$$). Clearly the very first character of BWT($v\mathtt{c}\$$) is the new character $\mathtt{c}$, adding at most one run. □

In general, appending, deleting, or substituting with a symbol that is not the smallest of the alphabet can increase the number of runs of a word by $\Theta(\sqrt{n})$.

Proposition 13. *There exists an infinite family of words such that: (i)* $r_\$(w\mathtt{b}) - r_\$(w) = \Theta(\sqrt{n})$; *(ii)* $r_\$(\widehat{w}) - r_\$(w) = \Theta(\sqrt{n})$; *(iii)* $r_\$(\widehat{w}\mathtt{a}) - r_\$(w) = \Theta(\sqrt{n})$.

Proof. Such a family is composed of the words $w_k\mathtt{b}$ with $k > 5$. The proof can be derived by using the results summarized in Table 2 and Lemma 5. □

6 Conclusion

In this paper, we investigate how a single edit operation on a word (insertion, deletion or substitution of a character) can affect the number of runs of the BWT of the word itself. Our contribution is twofold. On the one hand, we prove that $\Omega(\log n)$ is a lower bound whatever edit operation has been applied, by exhibiting infinite families of words for which each edit operation increases the number of runs by a multiplicative $\Theta(\log n)$-factor. This also proves that the upper bound $\mathcal{O}(\log n \log r)$ on the multiplicative sensitivity of r shown in [1] is tight for each edit operation. On the other hand, we improve the best known lower bound for the additive sensitivity of r, by showing an infinite family of words in which insertion, deletion and substitution of a character increase r by a $\Theta(\sqrt{n})$ additive factor. The tightness of the upper bound $\mathcal{O}(r \log r \log n)$ for the additive sensitivity of r proved in [1] is still an open question.

Funding Information. CU is partially funded by ANID national doctoral scholarship - 21210580. MS is partially funded by the PNRR project ITSERR (CUP B53C22001770006) and by the INdAM-GNCS Project (CUP E55F22000270001).

References

1. Akagi, T., Funakoshi, M., Inenaga, S.: Sensitivity of string compressors and repetitiveness measures. Inf. Comput. **291**, 104999 (2023)
2. Bannai, H., Gagie, T.: Refining the r-index. Theor. Comput. Sci. **812**, 96–108 (2020)
3. Borel, J., Reutenauer, C.: On Christoffel classes. RAIRO Theor. Inf. Appl. **40**(1), 15–27 (2006)
4. Brlek, S., Frosini, A., Mancini, I., Pergola, E., Rinaldi, S.: Burrows-wheeler transform of words defined by morphisms. In: Colbourn, C.J., Grossi, R., Pisanti, N. (eds.) IWOCA 2019. LNCS, vol. 11638, pp. 393–404. Springer, Cham (2019). https://doi.org/10.1007/978-3-030-25005-8_32
5. Burrows, M., Wheeler, D.J.: A block-sorting lossless data compression algorithm. Technical report, DIGITAL System Research Center (1994)
6. Castiglione, G., Restivo, A., Sciortino, M.: On extremal cases of Hopcroft's algorithm. Theoret. Comput. Sci. **411**(38–39), 3414–3422 (2010)
7. de Luca, A.: Sturmian words: structure, combinatorics, and their arithmetics. Theor. Comput. Sci. **183**(1), 45–82 (1997)
8. de Luca, A., Mignosi, F.: Some combinatorial properties of Sturmian words. Theor. Comput. Sci. **136**(2), 361–385 (1994)
9. Ferragina, P., Manzini, G.: Opportunistic data structures with applications. In: FOCS, pp. 390–398. IEEE Computer Society (2000)
10. Frosini, A., Mancini, I., Rinaldi, S., Romana, G., Sciortino, M.: Logarithmic equal-letter runs for BWT of purely morphic words. In: Diekert, V., Volkov, M. (eds.) Developments in Language Theory. DLT 2022. Lecture Notes in Computer Science, vol. 13257, pp. 139–151. Springer, Cham. https://doi.org/10.1007/978-3-031-05578-2_11
11. Gagie, T., Navarro, G., Prezza, N.: Optimal-time text indexing in BWT-runs bounded space. In: Czumaj, A. (ed.) SODA, pp. 1459–1477. SIAM (2018)

12. Gagie, T., Navarro, G., Prezza, N.: Fully functional suffix trees and optimal text searching in BWT-runs bounded space. J. ACM **67**(1), 1–54 (2020)
13. Giuliani, S., Inenaga, S., Lipták, Z., Prezza, N., Sciortino, M., Toffanello, A.: Novel results on the number of runs of the burrows-wheeler-transform. In: Bureš, T., et al. (eds.) SOFSEM 2021. LNCS, vol. 12607, pp. 249–262. Springer, Cham (2021). https://doi.org/10.1007/978-3-030-67731-2_18
14. Kempa, D., Kociumaka, T.: Resolution of the burrows-wheeler transform conjecture. Commun. ACM **65**(6), 91–98 (2022)
15. Knuth, D., Morris, J., Jr., Pratt, V.: Fast pattern matching in strings. SIAM J. Comput. **6**(2), 323–350 (1977)
16. Kociumaka, T., Navarro, G., Prezza, N.: Towards a definitive measure of repetitiveness. In: Kohayakawa, Y., Miyazawa, F.K. (eds.) LATIN 2021. LNCS, vol. 12118, pp. 207–219. Springer, Cham (2020). https://doi.org/10.1007/978-3-030-61792-9_17
17. Lagarde, G., Perifel, S.: Lempel-Ziv: a "one-bit catastrophe" but not a tragedy. In: SODA, pp. 1478–1495. SIAM (2018)
18. Lam, T.W., Li, R., Tam, A., Wong, S., Wu, E., Yiu, S.M.: High throughput short read alignment via Bi-directional BWT. In: BIBM, pp. 31–36. IEEE Computer Society (2009)
19. Langmead, B., Trapnell, C., Pop, M., Salzberg, S.L.: Ultrafast and memory-efficient alignment of short DNA sequences to the human genome. Genome Biol. **10**(3), R25 (2009)
20. Li, H., Durbin, R.: Fast and accurate long-read alignment with burrows-wheeler transform. Bioinformatics **26**(5), 589–595 (2010)
21. Lothaire, M.: Algebraic Combinatorics on Words. Cambridge University Press, Cambridge (2002)
22. Mäkinen, V., Navarro, G.: Succinct suffix arrays based on run-length encoding. In: Apostolico, A., Crochemore, M., Park, K. (eds.) CPM 2005. LNCS, vol. 3537, pp. 45–56. Springer, Heidelberg (2005). https://doi.org/10.1007/11496656_5
23. Mantaci, S., Restivo, A., Rosone, G., Sciortino, M., Versari, L.: Measuring the clustering effect of BWT via RLE. Theoret. Comput. Sci. **698**, 79–87 (2017)
24. Mantaci, S., Restivo, A., Sciortino, M.: Burrows-wheeler transform and Sturmian words. Inf. Process. Lett. **86**(5), 241–246 (2003)
25. Navarro, G.: Indexing highly repetitive string collections, part I: repetitiveness measures. ACM Comput. Surv. **54**(2), 1–31 (2022)
26. Sciortino, M., Zamboni, L.Q.: Suffix automata and standard Sturmian words. In: Harju, T., Karhumäki, J., Lepistö, A. (eds.) DLT 2007. LNCS, vol. 4588, pp. 382–398. Springer, Heidelberg (2007). https://doi.org/10.1007/978-3-540-73208-2_36
27. Seward, J.: 1996. https://sourceware.org/bzip2/manual/manual.html

The Domino Problem Is Undecidable on Every Rhombus Subshift

Benjamin Hellouin de Menibus[1,2(✉)], Victor H. Lutfalla[1,3], and Camille Noûs[1,4]

[1] Université Publique, Paris, France
[2] Université Paris-Saclay, CNRS, Laboratoire Interdisciplinaire des Sciences du Numérique, 91400 Orsay, France
hellouin@lisn.fr
[3] Normandie Univ, UNICAEN, ENSICAEN, CNRS, GREYC, 14000 Caen, France
[4] Laboratoire Cogitamus, Caen, France

Abstract. We extend the classical Domino problem to any tiling of rhombus-shaped tiles. For any subshift X of edge-to-edge rhombus tilings, such as the Penrose subshift, we prove that the associated X-Domino problem is Π_1^0-hard and therefore undecidable. It is Π_1^0-complete when the subshift X is given by a computable sequence of forbidden patterns.

Keywords: Rhombus tiling · decidability · Domino problem

1 Introduction

Tilings come, roughly speaking, in two families. Geometrical tilings are coverings of a space, usually the euclidean plane, by geometrical tiles without overlap; the constraints come from the geometry of the tiles. A famous example of geometrical tilings is the Penrose tilings [14]. Symbolic tilings are colourings of a discrete structure, usually $\mathbb{Z}^d$ for some d, whose constraints are given by forbidden patterns.

Both families have received a lot of attention for their dynamical and combinatorials properties. Specific areas of research on geometrical tilings are the study of their symmetries and their links with mathematical cristallography [3], whereas works on symbolic tilings have more links to computability and decidability theory. The seminal example is the Domino problem: given a set of colours and a set of forbidden patterns, is there a colouring of $\mathbb{Z}^2$ that satisfies those constraints? The proof by Berger [5] that this problem is undecidable shaped the whole domain of research.

There has been much work to extend classical results for the symbolic Domino problem on $\mathbb{Z}^2$ to other structures. An active area of research considers Domino problem on groups in order to relate properties of the group and of the tiling spaces; see [1] for a survey. Other considered extensions are Domino problems in self-similar structures to understand the limit between dimension 1 and 2 [4]

Supported by the ANR project C_SyDySi.

F. Drewes and M. Volkov (Eds.): DLT 2023, LNCS 13911, pp. 100–112, 2023.
https://doi.org/10.1007/978-3-031-33264-7_9

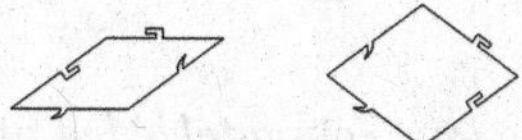

(a) Original definition of the Penrose tiles [14] with cuts and notches.

(b) Alternative definition with arrows on the edges [6].

Fig. 1. Definitions of Penrose tiles: adding symbols on the edges of the tiles allow for simple rhombus shapes.

or Domino problems inside a $\mathbb{Z}^2$-subshift to understand the effect of dynamical restrictions [2].

Coming back to geometrical tilings, complex examples such as the Penrose tilings were originally defined with jigsaw-type tiles with indentations on their edges (see Fig. 1). It can be restated as simple polygon tiles with symbols on their edges [6] with the condition that symbols must match. In essence these tilings are both symbolic and geometrical.

Similarly, given a set of shapes, we define symbolic tiles on those shapes by adding colours on the edges and study the induced symbolic-geometrical tilings. Let us consider the set of symbolic-geometrical tiles $\mathbf{T}$ of Fig. 2. Since the shapes are Penrose rhombuses, there are two natural questions regarding this tileset:

1. is there an infinite valid tiling of $\mathbb{R}^2$ with tiles in $\mathbf{T}$ up to translation?
2. is there an infinite valid tiling of $\mathbb{R}^2$ with tiles in $\mathbf{T}$ up to translation that projects to a geometrical Penrose tiling (that is, when removing the colours from the tiles)?

It is not hard to see that the first question is at least as hard as (and is in fact equivalent to) the classical Domino problem, which corresponds to the case where the input tiles are all the same rhombus. This motivates us to study the second question, where the geometrical (Penrose) subshift forces the use of the diferent rhombuses.

These are instances of the Domino problem in geometrical subshifts, which is the object of the present article. For any set of rhombuses $\mathcal{T}$ and a geometrical subshift X on these rhombuses, the X-`Domino` problem is defined as follows: given as input a set of tiles, that is, rhombuses from $\mathcal{T}$ with a colour on each edge, decide whether it is possible to tile the plane in such a way that:

1. tiles with a common edge have the same colour along the shared edge, and
2. the geometrical tiling (when colours are erased) is valid for X.

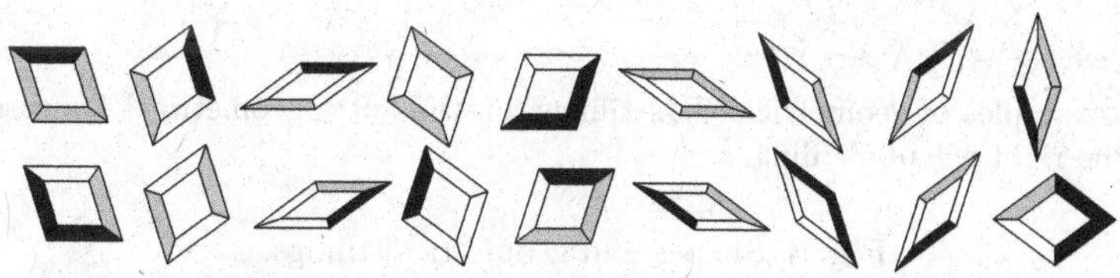

Fig. 2. An example of symbolic-geometrical tiles on Penrose rhombuses.

Our main result is the following:

Theorem 1. *Let X be a geometrical subshift given by a computable list of forbidden patterns. X*-`Domino` *is many-one equivalent to the classical Domino problem on $\mathbb{Z}^2$, that is, co-computably enumerable-complete, and thus undecidable.*

2 Definitions

2.1 Geometrical Tiling Spaces

Definition 1 (Shapes and patches). *We call* shape *a geometrical rhombus given as a pair of vectors $(\boldsymbol{u}, \boldsymbol{v})$ and a position p.*
We call shapeset *a finite set of shapes considered up-to-translation, see Fig. 3a. We call* patch *an edge-to-edge simply connected finite set of shapes,* i.e., *any two shapes are either disjoint, share a single common vertex or a full common edge, and there is no hole in the patch. We call* support *of a patch the union of its shapes. We call* pattern *a patch up to translation.*

Definition 2 (Tilings, full shift and subshifts). *Given a shapeset $\mathcal{T}$, we call $\mathcal{T}$-*tiling *an edge-to-edge covering of the euclidean plane without overlap by translates of the shapes in $\mathcal{T}$: see Fig. 3b.*

We call full shift *on $\mathcal{T}$, denoted by $X_\mathcal{T}$, the set of all $\mathcal{T}$-tilings.*

We call subshift *of $X_\mathcal{T}$ any subset X of $X_\mathcal{T}$ that is invariant by translation and closed for the tiling topology [15].*

Edge-to-edge rhombus tilings with finitely many shapes up to translation have *Finite Local Complexity* (FLC): that is, for any compact $C \subseteq \mathbb{R}^2$, there are finitely many patches whose support is included in C. The FLC hypothesis

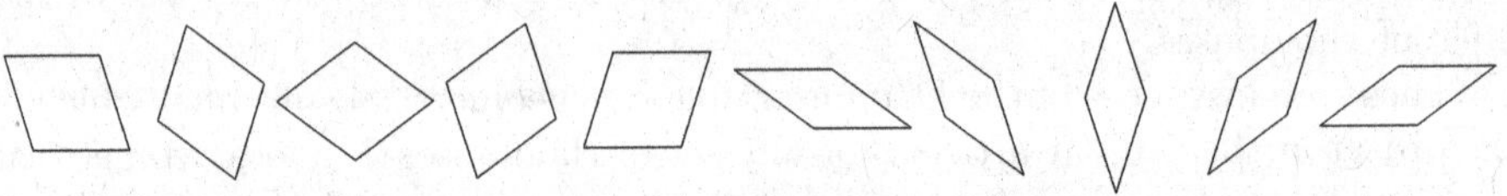

(a) An example of shapeset: the set $\mathcal{T}_{pen}$ of Penrose rhombuses up to translation.

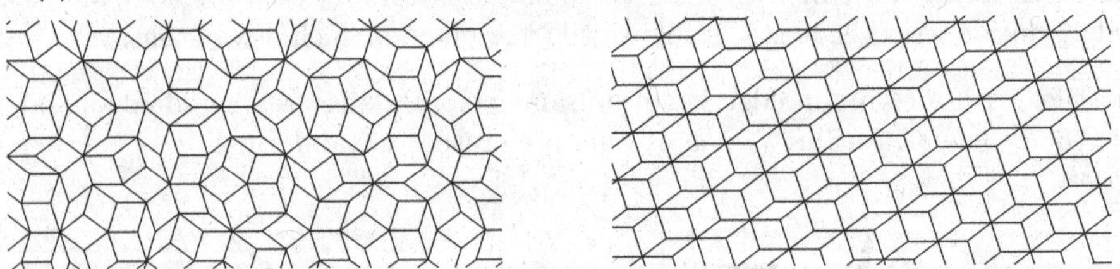

(b) Two examples of geometrical $\mathcal{T}_{pen}$-tilings: on the left a geometrical Penrose tiling and on the right a "cube"-tiling.

Fig. 3. Shapes and geometrical tilings.

appears a lot in the study of geometrical tilings and Delone sets; see [3]. In particular, FLC ensures that the tiling space shares most properties with standard $\mathbb{Z}^d$ tiling spaces, such as being compact for the usual tiling topology [12,15].

A subshift X can always be characterized by a countable (possibly infinite) set of forbidden patterns, that we denote as a sequence $\mathcal{F} := (f_n)_{n\in\mathbb{N}}$. In other words, X is the set of all tilings where no pattern in $\mathcal{F}$ appears. We say that the subshift X is *effective* when $\mathcal{F}$ is computable, *i.e.*, when there exists an algorithm that, on input n, outputs f_n.

We say that a pattern (or patch) has *minimal radius* r when its support contains a disk of radius r centered on a vertex of a rhombus of the pattern, and when removing any shape from the pattern would break that property.

A sequence $\mathcal{F}$ of forbidden patterns being fixed, we call locally-allowed patterns $\mathcal{A}(X)$ the set of finite patterns where no forbidden pattern appears, and rank-r locally-allowed patterns $\mathcal{A}_r(X)$ the set of patterns of minimal radius r that do not contain any of the first r forbidden patterns. Note that when X is an FLC subshift $\mathcal{A}_r(X)$ is finite for all r: indeed, there exists a constant d (maximum diameter of the shapes) such that the support of any minimal radius r pattern is included in a disk of radius $r + d$.

Note that patterns in $\mathcal{A}(X)$ may not be globally allowed in X (appear in no infinite tiling in X). They may even appear in no tiling of the full shift $X_{\mathcal{T}}$ if they contain a geometrical impossibility. Such patterns are called *deceptions* [7].

The interest of locally allowed patterns is that, as seen below in Lemma 2, they are computable from the list of forbidden patterns, whereas it is not the case for globally allowed patterns.

2.2 Symbolic-geometrical Tiling Spaces

Definition 3 (Symbols, tiles and tilesets). *A* tile *is a shape endowed with a colour on each edge, as seen in Fig. 4(a).*

Formally, given a finite set C whose elements are called colours and a rhombus shape r, we call r-Wang tile *or simply* r-tile *a quintuple (r, a_0, a_1, a_2, a_3) with $a_i \in C$. Formally, with $r = (\boldsymbol{u}, \boldsymbol{v}, p)$, the side $(p, p + \boldsymbol{u})$ has colour a_0, the side $(p + \boldsymbol{u}, p + \boldsymbol{u} + \boldsymbol{v})$ has colour a_1 and so on.*

Given a shapeset $\mathcal{T}$, we call $\mathcal{T}$-tile *a r-tile for some shape r that is in $\mathcal{T}$ up to translation. We call* $\mathcal{T}$-tileset *a finite set of $\mathcal{T}$-tiles, considered up to translation, such that each shape has at least a tile.*

Definition 4 (Colour erasing operator π). *We define the colour erasing operator π by:*

- *for any r-tile t, $\pi(t) := r$*
- *for a tiling x (or finite patch of tiles), $\pi(x) := \{\pi(t),\ t \in x\}$*
- *for a set of tilings X, $\pi(X) := \{\pi(x),\ x \in X\}$*

Definition 5 (Tiling). *Given a finite set of colours C and a $\mathcal{T}$-tileset $\mathbf{T}$, we call* $\mathbf{T}$-tiling *a tiling x such that $\pi(x) \in X_{\mathcal{T}}$ and such that any two tiles in x that share an edge have the same colour on their shared edge. See Fig. 4(b).*

We denote by $X_{\mathbf{T}}$ the subshift of all $\mathbf{T}$-tilings.

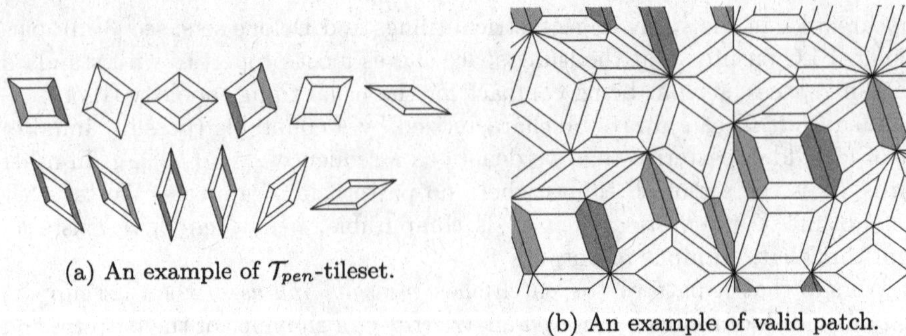

(a) An example of $\mathcal{T}_{pen}$-tileset.

(b) An example of valid patch.

Fig. 4. Rhombus Wang tiles.

A symbolic-geometrical subshift X is given by a set of shapes (geometrical constraints), a set of forbidden patterns (geometrical subshift) and colourings on the tiles (symbolic constraints). Even when the geometrical subshift is a full shift, geometrical and symbolic constraints can interact in interesting ways. For example, there is a choice of tiles on the Penrose rhombuses such that all valid tilings correspond to a geometrical Penrose tiling after erasing colours (in particular, no valid tiling use a single shape).

The definitions of minimal radius r patterns and locally allowed patterns extend naturally to symbolic-geometrical tilings.

2.3 Computability and Decidability

Definition 6. *A* decision problem *is a function* $A : dom(A) \to \{0,1\}$, *where* $dom(A)$ *is called the* input domain *of* A.

Definition 7. *A decision problem* A *is said to be decidable when there exists an algorithm (or Turing machine) that, given as input any* $x \in dom(A)$, *terminates and outputs* $A(x)$.

A weaker notion of computability for decision problems is the following:

Definition 8 (co-computably enumerable, Π_1^0). *A decision problem* A *is called* co-computably enumerable, *also known as* co-recursively enumerable, *when there exists a total computable function* $f : dom(A) \times \mathbb{N} \to \{0,1\}$ *such that:*

$$\forall x \in dom(A), \quad A(x) \Leftrightarrow \forall n \in \mathbb{N}, f(x,n)$$

Alternatively, a decision problem A *is co-computably enumerable if there is an algorithm that, on input* x, *terminates if and only if* $A(x)$ *is false.*

We denote by Π_1^0 *the class of co-computably enumerable problems.*

Π_1^0 is a class of the arithmetical hierarchy; see [11,13].

Definition 9 (Many-one reductions). *Given two decision problems* A *and* B, *we say that* A many-one reduces *to* B, *and write* $A \preccurlyeq_m B$, *when there exists a total computable function* $f : dom(A) \to dom(B)$ *such that* $A = B \circ f$.

Definition 10 (Π_1^0-hardness and Π_1^0-completeness). *A problem A is called Π_1^0-hard if $B \preccurlyeq_m A$ for any problem B in Π_1^0.*

A problem A is called Π_1^0-complete when it is both in Π_1^0 and Π_1^0-hard.

Notice that Π_1^0-hard problems are undecidable. The canonical example of a Π_1^0-complete problem is the co-halting problem, that is the problem of deciding whether a Turing Machine does not terminate on empty input in finite time.

Many-one reductions are a restrictive case of Turing reductions that are appropriate to study classes of decision problems such as Π_1^0, as the following classical Lemma shows:

Lemma 1 (Π_1^0-hardness). *Given two problems A and B such that $A \preccurlyeq_m B$,*

- *if B is Π_1^0, then A is Π_1^0.*
- *if A is Π_1^0-hard, then B is Π_1^0-hard.*

2.4 Domino Problems

In this paper, our goal is to prove the Π_1^0-hardness of a generalisation of the classical Domino problem, which is known to be Π_1^0-complete.

The classical Domino problem asks, given as input a finite set of Wang tiles, *i.e.*, square tiles with a colour on each edge, whether there exists an infinite valid tiling with these tiles: see Fig. 5.

Theorem 2 (Berger 1966 [5]). *The classical Domino problem is Π_1^0-complete.*

Berger's paper provides a many-one reduction to the co-halting problem. We extend this classical problem to rhombus-shaped Wang tiles.

Definition 11 (X-`Domino`). *Given a subshift X on $\mathcal{T}$, the* `Domino` *problem on X is defined as:*

Input *A finite set $\mathbf{T}$ of $\mathcal{T}$-tiles*
Output *Is there a $\mathbf{T}$-tiling $x \in X_{\mathbf{T}}$ such that $\pi(x) \in X$?*

The classical `Domino` problem is $X_{\{\square\}}$-`Domino`, that is, the domino problem on the full shift with a single shape (usually a square shape, but any single rhombus works). *Penrose*-`Domino` would be, given a set of tiles on Penrose rhombuses as in Fig. 4 looking for a Penrose tiling with matching edges.

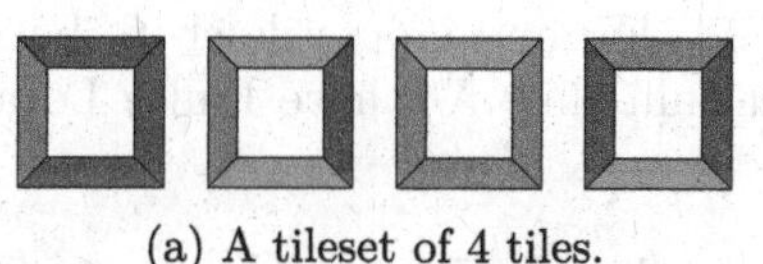

(a) A tileset of 4 tiles.

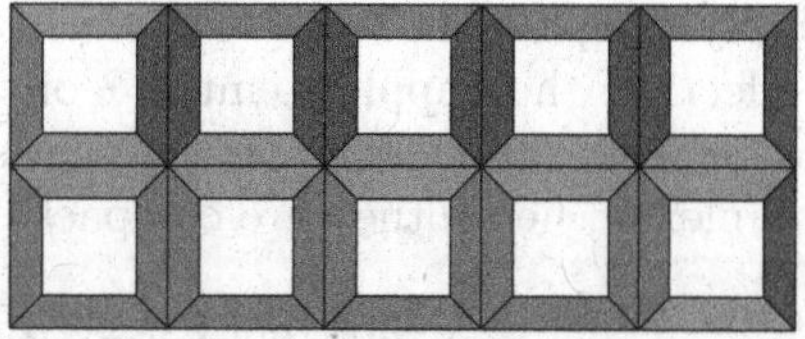

(b) A valid patch with these tiles.

Fig. 5. Tiles and tilings: a patch or tiling is called *valid* when any two adjacent tiles have the same colour on their shared edge.

3 Complexity of the Domino Problem on Rhombus Subshifts

3.1 The Domino Problem on Effective Rhombus Subshifts Is Π_1^0

Lemma 2. *Let $\mathcal{T}$ be a finite set of rhombus shapes. The following problem is computable:*

Input *An integer n, a finite list of forbidden patterns $\mathcal{F}$, and a tileset $\mathbf{T}$ of $\mathcal{T}$-tiles*
Output *The list $\mathcal{A}_n(\mathcal{F})$ of rank-n locally-allowed patterns, that is, minimal radius n patterns of $\mathbf{T}$-tiles that avoid all patterns in $\mathcal{F}$*

Proof. The algorithm is as follows:

1. By combinatorial exploration, try all possibilities and list all $\mathbf{T}$-patterns with minimal radius n.
2. Eliminate all listed patterns x such that the colour-erased pattern $\pi(x)$ contains some pattern in $\mathcal{F}$.
3. Output the remaining patterns.

In Point 1, remember that the set of edge-to-edge tilings on a fixed finite set of rhombus tiles have finite local complexity, so this process terminates in finite time.

Proposition 1 (X-`Domino` $\in \Pi_1^0$)**.** *For any shapeset $\mathcal{T}$ and any effective subshift X on $\mathcal{T}$, X-`Domino` is co-computably enumerable.*

Recall that X is called effective when it is defined by a computable sequence $\mathcal{F}$ of forbidden patterns. Note that, if X is not effective, then X-`Domino` is Π_1^0 when provided with an enumeration of $\mathcal{F}$ as oracle.

This is essentially the same proof as for the classical Domino problem.

Proof . The following problem, called disk-tiling-X, is decidable:

Input A finite set $\mathbf{T}$ of $\mathcal{T}$-tiles and an integer n
Output Is there a valid (finite) patch x with tiles in $\mathbf{T}$ such that $\pi(x)$ is a rank n locally allowed pattern of X, *i.e.*, $\pi(x) \in \mathcal{A}_n(X)$?

Simply compute the first n forbidden patterns $\mathcal{F}_n$, which is possible because X is effective, then apply Lemma 2 on $(n, \mathcal{F}_n, \mathbf{T})$. For any input tileset $\mathbf{T}$, both the geometrical subshift X and the symbolic full shift $X_{\mathbf{T}}$ have Finite Local Complexity, hence they are compact [15], and

$$X\text{-}\mathtt{Domino}(\mathbf{T}) \Leftrightarrow \forall n \in \mathbb{N},\ \text{disk-tiling-}X(\mathbf{T}, n).$$

Remark that disk-tiling-$X(\mathbf{T}, n)$ is true when there exists a rank n locally allowed pattern, that is, a pattern of minimal radius n that avoids the first n forbidden patterns $\mathcal{F}$ of X with tileset $\mathbf{T}$. If $\forall n \in \mathbb{N}$, disk-tiling-$X(\mathbf{T}, n)$, there exists a sequence $(p_n)_{n\in\mathbb{N}}$ with $\pi(p_n) \in \mathcal{A}_n(X)$. Since the radius of the patches tends to infinity, by compactness there exists a limit tiling x to which a subsequence converges. Now remark that $\pi(x) \in X$ because it avoids all forbidden patterns in $\mathcal{F}$. Indeed, for any k, the kth forbidden pattern does not appear in any $\pi(p_n)$ for $n \geq k$, so it does not appear in $\pi(x)$. Since disk-tiling-X is computable, we indeed have domino-$X \in \Pi_1^0$.

If X is a full shift on some shapeset $\mathcal{T}$, it is easy to see that the corresponding Domino problem is Π_1^0-hard by reduction to the classical version: given a finite set $\mathbf{T}$ of square Wang tiles, choose an arbitrary shape in $\mathcal{T}$ and colour it like $\mathbf{T}$, and colour every other shape with four new different *fresh* colours (so that any valid tiling may only use the first shape). In the rest of the paper, we extend this idea to work on an arbitrary subshift X.

3.2 The Domino Problem on Shape Uniformly Recurrent Subshifts Is Π_1^0-hard

In this section, we use the concept of *chains* of rhombuses [10] (also called *ribbons* [16]) to embed $\mathbb{Z}^2$ in rhombus tilings.

Definition 12 (Chains of rhombuses) *We call* chain of rhombuses *a bi-infinite sequence of rhombuses that share an edge direction; see Fig. 6.*

A chain of rhombuses has a normal vector $\boldsymbol{v}$: the direction of the common edge.

Lemma 3 (Occurrences of a rhombus, [10]). *In an edge-to-edge rhombus tiling, each rhombus of edge directions $\boldsymbol{u}$ and $\boldsymbol{v}$ is the intersection of two chains of normal vectors $\boldsymbol{u}$ and $\boldsymbol{v}$, and any intersection of two such chains is such a rhombus. Moreover, any two chains can cross at most once. See Fig. 6.*

As a consequence, two chains of same normal vector cannot cross, otherwise there would be an impossible flat rhombus at the intersection. Such chains are therefore called parallel.

Lemma 4 (Uniform monotonicity, [10]). *Given a finite shapeset $\mathcal{T}$, let θ_{min} be the smallest angle in a rhombus of $\mathcal{T}$. For any $\mathcal{T}$-tiling x, for any rhombus r appearing in a chain c of normal vector $\boldsymbol{u}$, the chain c is outside the cone centered in r and of half-angle θ_{min} along $\boldsymbol{u}$; see Fig. 7.*

Overall, an edge-to-edge rhombus tiling can be decomposed as d sets of parallel chains of rhombuses where d is the number of edge directions. Given an edge direction $\boldsymbol{u}$, the $\boldsymbol{u}$ chains can be indexed by either $\mathbb{Z}$, $\mathbb{N}$, $-\mathbb{N}$ or a finite integer interval in such a way that, starting from any position and moving along $\boldsymbol{u}$ one crosses the $\boldsymbol{u}$ chains in increasing order.

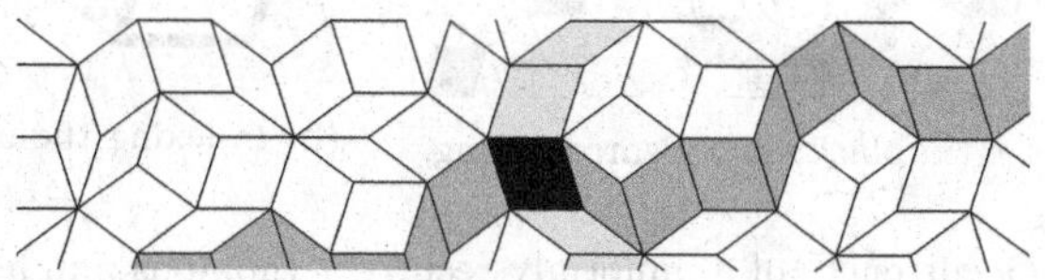

Fig. 6. In an edge-to-edge rhombus tiling, each rhombus (in black) is at the intersection of two chains of rhombuses (in shades of grey). (Color figure online)

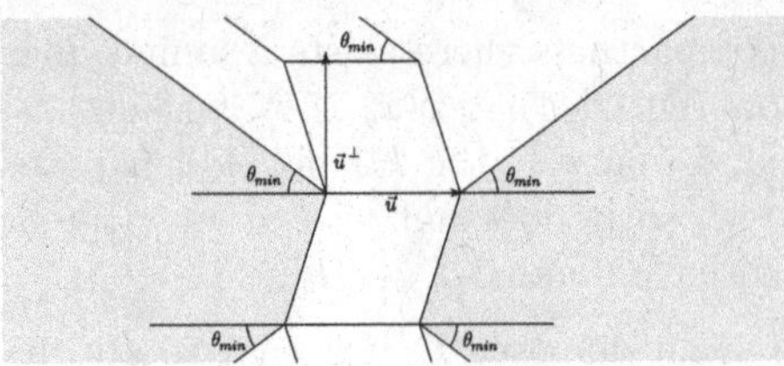

Fig. 7. The chain c is outside the grey cones left and right of the rhombus r.

Definition 13 (Shape uniform recurrence). *Given a rhombus shape r and a tiling x we say that r is uniformly recurrent in x, or that x is r-uniformly recurrent, if r appears in any disk of radius R in x for some R.*

A subshift X is called r-uniformly recurrent when every tiling $x \in X$ is r-uniformly recurrent.

A tiling (resp. a subshift) is called shape uniformly recurrent when it is r-uniformly recurrent for every shape r that appears in it.

Note that this is much weaker than the usual uniform recurrence, which holds for every pattern instead of a single shape.

The following lemma describes the structure shown in Fig. 8a which was already observed on Penrose tilings in [8, Chapter 11] [9, Chapter 6] and used to define an aperiodic tileset of 24 classical Wang tiles.

Lemma 5. *Let x be a edge-to-edge rhombus tiling and r a shape that is uniformly recurrent in x. The occurrences of r in x can be indexed by coordinates in $\mathbb{Z}^2$ such that two consecutive occurrences of r along a chain have adjacent $\mathbb{Z}^2$ coordinates; see Fig. 8b.*

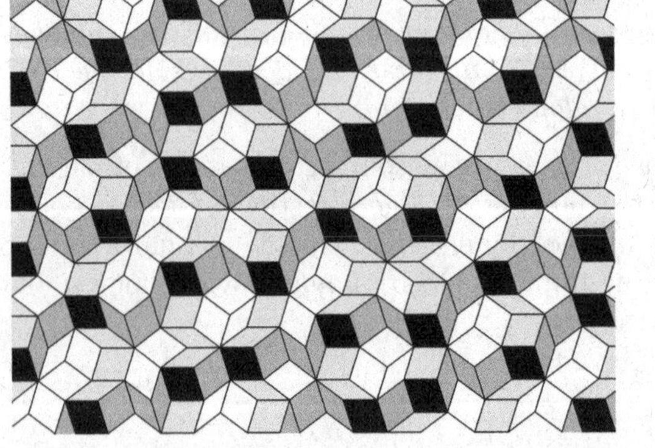

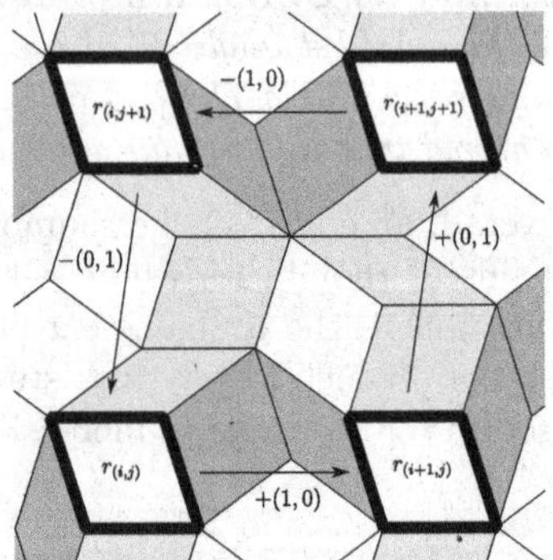

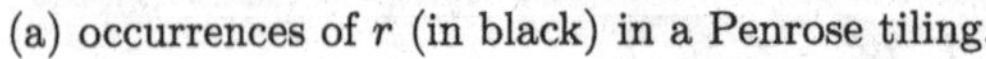

(a) occurrences of r (in black) in a Penrose tiling. (b) Indexing the occurrences of r.

Fig. 8. Occurrences of a uniformly recurrent rhombus r in a tiling.

Proof. Let x be a tiling in X and r a uniformly recurrent shape in x. Denote by $\boldsymbol{u}$ and $\boldsymbol{v}$ the two edge directions of the rhombus r. As explained in Lemma 3, the occurrences of r are exactly the intersections of a $\boldsymbol{u}$ chain and a $\boldsymbol{v}$ chain.

$\boldsymbol{u}$ chains cannot cross (Lemma 3) and are uniformly monotonous (Lemma 4), so they can be indexed by either $\mathbb{Z}$, $\mathbb{N}$, $-\mathbb{N}$ or a finite integer interval. Since r is uniformly recurrent, only the case of $\mathbb{Z}$ is possible. Indeed, there exists R such that any disk of radius R in x contains an occurrence of r, and in particular intersects a $\boldsymbol{u}$ chain. So, starting from any position, one finds arbitrarily many $\boldsymbol{u}$ chains in both the $\boldsymbol{u}$ and $-\boldsymbol{u}$ directions. The same holds for $\boldsymbol{v}$.

Denote $r_{(i,j)}$ the occurrence of r at the intersection of the ith $\boldsymbol{u}$ chain and the jth $\boldsymbol{v}$ chain. By definition of the indexing of chains, starting from occurrence $r_{(i,j)}$ and going along a $\boldsymbol{u}$ chain, the next occurrence of r is $r_{(i+1,j)}$, see Fig. 8b.

Proposition 2. *Let X be a nonempty subshift of edge-to-edge rhombus tilings that is r-uniformly recurrent for some shape r. X-*`Domino` *is Π_1^0-hard.*

Proof. We proceed by many-one reduction to the classical Domino problem which is known to be Π_1^0-complete [5].

Let $\mathcal{T}$ be the shapeset on which X is defined. Define the reduction φ_r as follows, where r is the uniformly recurrent shape. We are given as input a finite set of square tiles $\mathbf{T}_{wang}$ on the set of colours S_{wang}. Define a tileset $\mathbf{T}_{rhombus} = \varphi_r(\mathbf{T}_{wang})$ on colours $S_{rhombus} := S_{wang} \cup \{\text{blank}\}$, where blank is a fresh colour, as $\mathbf{T}_{rhombus} := \mathbf{T}_{coding} \cup \mathbf{T}_{link} \cup \mathbf{T}_{neutral}$ (see Fig. 9) with:

1. $\mathbf{T}_{coding}$ is a copy of $\mathbf{T}_{wang}$ on the shape r. In other words, for each tile $(a_0, a_1, a_2, a_3) \in \mathbf{T}_{wang}$, $\mathbf{T}_{coding}$ contains a tile (r, a_0, a_1, a_2, a_3).
2. $\mathbf{T}_{link}$ is a set of rhombus tiles that link occurrences of r and transmit the colours. Formally, for each shape $r' \neq r$ such that r and r' share exactly one edge direction $\boldsymbol{u}$, and for each colour $a \in S_{wang}$, $\mathbf{T}_{link}$ contains a tile of shape r' with colour a on both $\boldsymbol{u}$ edges and blank on other edges.
3. $\mathbf{T}_{neutral}$ completes the tileset with blank colour. Formally, for each shape r' that shares no edge direction with r, $\mathbf{T}_{neutral}$ contains one tile of shape r' with blank colour on each edge. See Fig. 9.

We prove that $\mathbf{T}_{wang}$ admits a valid tiling of $\mathbb{Z}^2$ if and only if $\varphi_r(\mathbf{T}_{wang})$ admits a valid tiling x such that $\pi(x) \in X$.

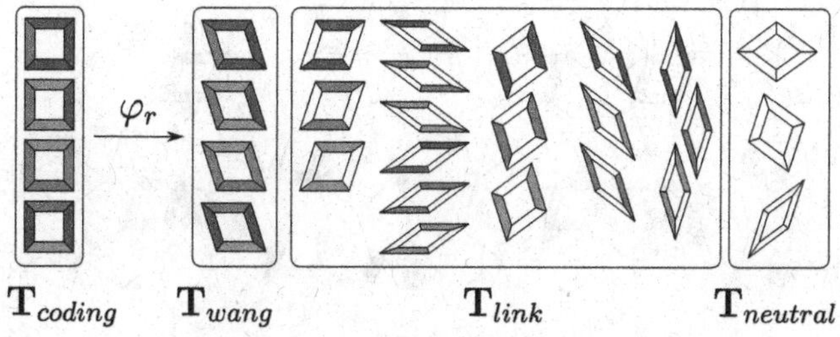

Fig. 9. The reduction φ_r.

Assume that $\mathbf{T}_{wang}$ admits a valid x_{wang} tiling of $\mathbb{Z}^2$. Let us pick $x \in X$. By hypothesis, r is uniformly recurrent in x. We colour x as follows (see Fig. 10):

Coding tiles: index occurrences of r as $\{r_{(i,j)} | (i,j) \in \mathbb{Z}^2\}$ as explained in Lemma 5. Copy the tiles from x_{wang} to the occurrences of r, *i.e.*, if at position (i,j) in x_{wang} there is a (a_0, a_1, a_2, a_3) tile, then colour $r_{(i,j)}$ as the coding tile $(r_{(i,j)}, a_0, a_1, a_2, a_3)$.

Linker tiles: by construction, the north colour of $r_{(i,j)}$ is equal to the south colour of $r_{(i,j+1)}$. We put linkers of that colour along that portion of chain. We proceed similarly for east-west links.

Neutral tiles: remaining tiles share no edge direction with r, so they must be neutral tiles.

The converse also holds: if there exists a valid tiling x on $\varphi_r(\mathbf{T}_{wang})$, since r is uniformly recurrent its occurrences can be indexed by $\mathbb{Z}^2$ (Lemma 5). By construction of the linker tiles, the coding tiles $r_{(i,j)}$ correspond to a valid $\mathbf{T}_{wang}$ tiling in $\mathbb{Z}^2$.

3.3 The Domino Problem on Any Rhombus Subshift Is Π_1^0-hard

In this section, for any subshift X, we build a subshift X' that has a uniformly recurrent shape and such that X'-`Domino` $\preccurlyeq_m X$-`Domino`.

Definition 14 (Restriction). *Given any rhombus subshift X on shapeset $\mathcal{T}$, and a subset $\mathcal{T}' \subset \mathcal{T}$, define $\rho_{\mathcal{T}'}(X)$ as the restriction of X to the configurations that contain only shapes in $\mathcal{T}'$, that is, $\rho_{\mathcal{T}'}(X) := X \cap X_{\mathcal{T}'}$.*

Lemma 6. *Let X be a nonempty subshift on a shapeset $\mathcal{T}$. For any $r \in \mathcal{T}$, either r is uniformly recurrent in every $x \in X$, or the restriction $\rho_{\mathcal{T}\setminus\{r\}}(X)$ is nonempty.*

Proof. This proof, once again, comes from finite local complexity and therefore compactness of edge-to-edge rhombus tilings.

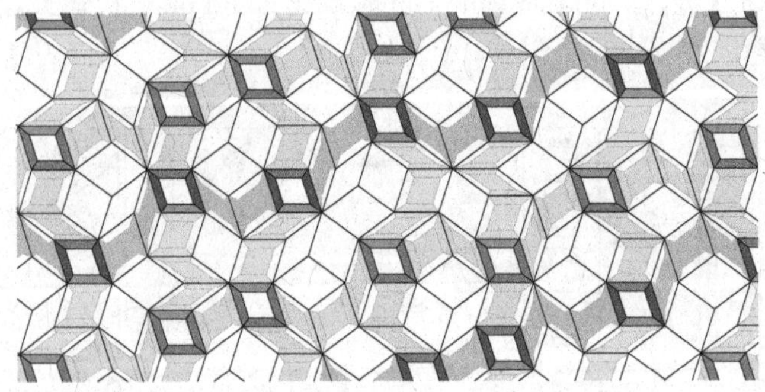

Fig. 10. A valid tiling x with tileset $\varphi_r(\mathbf{T}_{wang})$. Link tiles and neutral tiles have been visually modified to improve readability.

If r is not uniformly recurrent in every $x \in X$, by definition there exist arbitrarily large patterns in X that do not contain r. By compactness there exists an infinite tiling in X containing no r. Hence $\rho_{\mathcal{T}\setminus\{r\}}(X)$ is nonempty.

Lemma 7. *Let X be a nonempty subshift on a finite set of rhombuses $\mathcal{T}$. There is a subset $\mathcal{T}' \subseteq \mathcal{T}$ such that $\rho_{\mathcal{T}'}(X) \neq \emptyset$, and there exists $r \in \mathcal{T}'$ which is uniformly recurrent in every configuration $x \in \rho_{\mathcal{T}'}(X)$.*

Proof. We prove this by induction on the number of shapes, *i.e.*, on $\#\mathcal{T}$.

If $\mathcal{T} = \{r\}$, take $\mathcal{T}' = \mathcal{T}$ and r is uniformly recurrent in any tiling $x \in X$.

Assume the result holds for at most n shapes and $\#\mathcal{T} = n+1$. Pick some $r \in \mathcal{T}$: by Lemma 6 either r is uniformly recurrent in all tilings $x \in X$ or $\rho_{\mathcal{T}\setminus\{r\}}(X)$ is nonempty. In the first case, we conclude with $\mathcal{T}' = \mathcal{T}$. In the second case, we apply the induction hypothesis on $\rho_{\mathcal{T}\setminus\{r\}}(X)$, obtaining some $\mathcal{T}' \subseteq \mathcal{T} \setminus \{r\}$ and a tile $r' \in \mathcal{T}'$ such that r' is uniformly recurrent in $\rho_{\mathcal{T}'}(X)$.

Theorem 3. *Let X be a nonempty subshift of edge-to-edge rhombus tilings. X-`Domino` is Π_1^0-hard.*

This result, with Proposition 1, implies that X-`Domino` is Π_1^0-complete for effective subshifts X.

Proof. By Lemma 7 there exists a subset of the shapeset $\mathcal{T}' \subset \mathcal{T}$ such that the subshift $X' := \rho_{\mathcal{T}'}(X)$ has a uniformly recurrent tile t. By Proposition 2, X'-`Domino` is Π_1^0-hard. We now show that X'-`Domino` $\preccurlyeq_m$ X-`Domino` so that X-`Domino` is also Π_1^0-hard. The many-one reduction $\varphi_{\mathcal{T}'}$ from X'-`Domino` to X-`Domino` is defined as follows: given a $\mathcal{T}'$-wang tileset $\mathbf{T}'$ we define $\varphi_{\mathcal{T}'}(\mathbf{T}') := \mathbf{T}' \cup \mathbf{T}_{fresh}$ where $\mathbf{T}_{fresh}$ contains a tile with a *fresh* colour on each edge for each shape in $\mathcal{T} \setminus \mathcal{T}'$; see Fig. 11.

Remark that there is no reason that choosing the suitable $\mathcal{T}'$ given $\mathcal{T}$ and X (as an enumeration of forbidden patterns) can be done in a computable manner. It only matters that, for a fixed $\mathcal{T}'$, $\varphi_{\mathcal{T}'}$ is computable, which is easily seen with the above definition.

This reduction $\varphi_{\mathcal{T}'}$ is well defined as we have

$$X'\text{-}\mathtt{Domino}(\mathbf{T}') \Leftrightarrow X\text{-}\mathtt{Domino}(\varphi_{\mathcal{T}'}(\mathbf{T}')).$$

The implication holds because $X' \subset X$ and $\mathbf{T}' \subset \varphi_{\mathcal{T}'}(\mathbf{T}')$, so a $\mathbf{T}'$-tiling that projects by erasing colours in X' is also a $\varphi_{\mathcal{T}'}(\mathbf{T}')$-tiling that projects in X.

The converse holds because the tiles in $\mathbf{T}_{fresh}$ cannot appear in $\varphi_{\mathcal{T}'}(\mathbf{T}')$-tilings because they have a fresh colour on each side so no tile can be placed next to it. So a $\varphi_{\mathcal{T}'}(\mathbf{T}')$-tiling x is actually a $\mathbf{T}'$-tiling. Since x contains only tiles in $\mathbf{T}'$, $\pi(x)$ contains only shapes in $\mathcal{T}'$ so $\pi(x) \in \rho_{\mathcal{T}'}(X) = X'$.

Fig. 11. The reduction from a $\mathcal{T}'$-tileset to a $\mathcal{T}$-tileset : in the left box the original $\mathcal{T}'$-tileset, on the right the tiles with fresh colours $\mathbf{T}_{fresh}$.

References

1. Aubrun, N., Barbieri, S., Jeandel, E.: About the domino problem for subshifts on groups. In: Berthé, V., Rigo, M. (eds.) Sequences, Groups, and Number Theory. TM, pp. 331–389. Springer, Cham (2018). https://doi.org/10.1007/978-3-319-69152-7_9
2. Aubrun, N., Sablik, M., Esnay, J.: Domino problem under horizontal constraints. In: STACS 2020, (2020). https://doi.org/10.4230/LIPIcs.STACS.2020.26
3. Baake, M., Grimm, U.: Aperiodic Order: A Mathematical Invitation. Cambridge University Press, Cambridge (2013)
4. Barbieri, S., Sablik, M.: The domino problem for self-similar structures. In: Beckmann, A., Bienvenu, L., Jonoska, N. (eds.) CiE 2016. LNCS, vol. 9709, pp. 205–214. Springer, Cham (2016). https://doi.org/10.1007/978-3-319-40189-8_21
5. Berger, R.: The Undecidability of the Domino Problem. American Mathematical Soc., Providence (1966)
6. de Bruijn, N.G.: Algebraic theory of Penrose's non-periodic tilings of the plane. I. In: Indagationes Mathematicae (Proceedings), January 1981. https://doi.org/10.1016/1385-7258(81)90016-0
7. Dworkin, S., Shieh, J.-I.: Deceptions in quasicrystal growth. Commun. Math. Phys, March 1995. https://doi.org/10.1007/BF02101553
8. Grünbaum, B., Shephard, G.C.: Tilings and Patterns. Courier Dover Publications, New York (1987)
9. Jang, H.: Directional Expansiveness. PhD thesis, George Washington University (2021)
10. Kenyon, R.: Tiling a polygon with parallelograms. Algorithmica, April 1993. DOIurlhttps://doi.org/10.1007/BF01228510
11. Kozen, D.C.: Theory of Computation. Springer, May 2006. https://doi.org/10.1007/1-84628-477-5
12. Lutfalla, V.H.: Geometrical tilings : Distance, topology, compactness and completeness (2022). https://arxiv.org/abs/2208.14922
13. Monin, B., Patey, L.: Calculabilité: Degrés Turing, Théorie Algorithmique de l'aléatoire, Mathématiques à Rebours. Hypercalculabilité, Calvage et Mounet (2022)
14. Penrose, R.: The role of aesthetics in pure and applied mathematical research. Bulletin of the Institute of Mathematics and its Applications (1974)
15. Robinson, E.A.: Symbolic dynamics and tilings of $\mathbb{R}^d$. In: Proceedings of Symposia in Applied Mathematics, vol. 60, pp. 81–120 (2004)
16. Senechal, M.: Quasicrystals and Geometry. Cambridge University Press, Cambridge (1996)

Synchronization of Parikh Automata

Stefan Hoffmann(✉)

Fachbereich 4 - Abteilung Informatikwissenschaften, Universität Trier, Trier, Germany
hoffmanns.tcs@gmail.com

Abstract. Synchronizability of automata has been investigated and extended to various models. Here, we investigate notions of synchronizability for Parikh automata, i.e., automata with counters that are checked against a semilinear set at the end of the computation. We consider various notions of synchronizability (or directability) for Parikh automata and show that they give decidable and PSPACE-complete problems. We then show that for deterministic and complete Parikh automata on letters the synchronization problems are NP-complete over a binary alphabet and solvable in polynomial time for a unary alphabet when the dimension is fixed and the semilinear set is encoded in unary. For a binary encoding the problem remains NP-hard even for unary two-state deterministic and complete Parikh automata of dimension one.

Keywords: Synchronization · Parikh Automata

1 Introduction

The classical *synchronization problem* asks whether a given automaton admits a *synchronizing (or reset) word* that drives all states into a single state. The idea of bringing an automaton to a well-defined state by reading a sequence of inputs, starting from any state, can be seen as implementing a software reset. For more specific applications of synchronization to model checking, verification or robotics we refer to [32,34].

Synchronization was firstly introduced for complete and deterministic automata [5], but has since been extended to different automata models: partial automata [29], nondeterministic automata [25] (called directability in this context), weighted and timed automata [9], register automata [2,31], nested word automata [7], Markov decision processes [11], probabilistic automata [10,26] and push-down automata [3,13,14]. Also variations as synchronization on a budget were introduced [16].

While this problem is NL-complete (and hence solvable in polynomial time) for deterministic and complete finite automata [34], more general notions of synchronization for nondeterministic or partial automata [29] yield PSPACE-complete problems, or even undecidable problems as for push-down automata [14].

Here, we define and investigate notions of synchronization for *Parikh automata.* Parikh automata were introduced in [27] as an automaton model corresponding to the weak second-order logic of one successor (WS1S) with cardi-

F. Drewes and M. Volkov (Eds.): DLT 2023, LNCS 13911, pp. 113–127, 2023.
https://doi.org/10.1007/978-3-031-33264-7_10

nality constraints. These automata have computable Parikh image, which yields that the emptiness problem is decidable for languages recognized by Parikh automata.

Table 1 summarizes the complexities of the introduced notions of synchronization for different types of Parikh automata.

Table 1. Overview of the results. See the main text for definitions. For the result listed in the last three rows on the left side, we assume the semilinear set is encoded in unary, the dimension is fixed and the number of period vectors is bounded.

No restriction: All notions decidable and PSPACE-c, (Theorem 5.2 & 4.1).	
on letters, deterministic & complete	on letters & deterministic
$\|\Sigma\| = 1$ P, Theorem 5.5	PSPACE-c even for dim. zero as D_i-direct., $i \in \{1, 2, 3\}$, is PSPACE-c for (not necessarily complete) det. aut. [29]
$\|\Sigma\| = 2$ NP-c, Theorem 5.4	
$\|\Sigma\| \geq 3$ PSPACE-c, Theorem 5.2 & 4.1	

2 Preliminaries

General Notions. By Σ we denote a *finite alphabet.* By Σ^* we denote the *set of all words*, i.e., finite sequences, with concatenation as an operation. If $u \in \Sigma^*$, by $|u|$ we denote the *length* of u. The *empty word*, i.e., the word of length zero, is denoted by ε. Subsets of Σ^* are called *languages.* We set $\mathbb{N}_0 = \{0, 1, 2, \ldots\}$. For a set X, we denote by $\mathcal{P}(X) = \{Y \mid Y \subseteq X\}$ the *power set (of X).* The alphabet is called *unary* (resp *binary, ternary*) if $|\Sigma| = 1$ (resp. $|\Sigma| = 2$, $\Sigma| = 3$).

We assume the reader to have some basic knowledge in computational complexity theory and formal language theory, as contained, e.g., in [22]. For instance, we make use of regular expressions to describe languages. We also make use of complexity classes like NL, P, NP, PSPACE or 2-EXPSPACE (double exponential space) and we use log-space many-one reductions.

Automata. A *(finite and nondeterministic word) automaton* is a quintuple $\mathcal{A} = (Q, \Sigma, \delta, q_0, F)$ where Q is the *finite state set*, Σ the *input alphabet*, $\delta \subseteq Q \times \Sigma \times Q$ the *transition relation*, q_0 the *start (or initial) state* and $F \subseteq Q$ the set of *final states.*

A *path (or run) in $\mathcal{A}$ of length n* is a sequence of tuples $(p_i, x_{i+1}, p_{i+1}) \in \delta$ with $i \in \{0, \ldots, n-1\}$. A run is called *accepting* if $p_0 = q_0$ and $p_n \in F$. The *label of the run* is the word $x_1 \cdots x_n$. The language $L(\mathcal{A})$ *recognized* (or *accepted*) by $\mathcal{A}$ is the set of all words that label accepting runs in $\mathcal{A}$.

The transition relation can be seen as a function $\delta \colon Q \times \Sigma \to \mathcal{P}(Q)$. We extend it to a function $\hat{\delta} \colon Q \times \Sigma^* \to \mathcal{P}(Q)$ by $\hat{\delta}(q, \varepsilon) = q$ and $\hat{\delta}(q, ux) = \delta(\hat{\delta}(q, u), x)$. We also denote the extension $\hat{\delta}$ by δ in the following. Furthermore, for $S \subseteq Q$

and $u \in \Sigma^*$, we set $\delta(S, u) = \bigcup_{q \in S} \delta(q, u)$. If $|\delta(q, x)| \leq 1$ for all $q \in Q$ and $x \in \Sigma$, then $\mathcal{A}$ is called *deterministic*; if $|\delta(q, x)| \geq 1$ for all $q \in Q$ and $x \in \Sigma$, then $\mathcal{A}$ is called *complete*, see [22].

Parikh Images and Semilinear Sets. Let $u \in \Sigma^*$ and $a \in \Sigma$. By $|u|_a$ we denote the number of *occurrences of the letter* a in u. Instead of fixing an order on Σ and working with $\mathbb{N}_0^{|\Sigma|}$, we identify vectors and points with functions $\mathbb{N}_0^{\Sigma} = \{f \mid f \colon \Sigma \to \mathbb{N}_0\}$. The *Parikh mapping* $\psi \colon \Sigma^* \to \mathbb{N}_0^{\Sigma}$ is given by $\psi(u)(a) = |u|_a$.

By $\vec{0} \in \mathbb{N}_0^{\Sigma}$ we denote the *zero vector* (or *point*), i.e., $\vec{0}(a) = 0$ for each $a \in \Sigma$.

We denote points (and vectors) by tuples (row vectors) and identify row vectors, columns vectors and points, i.e., we do not distinguish them notationwise, as there is not much of a difference in usage in our contexts. So we write $A\vec{v}$ here and in the following for a matrix product that would usually be written as $A\vec{v}^T$.

If $\Sigma = \{a, b\}$, we set $\varphi(a) = (1, 0)$, $\varphi(b) = (0, 1)$.

A subset of $\mathbb{N}_0^d$ $(d \geq 0)$ of the form $\{\vec{v}_0 + c_1\vec{v}_1 + \ldots + c_n\vec{v}_n \mid c_1, \ldots, c_n \in \mathbb{N}_0\}$ for $\vec{v_i} \in \mathbb{N}_0^d$, $i \in \{0, 1, \ldots, n\}$, is called *linear*. The vectors $\vec{v}_1, \ldots, \vec{v}_n$ are called the *period vectors*, and $\vec{v}_0$ the *offset (or base) vector* of the linear set. A *semilinear set* is a finite union of linear sets. A linear set can be encoded as a finite list of vectors, and a semilinear set as a finite list of such lists of vectors. We call such an encoding a *representation by generators*, where the numbers appearing in the components of the vectors can be encoded in *binary* or in *unary*.

For $\vec{v}, \vec{u} \in \mathbb{N}_0^d$ with $\vec{v} = (v_1, \ldots, v_d)$ and $\vec{u} = (u_1, \ldots, u_d)$, we write $\vec{v} \leq \vec{u}$ if $v_i \leq u_i$ for all $i \in \{1, \ldots, d\}$.

For a $\vec{v} \in \mathbb{Z}^d$ and $A \in \mathbb{Z}^{d \times n}$ with $\vec{v} = (v_1, \ldots, v_d)$, $A = (a_{ij})_{1 \leq i \leq d, 1 \leq j \leq n}$ we set $||\vec{v}|| = \max\{|v_i| : i \in \{1, \ldots, d\}\}$ and $||A|| = \max\{|a_{ij}| \mid i \in \{1, \ldots, d\}, j \in \{1, \ldots, n\}\}$.

We need the following result from [6] implied by results of Pottier [30] and Domenjoud [8].

Proposition 2.1 ([6]). *Let $A \in \mathbb{Z}^{d \times n}$ and $\vec{c} \in \mathbb{Z}^d$. There exist offset vectors $\vec{v}_{0,1}, \ldots, \vec{v}_{0,k} \in \mathbb{Z}^d$ and a common set of period vectors $\vec{v}_1, \ldots, \vec{v}_m \in \mathbb{Z}^d$ with*

$$\{\vec{x} \in \mathbb{N}_0^n \mid A\vec{x} = \vec{c}\} = \bigcup_{i=1}^{k} \{\vec{v}_{0,i} + c_1 \cdot \vec{v}_1 + \ldots + c_m \cdot \vec{v}_m \mid c_1, \ldots, c_m \in \mathbb{N}_0\}$$

such that $k \leq (n+1)^d$, $m \leq n^d$, $||\vec{v}_{0,i}|| \leq ((n+1) \cdot ||A|| + ||\vec{c}|| + 1)^d$ for all $i \in \{1, \ldots, k\}$ and $||\vec{v}_j|| \leq (n \cdot ||A|| + 1)^d$ for all $j \in \{1, \ldots, m\}$.

See also the related work [28, Chap. 7.3] for results on the descriptional complexity of Parikh images of regular languages.

We also need the following common result.

Lemma 2.2. *Let $C \subseteq \mathbb{N}_0^{\Sigma}$ be a semilinear set given by generators encoded with n bits. There exists a word automaton $\mathcal{A}$ with at most 2^n states for the binary encoding of C or n states and computable in polynomial time for the unary encoding of C such that $\psi(L(\mathcal{A})) = C$.*

Synchronizable Automata. If $\mathcal{A} = (Q, \Sigma, \delta, q_0, F)$ is a deterministic and complete automaton, then it is *synchronizable* if $\bigcup_{s \in Q} \bigcap_{q \in Q} L(\mathcal{A}_{q,\{s\}}) \neq \emptyset$ where we set $\mathcal{A}_{q,\{s\}} = (Q, \Sigma, \delta, q, \{s\})$. However, for non-deterministic automata $\mathcal{A} = (Q, \Sigma, \delta, q_0, F)$, there are different notions of synchronizability [25,29]. A word $u \in \Sigma^*$ is called

1. D_1*-directing* if $\delta(q, u) \neq \emptyset$ for all $q \in Q$ and $|\delta(Q, u)| = 1$;
2. D_2*-directing* if $\delta(q, u) = \delta(Q, u)$ for all $q \in Q$;
3. D_3*-directing* if $\bigcap_{q \in Q} \delta(q, u) \neq \emptyset$.

Then $\mathcal{A}$ is said to be D_i*-directable* if it possesses a D_i-directing word for $i \in \{1, 2, 3\}$.

It is PSPACE-complete to decide whether a given nondeterministic automaton is D_i-directable for each $i \in \{1, 2, 3\}$ [29].

Furthermore, for hardness proofs, we need the following results from [12,20], stating that if we want to know whether a deterministic and complete automaton admits a synchronizing word coming from a specific set, then this can give PSPACE-complete or NP-complete problems.

Theorem 2.3 ([12,20]). *Deciding whether a deterministic and complete word automaton admits a synchronizing word in* $\{a, b\}^* c\{a, b\}^*$ *with* $\Sigma = \{a, b, c\}$ *is PSPACE-complete [12]. Furthermore, deciding whether a deterministic and complete word automaton admits a synchronizing word in* $b^*ab^*ab^*$ *with* $\Sigma = \{a, b\}$ *is* NP-*complete [20].*

Parikh Automata. Let $D \subseteq \mathbb{N}_0^d$ be finite. A *Parikh automaton* of dimension d over Σ is a pair $(\mathcal{A}, C)$ where $\mathcal{A} = (Q, \Sigma \times D, \delta, q_0, F)$ is a finite automaton with alphabet $\Sigma \times D$ and $C \subseteq \mathbb{N}_0^d$ is a semilinear set. The *recognized* (or *accepted*) language is

$$L(\mathcal{A}, C) = \{u_1 u_2 \cdots u_n \in \Sigma^* \mid \exists \vec{c} \in C\ \exists \vec{c}_1, \vec{c}_2, \ldots, \vec{c}_n \in D : \\ \vec{c} = \vec{c}_1 + \vec{c}_2 + \ldots + \vec{c}_n \wedge (u_1, \vec{c}_1)(u_2, \vec{c}_2) \cdots (u_n, \vec{c}_n) \in L(\mathcal{A})\}.$$

A Parikh automaton is *deterministic* if for every state $q \in Q$ and $a \in \Sigma$ there exists at most one pair $(q', \vec{v}) \in Q \times D$ such that $(q, (a, \vec{v}), q') \in \delta$. The Parikh automaton is called *complete*, if there exists at least one such tuple.

With every Parikh automaton $(\mathcal{A}, C)$ we associate a *word automaton* $\mathcal{A}^{\text{WORD}}$ by forgetting about the additional component D in the alphabet. More formally, set $\mathcal{A}^{\text{WORD}} = (Q, \Sigma, \delta_{\text{WORD}}, q_0, F)$ with

$$\delta_{\text{WORD}} = \{(p, x, q) \mid \exists \vec{v} \in D : (p, (x, \vec{v}), q) \in \delta\}.$$

A *Parikh automaton on letters* is a Parikh automaton $(\mathcal{A}, C)$ such that if $(q, (a, \vec{u}), q')$ and $(p, (a, \vec{v}), p')$ are two transitions in $\mathcal{A}$ with $a \in \Sigma$, then $\vec{u} = \vec{v}$, i.e., the letter determines the added vector. This class was introduced in [4]. We note that Parikh automata on letters are defined in analogy to *visibly push-down automata* [1]. Fig. 1 shows an example of a Parikh automaton on letter.

3 Notions of Synchronization

Here, we introduce several notions of synchronization. A Parikh automaton $(\mathcal{A}, C)$ is called:

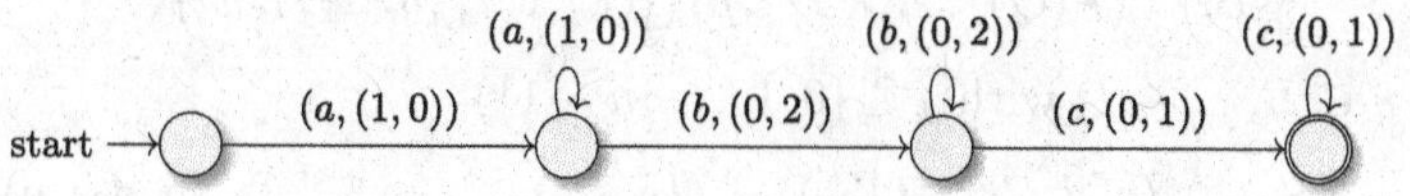

Fig. 1. A deterministic Parikh automaton on letters with semilinear set $\{(n,n) \in \mathbb{N}_0^2 \mid n \geq 1\}$ for the language $\{a^r b^s c^t \mid r = 2s + t, s, t \geq 1, r \geq 3\}$.

∀∃-D$_1$-directable: There exists a word $u \in \Sigma^*$ with $u = u_1 \cdots u_n$, $u_i \in \Sigma$, and $s \in Q$ such that for each $q \in Q$ there exist $\vec{c}_1, \vec{c}_2, \ldots, \vec{c}_n \in D$ such that $\delta(q, (u_1, \vec{c}_1) \cdots (u_n, \vec{c}_n)) = \{s\}$ and $\vec{c}_1 + \ldots + \vec{c}_n \in C$.

∃∀-D$_1$-directable: There exists a word $u \in \Sigma^*$ with $u = u_1 \cdots u_n$, $u_i \in \Sigma$, $s \in Q$ and $\vec{c}_1, \vec{c}_2, \ldots, \vec{c}_n \in D$ with $\vec{c}_1 + \ldots + \vec{c}_n \in C$ such that for each $q \in Q$ we have $\delta(q, (u_1, \vec{c}_1) \cdots (u_n, \vec{c}_n)) = \{s\}$.

∀∃-D$_2$-directable: There exists $u \in \Sigma^*$ with $u = u_1 \cdots u_n$, $u_i \in \Sigma$, such that for each $q \in Q$ there exist $\vec{c}_{q,1}, \vec{c}_{q,2}, \ldots, \vec{c}_{q,n} \in D$ such that $\vec{c}_{q,1} + \vec{c}_{q,2} + \cdots + \vec{c}_{q,n} \in C$ and for each $q', q'' \in Q$ we have $\delta(q', w') = \delta(q'', w'')$ where $w' = (u_1, \vec{c}_{q',1})(u_2, \vec{c}_{q',2}) \cdots (u_n, \vec{c}_{q',n})$ and $w'' = (u_1, \vec{c}_{q'',1})(u_2, \vec{c}_{q'',2}) \cdots (u_n, \vec{c}_{q'',n})$.

∃∀-D$_2$-directable: There exists $u \in \Sigma^*$ with $u = u_1 \cdots u_n$, $u_i \in \Sigma$, and counter values $\vec{c}_1, \vec{c}_2, \ldots, \vec{c}_n \in D$ with $\vec{c}_1 + \vec{c}_2 + \ldots + \vec{c}_n \in C$ such that for each $q \in Q$ we have $\delta(Q, w) = \delta(q, w)$ where $w = (u_1, \vec{c}_1)(u_2, \vec{c}_2) \cdots (u_n, \vec{c}_n)$.

∀∃-D$_3$-directable: $\bigcup_{s \in Q} \bigcap_{q \in Q} L(\mathcal{A}_{q,\{s\}}, C) \neq \emptyset$.

∃∀-D$_3$-directable: There exists a word $u \in \Sigma^*$ with $u = u_1 \cdots u_n$, $u_i \in \Sigma$, $s \in Q$ and $\vec{c}_1, \vec{c}_2, \ldots, \vec{c}_n \in D$ with $\vec{c}_1 + \ldots + \vec{c}_n \in C$ such that for each $q \in Q$ we have $s \in \delta(q, (u_1, \vec{c}_1) \cdots (u_n, \vec{c}_n))$.

For Parikh automata without counter values (considered as word automata) these notions reduce to the corresponding notion of D_i-directability. They are generalizations of these notions, where we either allow the counter values for each state to be different, or we demand a common sequence of counter values that work for all states (see Sect. 6 for a short discussion of other possible notions).

4 Decidability

All notions of synchronization give PSPACE-hard problem, as D_i-directability of word automata is PSPACE-complete [29].

Here, we always encode the semilinear set by generators.

Theorem 4.1. *It is decidable and in PSPACE whether a given n-state Parikh automaton is ∀∃-D_i-directing or ∃∀-D_i-directing for each $i \in \{1, 2, 3\}$ when the semilinear set is given by generators in binary or unary encoding and a shortest directing word for all notions of directability has length at most $r \cdot 2^{n^2}$ (unary encoding) or 2^{r+n^2} (binary encoding) where r is the number of bits to encode the semilinear set.*

Proof (Sketch). Let $(\mathcal{A}, C)$ be a Parikh automaton with $\mathcal{A} = (Q, \Sigma \times D, \delta, q_0, F)$ and $Q = \{q_0, \ldots, q_{n-1}\}$.

<u>∀∃-D$_i$-directing for $i \in \{1, 2, 3\}$</u>: Consider the Parikh automata

$$\mathcal{B}_i = (\mathcal{P}(Q)^n, \Sigma \times D^n, \mu, (\{q_0\}, \ldots, \{q_{n-1}\}), E_i)$$

with, for $a \in \Sigma$, $d_j \in D$ with $j \in \{0, 1, \ldots, n-1\}$),

$$\mu((S_0, \ldots, S_{n-1}), (a, d_0, \ldots, d_{n-1})) = (\{q \in Q \mid \exists s \in S_0 : (s, (a, d_0), q) \in \delta\}, \\ \ldots, \{q \in Q \mid \exists s \in S_{n-1} : (s, (a, d_{n-1}), q) \in \delta\})$$

and $E_1 = \{(\{s\}, \{s\}, \ldots, \{s\}) \mid s \in Q\}$, $E_2 = \{(S, S, \ldots, S) \mid \emptyset \neq S \subseteq Q\}$ and $E_3 = \{(S_0, S_1, \ldots, S_{n-1}) \mid \bigcap_{i=0}^{n-1} S_i \neq \emptyset\}$. Also, define the semilinear set $C' = \{(\vec{v}_0, \vec{v}_1, \ldots, \vec{v}_{n-1}) \mid \vec{v}_i \in C\} = C^n \subseteq \mathbb{N}_0^{dn}$. Then, it is easy to check that

$$L((\mathcal{B}_i, C')) = \{u \in \Sigma^* \mid u \text{ is } \forall\exists\text{-D}_i\text{-directing for } (\mathcal{A}, C)\}$$

for $i \in \{1, 2, 3\}$.

Let r be the number of bits used to encode the semilinear set C' (which is polynomially dependent on the number of bits used for C). By using an automaton-representation of C and the pigeonhole principle on the accepting runs of the above automaton, we can show that a shortest word in $L((\mathcal{B}_i, C))$ has length at most 2^{r+n^2}. Note that $\mathcal{B}_i$ has 2^{n^2} states.

The automata $\mathcal{B}_i$ can be simulated in non-deterministic polynomial space $O(r \cdot n^2)$ by a "guess-on-the-fly" approach. A non-deterministic machine stores a state vector $(S_0, \ldots, S_{n-1})$ of $\mathcal{B}_i$, which consists of n subsets from Q, guesses an input symbol $(a, \vec{d_0}, \ldots, \vec{d_{n-1}}) \in \Sigma \times D^n$ and updates the state vector to $\mu((S_0, \ldots, S_{n-1}), (a, \vec{d_0}, \ldots, \vec{d_{n-1}}))$. Additionally, we maintain n counters with values in $\mathbb{N}_0^d$, each initialized to the zero vector, where we update the i-th counter by adding $\vec{d_i}$ to it. The machine starts with $(\{q_0\}, \ldots, \{q_{n-1}\})$ and repeats the above steps $2^r \cdot 2^{n^2}$ times (which can be ensured with an additional counter using only polynomial space). Hence, every computation path terminates and the counter values only need polynomial space when using a binary encoding (and so also for a unary encoding). At the end, we check if the counter values are contained in C', which can be done in non-deterministic polynomial time [15, Proposition III.2]. By Savitch's theorem [33] we can deduce that the problem is contained in PSPACE.

<u>∃∀-D$_i$-directing for $i \in \{1, 2, 3\}$</u>: Construct the same Parikh automata $\mathcal{B}_i$ but let $C'' = \{(\vec{v}, \vec{v}, \ldots, \vec{v}) \mid \vec{v} \in C\}$. We have

$$L((\mathcal{B}_i, C'')) = \{u \in \Sigma^* \mid u \text{ is } \exists\forall\text{-D}_i\text{-directing for } (\mathcal{A}, C)\}$$

for $i \in \{1, 2, 3\}$. Then perform the same steps as before. □

5 Deterministic Complete Parikh Automata on Letters

For deterministic and complete Parikh automata on letters, all notions of directability coincide and, in this case, we simply say the Parikh automaton is *synchronizing*.

For the results in this section, we make the following additional assumptions about the semilinear set.

1. The semilinear set is encoded in *unary*.
2. Each linear part has a bounded number of period vectors.
3. The dimension d is fixed.

We show the following on the complexity to decide synchronizability.

1. PSPACE-complete for $|\Sigma| \geq 3$ (Sect. 5.2).
2. NP-complete for a binary alphabet (Sect. 5.3).
3. In polynomial time for a unary alphabet (Sect. 5.4).

But first (Sect. 5.1) we start with a result on simplifying the semilinear set.

5.1 Simplifying the Semilinear Set

The following simplification of the semilinear set was stated in [4, Proposition 5.3]. However, we need additional claims about polynomial time computability, which we state and prove here.

Proposition 5.1. *Suppose $(\mathcal{A}, C)$ is a Parikh automaton on letters. Then, there exists a semilinear set $\overline{C} \subseteq \mathbb{N}_0^{\Sigma}$ such that, for all $w \in \Sigma^*$ and all $q, q' \in Q$,*

$$w \in L(\mathcal{A}_{q,\{q'\}}, C) \Leftrightarrow \psi(w) \in \overline{C} \wedge w \in L(\mathcal{A}_{q,\{q'\}}^{WORD}).$$

Furthermore, a representation of $\overline{C}$ by generators encoded in unary is computable in polynomial time if the following additional assumptions on the input are met:

1. *C is specified by generators encoded in unary,*
2. *the alphabet is fixed,*
3. *the dimension of $(\mathcal{A}, C)$ is fixed,*
4. *the number of period vectors specifying each linear set from C is bounded by a fixed number.*

Proof (Sketch). Without loss of generality, we can suppose that for every letter $a \in \Sigma$ there exists at least one transition $(q, (a, \vec{u}), q')$ in $\mathcal{A}$. Define $\varphi \colon \mathbb{N}_0^{\Sigma} \to \mathbb{N}_0^d$ by $\varphi(\psi(a)) = \vec{u}$ for some arbitrarily chosen transition $(q, (a, \vec{u}), q')$ in $\mathcal{A}$. As $\mathcal{A}$ is a Parikh automaton on letters, this is well-defined. Then, extend to all of $\mathbb{N}_0^{\Sigma}$ by setting $\varphi(\vec{v}) = \sum_{a \in \Sigma} \vec{v}(a) \cdot \varphi(\psi(a))$ for $\vec{v} \in \mathbb{N}_0^{\Sigma}$. Set $\overline{C} = \varphi^{-1}(C)$.

For given sets $B, P \subseteq \mathbb{Z}^d$ with $P = \{\vec{v_1}, \ldots, \vec{v_n}\}$, we set

$$L(B, P) = \{\vec{b} + c_1 \cdot \vec{v_1} + \ldots + c_n \cdot \vec{v_n} \mid \vec{b} \in B \text{ and } c_1, \ldots, c_n \in \mathbb{N}_0\}$$

(this is a standard notation in the literature on semilinear sets; however it is similar to our notation for the accepted language of a Parikh automaton, but it only appears in this proof and hence this should not cause confusion).

Fix an ordering $\Sigma = \{a_1, \ldots, a_k\}$. The homomorphism $\varphi \colon \mathbb{N}_0^\Sigma \to \mathbb{N}_0^d$ can be given by a matrix $A \in \mathbb{Z}^{d \times |\Sigma|}$ where the i-th column is the vector $\varphi(\psi(a_i))$. Then $\varphi(\vec{v}) = A\vec{v}$ (recall that we identified row and column vectors, and so we write $A\vec{v}$ here and in the following for the matrix product that would usually be written as $A\vec{v}^T$). The set C is a finite union of linear sets, and as inverse images preserve unions, if we can compute the inverse image of a linear set in polynomial time under the stated condition, the claim follows. So, we can suppose C is a linear set given by $\vec{v_0}, \vec{v_1}, \ldots, \vec{v_n}$ with $C = \{\vec{v_0} + c_1\vec{v_1} + \ldots + c_n\vec{v_n} \mid c_1, \ldots, c_n \in \mathbb{N}_0\}$. We have

$$\vec{v} \in \overline{C} \Leftrightarrow \varphi(\vec{v}) \in C \Leftrightarrow \exists c_1, \ldots, c_n : \varphi(\vec{v}) = \vec{v_0} + c_1\vec{v_1} + \cdots + c_n\vec{v_n}.$$

Let B be the matrix whose first $|\Sigma|$ columns equal A and the next n columns equal $-\vec{v}_1, -\vec{v}_2, \ldots, -\vec{v}_n$, i.e. $B = (A|-\vec{v}_1|-\vec{v}_2| \cdots |-\vec{v}_n)$. Note that $B \in \mathbb{Z}^{d \times (|\Sigma|+n)}$ has non-positive entries. Then $\vec{v} = (x_1, \ldots, x_{|\Sigma|}) \in \overline{C}$ if and only if there exist $c_1, \ldots, c_n \in \mathbb{N}_0$ such that for the vector $\vec{u} = (x_1, \ldots, x_{|\Sigma|}, c_1, \ldots, c_n)$ we have $B\vec{u} = \vec{v_0}$. Let $Q \subseteq \mathbb{N}_0^{|\Sigma|+n}$ be the minimal elements in $\{\vec{x} \in \mathbb{N}_0^{|\Sigma|+n} \mid B\vec{x} = \vec{v_0}\}$ and $P \subseteq \mathbb{N}_0^{|\Sigma|+n}$ be the minimal elements in $\{\vec{x} \in \mathbb{N}_0^{|\Sigma|+n} \setminus \{\vec{0}\} \mid B\vec{x} = \vec{0}\}$. In the proof of [17, Theorem 6.1] it is shown that $\{\vec{x} \in \mathbb{N}_0^{|\Sigma|+n} \mid B\vec{x} = \vec{v_0}\} = L(Q, P)$. Let $\pi \colon \mathbb{N}_0^{|\Sigma|+n} \to \mathbb{N}_0^{|\Sigma|}$ be the homomorphism given by setting $\pi((x_1, \ldots, x_{|\Sigma|}, x_{|\Sigma|+1}, \ldots, x_{|\Sigma|+n})) = (x_1, \ldots, x_{|\Sigma|})$. Then

$$\overline{C} = \{(x_1, \ldots, x_{|\Sigma|}) \mid \exists c_1, \ldots, c_n \in \mathbb{N}_0 : \\ (x_1, \ldots, x_{|\Sigma|}, c_1, \ldots, c_n) \in L(Q, P)\} = \pi(L(Q, P)).$$

With Prop. 2.1 we can deduce that Q and P are computable in polynomial time, which gives the claim (the proof is left out due to space). □

Remark 1. The set $\overline{C}$ is not computable in polynomial time in general, even when the semilinear set C is encoded in unary. For example, consider the alphabet $\Sigma = \{a_1, b_1, a_2, b_2, \ldots, a_n, b_n\}$. Define the homomorphism $\varphi \colon \mathbb{N}_0^{2n} \to \mathbb{N}_0^n$ by setting, for $i \in \{1, \ldots, n\}$, the images $\varphi(\psi(a_i))$ and $\varphi(\psi(b_i))$ equal to the i-th standard unit vector that is one precisely at the i-entry and zero elsewhere. Then

$$\varphi^{-1}(\{(1, \ldots, 1)\}) = \{(v_1, u_1, v_2, u_2, \ldots, v_n, u_n) \in \mathbb{N}_0^{2n} \mid v_i + u_i = 1, i \in \{1, \ldots, n\}\}$$

contains precisely 2^n elements and so is not polynomial time computable.

Note that the homomorphism $\varphi \colon \mathbb{N}_0^\Sigma \to \mathbb{N}_0^d$ in the proof of Proposition 5.1 can be computed in polynomial time and we have

$$u \text{ synchronizes } (\mathcal{A}, C) \Leftrightarrow \varphi(\psi(u)) \in C \wedge u \in \bigcup_{s \in Q} \bigcap_{q \in Q} L(\mathcal{A}_{q,\{s\}}^{\text{WORD}}).$$

Hence, when we only need the above property or only have to query C by applying φ and ψ, then we do not have to compute $\overline{C}$.

5.2 Ternary Alphabet

For a ternary alphabet, or an alphabet containing at least three letters, the problem is still PSPACE-hard. This follows rather straightforward from Theorem 2.3.

Theorem 5.2. *Fix an alphabet Σ with $|\Sigma| \geq 3$. The problem of synchronization of deterministic and complete Parikh automata on letters is PSPACE-hard even when only the fixed semilinear constraint language $C = \{1\}$ is used.*

Proof. Let $\Sigma = \{a, b, c\}$ and $\mathcal{A} = (Q, \Sigma, \delta, q_0, F)$ be a deterministic and complete automaton. We reduce from the PSPACE-complete problem to decide whether $\mathcal{A}$ has a synchronizing word in $\{a,b\}^*c\{a,b\}^*$ (Thm 2.3).

Let $\mathcal{B} = (Q, \Sigma \times \{0,1\}, \mu, q_0, F)$ where

$$\mu = \{(p,(a,0),q) \mid (p,a,q) \in \delta\} \cup \\ \{(p,(b,0),q) \mid (p,b,q) \in \delta\} \cup \{(p,(c,1),q) \mid (p,c,q) \in \delta\}$$

and $C = \{1\}$. Then $(\mathcal{B}, C)$ is a deterministic and complete Parikh automaton on letters that has a synchronizing word if and only if $\mathcal{A}$ has a synchronizing word in $\{a,b\}^*c\{a,b\}^*$ (which is actually the same word for $(\mathcal{B}, C)$ and $\mathcal{A}$). □

5.3 Binary Alphabet

We show NP-completeness for a binary alphabet.

NP-hardness can be shown as in Theorem 5.2 but using Theorem 2.3 with the language $b^*ab^*ab^*$.

Theorem 5.3. *For a binary alphabet, the problem of synchronization for deterministic and complete Parikh automata on letters is* NP*-hard even when only the fixed semilinear constraint set $C = \{2\}$ is used.*

Next, we show that the problem with the mentioned restrictions is contained in NP.

Theorem 5.4. *For binary alphabets, the problem of synchronization for deterministic and complete Parikh automata on letters is* NP*-complete.*

Proof. By Theorem 5.3, we only have to show containment in NP. Let $(\mathcal{A}, C)$ be a Parikh automaton with n states. By Proposition 5.1 there exists a semilinear set $\overline{C} \subseteq \mathbb{N}_0^{\Sigma}$ computable in polynomial time such that

$$u \text{ synchronizes } (\mathcal{A}, C) \Leftrightarrow u \in \psi^{-1}(\overline{C}) \cap \bigcup_{s \in Q} \bigcap_{q \in Q} L(\mathcal{A}_{q,\{s\}}^{\text{WORD}}). \tag{1}$$

Write $\overline{C} = \bigcup_{i=1}^{n} \{\vec{v}_{i,0} + c_1 \cdot \vec{v}_{i,1} + \ldots + c_{n_i} \cdot \vec{v}_{i,n_i} \mid c_1, \ldots, c_{n_i} \in \mathbb{N}_0\}$, i.e., a finite union of linear sets with $\vec{v}_{i,j} \neq \vec{0}$ for all $i \in \{1, \ldots, n\}$ and $j \in \{1, \ldots, n_i\}$. We have three cases, and which is the case can be determined in polynomial time by inspection of the vectors $\vec{v}_{i,j}$.

1. The set $\overline{C}$ is finite, i.e., $\forall i \in \{1, \ldots, n\} : n_i = 0$.
2. There exists an infinite sequence $\vec{u_i} = (x_i, y_i) \in \overline{C}$ such that $x_i < x_{i+1}$ and $y_i < y_{i+1}$.
3. Neither of the previous cases applies. Then for each $i \in \{1, \ldots, n\}$ it holds that for all $j \in \{1, \ldots, n_i\}$ we have $\vec{v}_{i,j} \in \{c \cdot (1,0) \mid c > 0\}$ or for all $j \in \{1, \ldots, n_i\}$ we have $\vec{v}_{i,j} \in \{c \cdot (0,1) \mid c > 0\}$.

In the first case, we can nondeterministically guess a word $u \in \Sigma^*$ with $(|u|_a, |u|_b) \in \overline{C}$. This is possible in polynomial time as, with the stated assumptions, the magnitudes of the components of the vectors in $\overline{C}$ depend polynomially on those from C by Proposition 5.1. By Eqn. (1) this word synchronizes $(\mathcal{A}, C)$ if and only if it synchronizes $\mathcal{A}^{\text{WORD}}$. Checking if a given word synchronizes a word automaton can be done in polynomial time [20, Theorem 4].

In the second case, we claim that $(\mathcal{A}, C)$ has a synchronizing word if and only if $\mathcal{A}^{\text{WORD}}$ has a synchronizing word. The latter problem is solvable in polynomial time, as testing synchronizability of ordinary deterministic word automata can be done in polynomial time [34]. In fact, every synchronizing word for $(\mathcal{A}, C)$ also synchronizes $\mathcal{A}^{\text{WORD}}$ by simply forgetting about the counting register values. Conversely, suppose $u \in \Sigma^*$ synchronizes $\mathcal{A}^{\text{WORD}}$. By assumption there exist $i \in \mathbb{N}_0$ and $(x_i, y_i) \in \overline{C}$ such that $|u|_a \le x_i$ and $|u|_b \le y_i$. Then choose a word $v \in \Sigma^*$ such that $\psi(uv) = (x_i, y_i) \in \overline{C}$. The word uv synchronizes $\mathcal{A}^{\text{WORD}}$ as well, as we can append any word to a synchronizing word and the resulting word is still synchronizing (here we use that $\mathcal{A}^{\text{WORD}}$ is deterministic). So, by Eqn. (1), as $\psi(uv) \in \overline{C}$, the word uv synchronizes $(\mathcal{A}, C)$.

Now, for the last case. We show that we can test each linear set $\{\vec{v}_{i,0} + c_1\vec{v}_{i,1} + \ldots + c_{n_i}\vec{v}_{n_i,1}\}$ for $i \in \{1, \ldots, n\}$ in nondeterministic polynomial time. Fix some $i \in \{1, \ldots, n\}$. If $n_i = 0$, then the linear set equals $\{\vec{v}_{i,0}\}$ and we can argue as in the first case. Otherwise, without loss of generality, each $\vec{v}_{i,j}$ for $j \in \{1, \ldots, n_i\}$ is a non-zero multiple of $(1,0)$, i.e., $\vec{v}_{i,j} = c_j \cdot (1,0)$. Let $\vec{v}_{i,0} = (x, y)$ and $j \in \{1, \ldots, n_i\}$. Set $L_i = \psi^{-1}(\{\vec{v}_{i,0} + c_1 \cdot \vec{v}_{i,1} + \ldots + c_{n_i} \cdot \vec{v}_{i,n_i}\})$. Observe that for any synchronizing word u for $\mathcal{A}^{\text{WORD}}$ that belongs to L_i we have $|u|_b = y$. Conversely, if u is a synchronizing word for $\mathcal{A}^{\text{WORD}}$ with $|u|_b = y$, we can extend it with a word $v \in a^*$ such that $|uv|_a = x + d \cdot c_j$ for some $d \in \mathbb{N}_0$. The word uv synchronizes $\mathcal{A}^{\text{WORD}}$ as well and $uv \in L_i$. Consequently, there exists a synchronizing word u of $\mathcal{A}^{\text{WORD}}$ with $|u|_b = y$ if and only if there exists a synchronizing word of $\mathcal{A}^{\text{WORD}}$ in L_i. Next we show how to find such a word in nondeterministic polynomial time. An important part of our argument is the fact that we only have to search among "sufficiently short" words.

<u>Claim:</u> Let $N \ge 0$ be the number of states of $\mathcal{A}^{\text{WORD}}$. If there exists a synchronizing word $u \in \Sigma^*$ for $\mathcal{A}^{\text{WORD}}$ with $|u|_b = y$, then there exists a synchronizing word

$$a^{i_0} b a^{i_1} b a^{i_2} b \cdots b a^{i_y}$$

for $\mathcal{A}^{\text{WORD}}$ with $0 \le i_r < 2^N$ for all $r \in \{0, 1, \ldots, y\}$.

Proof of the Claim. Write

$$u = a^{j_0} b a^{j_1} b a^{j_2} b \cdots b a^{j_y}$$

for numbers $j_r \geq 0$ with $r \in \{0, 1, \ldots, |u|_b\}$. Consider the subsets

$$\delta_{\text{WORD}}(Q, a^{j_0}ba^{j_1}b\cdots a^{j_{r-1}}b), \delta_{\text{WORD}}(Q, a^{j_0}ba^{j_1}b\cdots a^{j_{r-1}}ba),$$
$$\delta_{\text{WORD}}(Q, a^{j_0}ba^{j_1}b\cdots a^{j_{r-1}}ba^2), \ldots, \delta_{\text{WORD}}(Q, a^{j_0}ba^{j_1}b\cdots a^{j_{r-1}}ba^{j_r}).$$

If $j_r \geq 2^N$, by the pigeonhole principle, we find $0 \leq i_r < 2^N$ such that

$$\delta_{\text{WORD}}(Q, a^{j_0}ba^{j_1}b\cdots a^{j_{r-1}}ba^{j_r}) = \delta_{\text{WORD}}(Q, a^{j_0}ba^{j_1}b\cdots a^{j_{r-1}}ba^{i_r})$$

For each $r \in \{0, 1, \ldots, y\}$ apply this argument if $j_r \geq 2^N$ and setting $i_r = j_r$ if $j_r < 2^N$. This gives the claim. *[End, Proof of the Claim.]*

By the above claim, it is sufficient to check if $\mathcal{A}^{\text{WORD}}$ has a synchronizing word of the form

$$a^{i_0}ba^{i_1}ba^{i_2}b\cdots ba^{i_{|u|_b}}$$

with $0 \leq i_r \leq 2^N$ for $r \in \{0, 1, \ldots, |u|_b\}$. By encoding each number i_r in binary we need at most N bits for each i_r. In total, we need at most $n \cdot y$ bits for all these numbers (recall that $\vec{v_{i,0}} = (x, y)$ and, as written above, x and y have size polynomial in the input). Hence they can be guessed nondeterministically using polynomial time. Verification can also be done in polynomial time: Let M be the transition matrix of the letter a in $\mathcal{A}^{\text{WORD}}$ (i.e., the matrix indexed by the states and with a one if the letter a goes from the row state to the column state and zero otherwise). By successively multiplying and squaring them (square-and-multiply algorithm), we can compute in polynomial time the result of applying the letter i_r many times for each $r \in \{0, 1, \ldots, |u|_b\}$. Afterwards, check that the resulting states can be appended to form paths ending in the same state, or alternatively multiply out the results with the transition matrix for b in between according to the guessed word and check that the resulting matrix contains a column of ones and zeros everywhere else (which is equivalent to the word being synchronizing in the word automaton $\mathcal{A}^{\text{WORD}}$).

Note that we needed the unary encoding of y here in guaranteeing that we have enough space and time to "layout" the b's in between the consecutive runs of a's. □

5.4 Unary Alphabet

For a unary alphabet we have a polynomial time algorithm.

Theorem 5.5. *For a unary alphabet, the problem of synchronization for deterministic and complete Parikh automata on letters is in* P.

Proof. Let $(\mathcal{A}, C)$ be a Parikh automaton as stated. By Proposition 5.1 we can compute (with the stated assumptions on the input) a semilinear set $\overline{C} \subseteq \mathbb{N}_0^{\Sigma}$ in polynomial time such that the set of synchronizing words is $\psi^{-1}(\overline{C}) \cap \bigcup_{s \in Q} \bigcap_{q \in Q} L(\mathcal{A}_{q,\{s\}}^{\text{WORD}})$. However, a unary deterministic automaton is synchronizable if and only if it has a single state with a self-loop and every path ultimately leads into this state. This can be checked in polynomial time for

$\mathcal{A}^{\text{WORD}}$, and if this is the case, then the set of synchronizing words is $a^n a^*$ for some n, where n is at most the number of states of $\mathcal{A}$.

Furthermore, note that as $|\Sigma| = 1$ we have $\overline{C} \subseteq \mathbb{N}_0$ and the Parikh mapping is an isomorphism. Hence, as it is encoded in unary, it directly corresponds to a regular language in Σ^* recognized by an automaton (see also Lemma 2.2) $\mathcal{B}$ computable in polynomial time. Then, we only have to test if $a^n a^* \cap L(\mathcal{B})$ is non-empty, which can be done in polynomial time. □

6 Conclusion

Let $\vec{v} \in \mathbb{N}_0^d$ and $C \subseteq \mathbb{N}_0^d$ be a semilinear set. Then for $\mathcal{A} = (\{s,t\}, \{a\} \times \{\vec{0}, \vec{v}\}, \delta, s, \{t\})$ with $\delta(s, (a, \vec{v})) = t$ and $\delta(t, (a, \vec{v})) = t$ we have that $\vec{v} \in C$ if and only if there exists a synchronizing word for $(\mathcal{A}, C)$. As the uniform membership problem for semilinear sets is NP-complete in dimension one and an encoding by generators in binary [23,24], we find that when using a binary encoding, the synchronization problem is NP-hard even for two-state deterministic and complete unary Parikh automata of dimension one not leaving much room to identify tractable instances (note that for the results from Sect. 5 we assumed a unary encoding).

Semilinear sets can be represented by Presburger formulae [18]. By a similar construction, the problem whether a given Presburger formula is true can be reduced to the synchronization problem. The complexity of deciding Presburger formulae sits between non-deterministic double exponential time and double exponential space 2-EXPSPACE [19]. However, the complexity of decision procedures varies greatly depending on which fragment of this logic we are considering [19] and, by quantifier elimination [19], every semilinear set can be represented by a quantifier-free formula with modular predicates. So, investigating the complexity landscape further for this representation, as well as exploring different fragments of Presburger arithmetic, are questions for future research.

Other notions of synchronization are conceivable. For example, there is a corresponding notion of *language synchronization* [21] asking whether there exists a word $w \in \Sigma^*$ such that for every $u, v, z \in \Sigma^*$ the word uwz is in $L(\mathcal{A}, C)$ if and only if vwz is in $L(\mathcal{A}, C)$. In case of languages accepted by word automata, this problem can be solved by computing a minimal automaton and testing whether there exists a word driving this automaton into a single state, regardless of the starting state. However, we do not know how determine whether there exists such a word for languages accepted by Parikh automata, as in general there exist infinitely many words u, v such that there exists z so that $uz \in L(\mathcal{A}, C)$ but $vz \notin L(\mathcal{A}, C)$. Similar difficulties are faced when not only considering the state, but the state together with an attainable counter value and demand to reset all these combinations of states and counter values. Tackling such (probably more difficult questions) is a problem left open by the present work.

Acknowledgement. I thank one anonymous referee for pointing out an error in a previous version of the statement of Proposition 5.1 (the example from Remark 1 is due to this referee). The notion of language synchronization was noted by a referee. I thank all referees for pointing to unclear formulations or making numerous comments that helped in simplifying proofs.

References

1. Alur, R., Madhusudan, P.: Visibly pushdown languages. In: Babai, L. (ed.) STOC 2004, pp. 202–211. ACM (2004). https://doi.org/10.1145/1007352.1007390
2. Babari, P., Quaas, K., Shirmohammadi, M.: Synchronizing data words for register automata. In: Faliszewski, P., Muscholl, A., Niedermeier, R. (eds.) MFCS 2016. LIPIcs, vol. 58, pp. 15:1–15:15. Schloss Dagstuhl (2016). https://doi.org/10.4230/LIPIcs.MFCS.2016.15
3. Balasubramanian, A.R., Thejaswini, K.S.: Adaptive synchronisation of pushdown automata. In: Haddad, S., Varacca, D. (eds.) 32nd International Conference on Concurrency Theory, CONCUR 2021, 24–27 August 2021, Virtual Conference. LIPIcs, vol. 203, pp. 17:1–17:15. Schloss Dagstuhl - Leibniz-Zentrum für Informatik (2021). https://doi.org/10.4230/LIPIcs.CONCUR.2021.17
4. Cadilhac, M., Finkel, A., McKenzie, P.: Affine Parikh automata. RAIRO Theor. Informatics Appl. **46**(4), 511–545 (2012). https://doi.org/10.1051/ita/2012013
5. Černý, J.: Poznámka k homogénnym experimentom s konečnými automatmi. Matematicko-fyzikálny časopis **14**(3), 208–216 (1964)
6. Chistikov, D., Haase, C.: The taming of the semi-linear set. In: Chatzigiannakis, I., Mitzenmacher, M., Rabani, Y., Sangiorgi, D. (eds.) ICALP 2016. LIPIcs, vol. 55, pp. 128:1–128:13. Schloss Dagstuhl (2016). https://doi.org/10.4230/LIPIcs.ICALP.2016.128
7. Chistikov, D., Martyugin, P., Shirmohammadi, M.: Synchronizing automata over nested words. Journal of Automata, Languages, and Combinatorics 24(2–4), 219–251 (2019). https://doi.org/10.25596/jalc-2019-219
8. Domenjoud, E.: Solving systems of linear diophantine equations: an algebraic approach. In: Tarlecki, A. (ed.) MFCS 1991. LNCS, vol. 520, pp. 141–150. Springer, Heidelberg (1991). https://doi.org/10.1007/3-540-54345-7_57
9. Doyen, L., Juhl, L., Larsen, K.G., Markey, N., Shirmohammadi, M.: Synchronizing words for weighted and timed automata. In: Raman, V., Suresh, S.P. (eds.) FSTTCS 2014. LIPIcs, vol. 29, pp. 121–132. Schloss Dagstuhl (2014). https://doi.org/10.4230/LIPIcs.FSTTCS.2014.121
10. Doyen, L., Massart, T., Shirmohammadi, M.: Infinite synchronizing words for probabilistic automata. In: Murlak, F., Sankowski, P. (eds.) MFCS 2011. LNCS, vol. 6907, pp. 278–289. Springer, Heidelberg (2011). https://doi.org/10.1007/978-3-642-22993-0_27
11. Doyen, L., Massart, T., Shirmohammadi, M.: The complexity of synchronizing markov decision processes. J. Comput. Syst. Sci. **100**, 96–129 (2019). https://doi.org/10.1016/j.jcss.2018.09.004
12. Fernau, H., Gusev, V.V., Hoffmann, S., Holzer, M., Volkov, M.V., Wolf, P.: Computational complexity of synchronization under regular constraints. In: Rossmanith, P., Heggernes, P., Katoen, J. (eds.) MFCS 2019. LIPIcs, vol. 138, pp. 63:1–63:14. Schloss Dagstuhl (2019). https://doi.org/10.4230/LIPIcs.MFCS.2019.63

13. Fernau, H., Wolf, P.: Synchronization of deterministic visibly push-down automata. In: Saxena, N., Simon, S. (eds.) FSTTCS 2020. LIPIcs, vol. 182, pp. 45:1–45:15. Schloss Dagstuhl (2020)
14. Fernau, H., Wolf, P., Yamakami, T.: Synchronizing deterministic push-down automata can be really hard. In: Esparza, J., Král', D. (eds.) MFCS 2020. LIPIcs, vol. 170, pp. 33:1–33:15. Schloss Dagstuhl (2020). https://doi.org/10.4230/LIPIcs.MFCS.2020.33
15. Figueira, D., Libkin, L.: Path logics for querying graphs: combining expressiveness and efficiency. In: LICS 2015, pp. 329–340. IEEE Computer Society (2015). https://doi.org/10.1109/LICS.2015.39
16. Fominykh, F.M., Martyugin, P.V., Volkov, M.V.: P(l)aying for synchronization. Int. J. Found. Comput. Sci. **24**(6), 765–780 (2013)
17. Ginsburg, S., Spanier, E.H.: Bounded ALGOL-like languages. Trans. Am. Math. Soc. **113**(2), 333–368 (1964)
18. Ginsburg, S., Spanier, E.H.: Semigroups, presburger formulas, and languages. Pacific J. Math. **16**(2), 285–296 (1966). pjm/1102994974
19. Haase, C.: A survival guide to Presburger arithmetic. ACM SIGLOG News **5**(3), 67–82 (2018). https://dl.acm.org/citation.cfm?id=3242964
20. Hoffmann, S.: Constrained synchronization and commutativity. Theor. Comput. Sci. **890**, 147–170 (2021). https://doi.org/10.1016/j.tcs.2021.08.030, https://www.sciencedirect.com/science/article/pii/S0304397521005077
21. Holzer, M., Jakobi, S.: On the computational complexity of problems related to distinguishability sets. Inf. Comput. **259**(2), 225–236 (2018). https://doi.org/10.1016/j.ic.2017.09.003
22. Hopcroft, J.E., Ullman, J.D.: Introduction to Automata Theory, Languages, and Computation. Addison-Wesley Publishing Company, Boton (1979)
23. Huynh, T.-D.: The complexity of semilinear sets. In: de Bakker, J., van Leeuwen, J. (eds.) ICALP 1980. LNCS, vol. 85, pp. 324–337. Springer, Heidelberg (1980). https://doi.org/10.1007/3-540-10003-2_81
24. Huynh, T.: The complexity of semilinear sets. J. Inf. Process. Cybern. (now J. Automata, Lang. Comb.) **18**(6), 291–338 (1982)
25. Imreh, B., Steinby, M.: Directable nondeterministic automata. Acta Cybernetica **14**(1), 105–115 (1999)
26. Kfoury, D.: Synchronizing sequences for probabilistic automata. Stud. Appl. Math. **49**, 101–103 (1970)
27. Klaedtke, F., Rueß, H.: Monadic second-order logics with cardinalities. In: Baeten, J.C.M., Lenstra, J.K., Parrow, J., Woeginger, G.J. (eds.) ICALP 2003. LNCS, vol. 2719, pp. 681–696. Springer, Heidelberg (2003). https://doi.org/10.1007/3-540-45061-0_54
28. (Lin), A.W.T.: Model Checking Infinite-State Systems: Generic and Specific Approaches. Ph.D. thesis, University of Edinburgh (2010)
29. Martyugin, P.: Computational complexity of certain problems related to carefully synchronizing words for partial automata and directing words for nondeterministic automata. Theor. Comput. Syst. **54**(2), 293–304 (2013). https://doi.org/10.1007/s00224-013-9516-6
30. Pottier, L.: Minimal solutions of linear diophantine systems?: bounds and algorithms. In: Book, R.V. (ed.) RTA 1991. LNCS, vol. 488, pp. 162–173. Springer, Heidelberg (1991). https://doi.org/10.1007/3-540-53904-2_94
31. Quaas, K., Shirmohammadi, M.: Synchronizing data words for register automata. ACM Trans. Comput. Logic **20**(2), 11:1–11:27 (2019). https://doi.org/10.1145/3309760

32. Sandberg, S.: 1 homing and synchronizing sequences. In: Broy, M., Jonsson, B., Katoen, J.-P., Leucker, M., Pretschner, A. (eds.) Model-Based Testing of Reactive Systems. LNCS, vol. 3472, pp. 5–33. Springer, Heidelberg (2005). https://doi.org/10.1007/11498490_2
33. Savitch, W.J.: Relationships between nondeterministic and deterministic tape complexities. J. Comput. Syst. Sci. **4**(2), 177–192 (1970). https://doi.org/10.1016/S0022-0000(70)80006-X
34. Volkov, M.V., Kari, J.: Černý's conjecture and the road colouring problem. In: Pin, J.É. (ed.) Handbook of Automata Theory, vol. I, pp. 525–565. European Mathematical Society Publishing House (2021)

Completely Distinguishable Automata and the Set of Synchronizing Words

Stefan Hoffmann(✉)

Fachbereich 4 - Abteilung Informatikwissenschaften, Universität Trier, Trier, Germany
hoffmanns.tcs@gmail.com

Abstract. An n-state automaton is synchronizing if it can be reset, or synchronized, into a definite state. The set of input words that synchronize the automaton is acceptable by an automaton of size $2^n - n$. Shortest paths in such an accepting automaton correspond to shortest synchronizing words. Here, we introduce completely distinguishable automata, a subclass of the synchronizing automata. Being completely distinguishable is a necessary condition for a minimal automaton for the set of synchronizing words to have size $2^n - n$. In fact, as we show, it has size $2^n - n$ if and only if the automaton is completely distinguishable and has a completely reachable subautomaton that only missed at most one state. We give different characterizations of completely distinguishable automata. Then we relate these notions to graph-theoretical constructions and investigate the subclass of automata with simple idempotents (SI-automata). We show that for these automata the properties of synchronizability, complete distinguishability and complete reachability and the minimal automaton for the set of synchronizing word having $2^n - n$ states are equivalent when the transformation monoid contains a transitive permutation group. A related result from the literature about SI-automata is wrong, we discuss and correct that mistake here. Lastly, using the results on SI-automata, we show that deciding complete reachability, complete distinguishability and whether the minimal automaton for the set of synchronizing words has $2^n - n$ states are all NL-hard problems.

Keywords: completely reachable automata · completely distingusihable automata · synchronizing automata · set of synchronizing words

An automaton is *synchronizing* if there exists a word, called a *reset* or *synchronizing word*, driving it into the same state regardless of its current state. Synchronizing words correspond to a sequence of actions (or inputs) that allow an operator (or observer) to regain control by resetting the automaton into a well-defined state. This notion arises in different contexts and applications in coding theory [3,27,40], robotics [11,16,29], group theory [1], symbolic dynamics [37], model checking and verification [35]. See also the surveys [35,38,40].

F. Drewes and M. Volkov (Eds.): DLT 2023, LNCS 13911, pp. 128–142, 2023.
https://doi.org/10.1007/978-3-031-33264-7_11

The Černý conjecture states that every n-state synchronizing automaton admits a reset word of length at most $(n-1)^2$ [10,36] (see Fig. 1 for the Černý family of automata reaching that bound). It is one of the most longstanding open problems in automata theory. Despite named after Černý, it first appeared in print in a paper by Starke [36]. Černý seems to have brought this problem to general attention on a conference in 1969 (and often [10] is cited with respect to this conjecture). See [39] for further information on the origin of this conjecture. For algorithmic questions, the problem of computing a *shortest synchronizing word* is computationally hard [11,31,32], and also its length approximation remains hard [15]. We refer to [35,40] for details on algorithmic issues and the Černý conjecture.

With every automaton we can associate the set of synchronizing words. The set of synchronizing words can be described by an automaton that might be exponentially larger and the size of a minimal automaton for this set is a measure of its complexity. Using the minimal automaton, this can be extended to an operation on languages, giving a regularity-preserving operation and thus fitting into the framework of the descriptional or state complexity of operations [14,41]. A shortest synchronizing word is related to a path in this automaton, linking the structure of these automata to the Černý conjecture. Hardness of certain algorithmic questions is related to this exponential blow-up of automata for the set of synchronizing words. For example, the NP-hardness of computing shortest synchronizing words [11,32], the PSPACE-complete problem to decide whether two given automata have the same set of synchronizing words [28], the *Constrained Synchronization Problem* [13], which asks for a synchronizing word of a given input automaton contained in a given regular constraint language. In all these example, if automata for the set of synchronizing words are computable in polynomial time for a given class of input automata, then these problems are solvable in polynomial time for this class of input automata.

Here, we investigate conditions related to a maximal blow-up of minimal automata for the set of synchronizing words. The class of completely distinguishable automata (Sect. 2.1) is motivated as being a necessary condition. We then use this notion and complete reachability to characterize this maximal blow-up (Sect. 2.2). Then we investigate a subclass from the literature on synchronizing automata and connect the problem to graph-theoretical notions (Sect. 3). We conclude by investigating associated decision problems (Sect. 4).

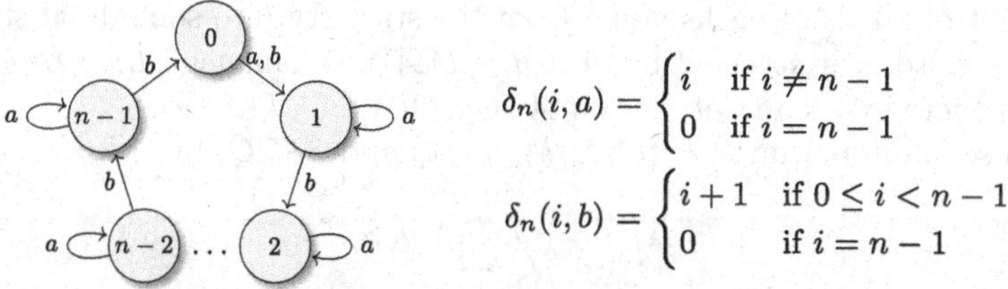

Fig. 1. The Černý family [9] of n-state automata $\mathcal{C}_n$ whose shortest synchronizing words have length $(n-1)^2$. A minimal automaton for the set of synchronizing words of $\mathcal{C}_n$ has $2^n - n$ states [28].

1 Preliminaries

By Σ we denote a *finite alphabet*, Σ^* denotes the *set of all words with concatenation* as operation. The *empty word* is denoted by ε. *Languages* are subsets of Σ^*. The *power set* of a set X is denoted by $\mathcal{P}(X)$. A *k-set* is a set with k elements, a *singleton set* is a 1-set. We assume some familiarity with formal language theory and complexity theory, e.g., we use regular expressions and the complexity classes NL, P, PSPACE. See the textbook [25] for more information.

Automata Theory: A *(deterministic and complete) finite semiautomaton* is a tuple $\mathcal{A} = (Q, \Sigma, \delta)$ (or just *automaton* for short) where Q is the *finite state set*, Σ the *input alphabet* and $\delta\colon Q \times \Sigma \to Q$ the *(totally defined) transition function*. We extend it to a function $\hat{\delta}\colon Q \times \Sigma^* \to Q$ by setting $\hat{\delta}(q, \varepsilon) = q$ and $\hat{\delta}(q, ux) = \delta(\hat{\delta}(q, u), x)$ for $u \in \Sigma^*$ and $x \in \Sigma$. For simplicity, we denote the extension $\hat{\delta}$ by δ as well.

A *trap state* is a state $q \in Q$ such that $\delta(q, x) = q$ for all $x \in \Sigma$.

Let $\mathcal{A} = (Q, \Sigma, \delta)$. The *power (semi-)automaton* is $\mathcal{P}(\mathcal{A}) = (\mathcal{P}(Q), \Sigma, \delta)$ where $\delta\colon \mathcal{P}(Q) \times \Sigma \to \Sigma$ is the *extension* of $\delta\colon Q \times \Sigma \to Q$ to subsets $\delta(S, x) = \{\delta(q, x) \mid q \in Q\}$ for $S \subseteq Q$ and $x \in \Sigma$

A subset $S \subseteq Q$ defines a *subautomaton* $\mathcal{A}_{|S} = (S, \Sigma, \delta_{|S})$ if $\delta(S, x) \subseteq S$ for all $x \in \Sigma$ where the transition function of $\mathcal{A}_{|S}$ is given by the transition function of $\mathcal{A}$ but restricted to S. Furthermore, for $\Gamma \subseteq \Sigma$ we let $\mathcal{A}_{|S,\Gamma} = (S, \Gamma, \delta_{|S,\Gamma})$ be the automaton $\mathcal{A}_{|S}$ but restricted to input symbols from Γ.

A state p is *reachable* from the state $q \in Q$ if there exists $u \in \Sigma^*$ such that $\delta(q, u) = p$. The automaton $\mathcal{A}$ is called *strongly connected* if every two states are reachable from each other.

Let $\mathcal{A} = (Q, \Sigma, \delta)$. Each word $u \in \Sigma^*$ *induces a function* $\delta_u\colon Q \to Q$ by $\delta_u(q) = \delta(q, u)$ for all $q \in Q$. The *transformation monoid of* $\mathcal{A}$ is $\mathcal{T}_M(\mathcal{A}) = \{\delta_u \mid u \in \Sigma^*\}$.

A *(deterministic and complete) finite automaton (as a language acceptor)* is a 5-tuple $\mathcal{A} = (Q, \Sigma, \delta, q_0, F)$ where (Q, Σ, δ) is a semiautomaton as defined before, q_0 is the *start state* and $F \subseteq Q$ is the *set of final states*. The language *accepted*, or *recognized*, by $\mathcal{A}$ is

$$L(\mathcal{A}) = \{u \in \Sigma^* \mid \delta(q_0, u) \in F\},$$

i.e., the set of labels of paths going from the start state to some final state (we also say a word u is *accepted* by $\mathcal{A}$ if $u \in L(\mathcal{A})$). A language $L \subseteq \Sigma^*$ is called *regular* if there exists an automaton $\mathcal{A}$ such that $L = L(\mathcal{A})$.

For a semiautomaton $\mathcal{A} = (Q, \Sigma, \delta)$, $q \in Q$ and $S \subseteq Q$, by

$$\mathcal{A}_{q,S} = (Q, \Sigma, \delta, q, S)$$

we denote the automaton (as a language acceptor) that results if we set the start state to q and the set of final states to S.

We call both types of automata – semiautomata and automata as languages acceptors – simply automata, as the context should make it clear what is meant. Furthermore, concepts defined for semiautomata carry over to automata as language acceptors.

Minimal Automata: For every regular language $L \subseteq \Sigma^*$ there exists an automaton with the minimal number of states accepting it that is unique up to renaming of the states and called the *minimal automaton* of L [25]. Let $\mathcal{A} = (Q, \Sigma, \delta, q_0, F)$ be an automaton. Two states $s, t \in Q$ are *distinguishable* if there exists $u \in \Sigma^*$ such that

$$(\delta(s,u) \in F \wedge \delta(t,u) \notin F) \vee (\delta(s,u) \notin F \wedge \delta(t,u) \in F). \tag{1}$$

The properties of distinguishability of states and the reachability of states both characterize the minimality of an automaton accepting a given language. This is a standard result from automata theory [25].

Theorem 1.1. *Let $\mathcal{B} = (S, \Sigma, \mu, s_0, E)$. The following are equivalent:*

1. *The automaton $\mathcal{B}$ has the least number of states such that $L(\mathcal{A}) = L(\mathcal{B})$.*
2. *Each state of $\mathcal{B}$ is reachable from s_0 and all pairs of states are distinguishable.*

Synchronizing Automata: Let $\mathcal{A} = (Q, \Sigma, \delta)$. A *synchronizing* (or *reset*) *word* $u \in \Sigma^*$ is a word such that $\delta(Q, u) = \{q\}$ for some $q \in Q$ (we say that u *synchronizes* $\mathcal{A}$ and that q is a *synchronizing state* of $\mathcal{A}$). By $\mathrm{Syn}(\mathcal{A})$ we denote the *the set of synchronizing words* of $\mathcal{A}$. The automaton is called *synchronizing* if $\mathrm{Syn}(\mathcal{A}) \neq \emptyset$.

Set $F = \{\{q\} \mid q \in Q\}$. As $\mathrm{Syn}(\mathcal{A}) = \{u \in \Sigma^* \mid \delta(Q, u) \in F\}$, we have $\mathrm{Syn}(\mathcal{A}) = L(\mathcal{P}(\mathcal{A})_{Q,F})$, i.e., the power automaton $\mathcal{P}(\mathcal{A})$ accepts $\mathrm{Syn}(\mathcal{A})$. The state $\emptyset$ can be discarded and the states in F can be merged into a single state to get an automaton with $2^n - 1 - (n-1) = 2^n - n$ states accepting $\mathrm{Syn}(\mathcal{A})$.

We call an n-state automaton $\mathcal{A}$ *sync-maximal* if the minimal automaton for $\mathrm{Syn}(\mathcal{A})$ has $2^n - n$ states.

Let $\mathcal{A} = (Q, \Sigma, \delta)$. We call two subsets of states $A, B \subseteq Q$ *distinguishable* in $\mathcal{A}$ if they are distinguishable in $\mathcal{P}(\mathcal{A})_{Q,F}$ according to Eq. (1) where $F = \{\{q\} \mid q \in Q\}$, i.e., there exists a word $u \in \Sigma^*$ (called a *distinguishing word*) such that

$$(|\delta(A,u)| = 1 \wedge |\delta(B,u)| \neq 1) \vee (|\delta(A,u)| \neq 1 \wedge |\delta(B,u)| = 1). \tag{2}$$

An automaton $\mathcal{A} = (Q, \Sigma, \delta)$ is called *completely subset-pair distinguishable* (or just *completely distinguishable* for short) if every two distinct non-empty and non-singleton subsets $A, B \subseteq Q$ are distinguishable in $\mathcal{A}$.

Applying Theorem 1.1 to $\mathrm{Syn}(\mathcal{A})$ and the power automaton, we can state the following characterization.

Theorem 1.2. *The automaton $\mathcal{A} = (Q, \Sigma, \delta)$ is sync-maximal if and only if the following two conditions hold true:*

1. *$\mathcal{A}$ is synchronizing, i.e., at least one singleton subset is reachable in $\mathcal{P}(\mathcal{A})$, and every non-empty subset with at least two states is reachable in $\mathcal{P}(\mathcal{A})$.*
2. *$\mathcal{A}$ is completely distinguishable.*

The automaton $\mathcal{A}$ is *completely reachable* if every non-empty subset of states is reachable in $\mathcal{P}(\mathcal{A})$ from Q, i.e., for each non-empty $A \subseteq Q$ there exists $u \in \Sigma^*$ such that $\delta(Q, u) = A$. These automata were introduced in [4].

In a completely reachable automaton for $p, q \in Q$ there exist $u, v \in \Sigma^*$ such that $\delta(Q, u) = \{p\}$ and $\delta(Q, v) = \{q\}$, which yields the next result.

Lemma 1.3. *If $\mathcal{A} = (Q, \Sigma, \delta)$ is completely reachable, then $\mathcal{A}$ is strongly connected and synchronizing.*

2 Completely Distinguishable Automata

First, we give other characterizations of complete distinguishability and discuss an erroneous statement from the literature (Sect. 2.1). Then we characterize sync-maximal automata in terms of complete distinguishability and complete reachability (Sect. 2.2).

2.1 Complete Distinguishability

First, completely distinguishable automata with at least three states are synchronizing (hence, mentioning synchronizability can be omitted from Theorem 1.2 for a characterization of sync-maximality for automata with at least three states). Automata with at most two states are always completely distinguishable.

Proposition 2.1. *A completely distinguishable automaton $\mathcal{A} = (Q, \Sigma, \delta)$ with $|Q| \geq 3$ is synchronizing.*

Next, we formulate various equivalent characterizations related to subset mapping properties. Item 6 of Theorem 2.2 is useful for an inductive reasoning.

Theorem 2.2. *Let $\mathcal{A} = (Q, \Sigma, \delta)$ with $|Q| \geq 3$. The following are equivalent:*

1. *$\mathcal{A}$ is completely distinguishable.*
2. *For every $A, B \subseteq Q$ with $A \subsetneq B$ and A non-empty there exists $u \in \Sigma^*$ such that $1 = |\delta(A, u)| < |\delta(B, u)|$.*
3. *For every non-empty $A \subseteq Q$ and $p \in Q \setminus A$ there exists $u \in \Sigma^*$ such that $\delta(p, u) \notin \delta(A, u)$ and $|\delta(A, u)| = 1$.*
4. *For every $A \subseteq Q$ with $|A| \geq 2$ and $p \in Q \setminus A$ there exists $u \in \Sigma^*$ such that $\delta(p, u) \notin \delta(A, u)$ and $|\delta(A, u)| < |A|$.*
5. *For every $q \in Q$ there exists $u \in \Sigma^*$ such that $1 = |\delta(Q \setminus \{q\}, u)|$ and $\delta(q, u) \notin \delta(Q \setminus \{q\}, u)$.*
6. *For every two distinct $A, B \subseteq Q$ with $\min\{|A|, |B|\} \geq 2$ there exists an element $u \in \Sigma^*$ such that $3 \leq |\delta(A, u)| + |\delta(B, u)| < |A| + |B|$ and $\delta(A, u) \neq \delta(B, u)$.*

Proof (sketch). Due to space we only proof the implication 5 ⇒ 6. Let $A, B \subseteq Q$ be distinct with $\min\{|A|, |B|\} \geq 2$. Without loss of generality, suppose there exists $q \in B \setminus A$. By assumption we find $u \in \Sigma^*$ such that $|\delta(Q \setminus \{q\}, u)| = 1$ and $\delta(q, u) \notin \delta(Q \setminus \{q\}, u)$. As $A \subseteq Q \setminus \{q\}$ we have $|\delta(A, u)| = 1$. So $|\delta(A, u)| + |\delta(B, u)| < |A| + |B|$. Furthermore, as $(Q \setminus \{q\}) \cap B \neq \emptyset$ and $q \in B$ we have

$$\delta(B, u) = \delta(Q, u) = \delta(Q \setminus \{q\}, u) \cup \{\delta(q, u)\}$$

and this set is a 2-set. So $3 \leq |\delta(A, u)| + |\delta(B, u)|$ and $\delta(A, u) \neq \delta(B, u)$. □

Remark 1. In [21] it was claimed that all pairs of non-empty and non-singleton subsets are distinguishable if and only if this is true for all pairs of distinct 2-sets. This is wrong. See the counterexamples in Fig. 2. The inductive "proof" in [21] missed a case (we note that this is related to the condition $\delta(A, u) \neq \delta(B, u)$ in Theorem 2.2 (Item 6)). The main result of [22] relied on that false result. However, it is still true, in Sect. 3 (Theorem 3.9) we give a proof using a more general result about SI-automata. Unfortunately, the result from [23] claiming that a strongly connected SI-automaton is completely reachable if and only if it is sync-maximal is false. However, Theorem 3.7 shows that strong connectivity of a special subautomaton implies this equivalence.

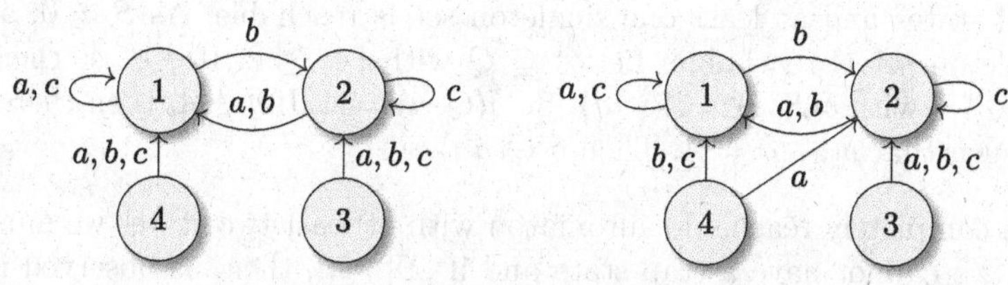

Fig. 2. Two automata that are not completely distinguishable, but where all pairs of distinct 2-sets are distinguishable. Left: $\{1, 2, 3\}$ and $\{3, 4\}$ are not distinguishable (every letter maps both to $\{1, 2\}$); Right: $\{1, 2, 3, 4\}$ is not distinguishable from $\{1, 2, 3\}$.

2.2 Sync-Maximal Automata

In [23] various necessary conditions were listed for sync-maximality. Here, we give a more precise statement, in fact a characterization in terms of complete distinguishability and the complete reachability of a subautomaton.

Observe that for $\mathcal{A} = (Q, \Sigma, \delta)$ the set of synchronizing states $S = \{q \in Q \mid \exists u \in \Sigma^* : \delta(Q, u) = \{q\}\}$ defines a strongly connected subautomaton $\mathcal{A}_{|S}$.

Theorem 2.3. *Let $n \geq 3$ and $\mathcal{A} = (Q, \Sigma, \delta)$ be a n-state automaton with set of synchronizing states $S \subseteq Q$. Then $\mathcal{A}$ is sync-maximal if and only if $\mathcal{A}$ is completely distinguishable and one of the following is true:*

1. *$\mathcal{A}$ is completely reachable or*
2. *the following two conditions hold:*
   (a) *$Q = S \cup \{q\}$ for some non-synchronizing state $q \notin S$ and $\mathcal{A}_{|S,\Gamma}$ with $\Gamma = \{x \in \Sigma \mid \delta(q, x) = q\}$ is completely reachable and*
   (b) *there exists $a \in \Sigma \setminus \Gamma$ such that $\delta(Q, a) = S$.*

Proof. As S defines a subautomaton, for every $u \in \Sigma^*$ we have $S \cap \delta(Q, u) \neq \emptyset$. So, no subset of $Q \setminus S$ is reachable in $\mathcal{P}(\mathcal{A})$. This implies $|Q \setminus S| \leq 1$ as every subset with at least two states is reachable in $\mathcal{A}$. If $Q = S$, then $\mathcal{A}$ is completely reachable as $\operatorname{Syn}(\mathcal{A}) \neq \emptyset$ with $n \geq 3$ implies that at least one singleton state set is reachable, and then by strong connectivity all singleton state sets are reachable.

So, suppose $Q \setminus S = \{q\}$ for some $q \in Q$. Let $A \subseteq S$ be non-empty. As $\mathcal{A}$ is sync-maximal, there exists $u \in \operatorname{Syn}(\mathcal{A})$. As $\delta(Q, u) \subseteq S$ and $\mathcal{A}$ is not strongly connected (otherwise $Q = S$), there are only outgoing transitions or self-loops at the state q. By sync-maximality, there exists $u \in \Sigma^*$ such that $\delta(Q, u) = A \cup \{q\}$. This implies $u \in \Gamma^*$ and so $\delta(S, u) = A$. Lastly, as $|S| \geq 2$, there exists $v \in \Sigma^*$ such that $\delta(Q, v) = S$. Now, as we cannot go back to the state q, for a shortest such word we can deduce that $v = a$ for some $a \in \Sigma \setminus \Gamma$. Conversely, if $\mathcal{A}$ is completely reachable and completely distinguishable, then $\mathcal{A}$ is clearly sync-maximal. Now, suppose $\mathcal{A}$ is completely distinguishable with $q \in Q$, $S \subseteq Q$, Γ and $a \in \Sigma \setminus \Gamma$ as in the statement. It remains to show that every non-singleton subset of states and at least one singleton set is reachable. As $S \neq \emptyset$, at least one singleton set is reachable. Let $A \subseteq Q$ with $|A| \geq 2$. If $q \in A$, then there exists $u \in \Gamma^*$ with $\delta(S, u) = A \setminus \{q\}$. So $\delta(Q, u) = A$. If $q \notin A$, then there exists $u \in \Gamma^*$ such that $\delta(S, u) = A$. Then $\delta(Q, au) = A$. □

For a completely reachable automaton with at least two states we must have $|\Sigma| \geq 2$, we cannot have a trap state and if $|\Sigma| = 2$, then, as observed in [4,6, 8,21], one letter must cyclically permute the states. From this and the previous result we can derive the same restrictions for sync-maximal automata if $Q = S$, or, if $Q \neq S$, that $|\Sigma| \geq 3$, we must have two distinct letters $a, b \in \Sigma$ with $|\delta(Q, a)| = |\delta(Q, b)| = |Q| - 1$ and for $|\Sigma| = 3$ one letter from Γ cyclically permutes the set S.

3 SI-Automata

A *permutation group* on a domain X is a subset of permutations on X closed under function composition. A permutation group G is called *transitive* if for every two $x, y \in X$ there exists $g \in G$ such that $g(x) = y$.

Let $\mathcal{A} = (Q, \Sigma, \delta)$. A word (resp. transformation $t \colon Q \to Q$) $u \in \Sigma^*$ is a *simple idempotent* if there exists a unique $p \in Q$ such that $\delta(p, u) \neq p$ (resp. $t(p) \neq p$) and $\delta(q, u) = q$ (resp. $t(q) = q$) for all $q \in Q \setminus \{p\}$. A word $u \in \Sigma^*$ is a *permutational word* if $\delta(Q, u) = Q$. A word u (resp. transformation $t \colon Q \to Q$) has *defect one* if $|\delta(Q, u)| = |Q| - 1$ (resp. $|t(Q)| = |Q| - 1$).

We call $\mathcal{A}$ an *automaton with simple idempotents (SI-automaton)* if every letter is either a simple idempotent or a permutational letter. We set $\Sigma_I = \{a \in \Sigma \mid a \text{ is a simple idempotent.}\}$ and $\Sigma_G = \{a \in \Sigma \mid a \text{ is a permutational letter.}\}$. SI-automata were introduced by Rystsov [33,34] and he has shown that a shortest synchronizing word has length at most $2(|Q|-1)^2$ for these automata. The Černý automata (Fig. 1) are SI-automata.

3.1 The Rystsov Graph and Related Constructions

A tuple $G = (V, E)$ is a *(directed) graph* where V is the set of *vertices* (or *nodes*) and $E \subseteq V \times V$ the set of *edges*. A sequence of vertices $v_0, v_1, \ldots, v_n \in V$ in $G = (V, E)$ is a *path* (of length $n \geq 1$) from v_0 to v_n if $(v_i, v_{i+1}) \in E$ for all $i \in \{0, 1, \ldots, n-1\}$ and all vertices are distinct, i.e., $|\{v_0, v_1, \ldots, v_n\}| = n + 1$. Let $k \geq 1$. The graph G is *strongly k-connected* if for any two vertices u, v there exist at least k paths from u to v such that for two such paths $v_0, v_1, \ldots, v_n$ and $u_0, u_1, \ldots, u_m$ we have $\{v_1, \ldots, v_{n-1}\} \cap \{u_1, \ldots, u_{m-1}\} = \emptyset$ and $v_0 = u_0 = u$, $v_n = u_m = v$. A strongly 1-connected graph is also just called *strongly connected.* The graph G is *vertex-transitive* if for every two vertices $v_1, v_2 \in V$ there exists a permutation $\pi \colon V \to V$ with $\pi(v_1) = v_2$ and such that, for all $u_1, u_2 \in V$, we have

$$(u_1, u_2) \in E \Leftrightarrow (\pi(u_1), \pi(u_2)) \in E.$$

In a vertex transitive graph, for $v_1, v_2 \in V$ we have $|\{v_3 \mid (v_1, v_3) \in E\}| = |\{v_3 \mid (v_2, v_3) \in E\}|$, and this number is called the *outdegree* of G.

The next result is due to Hamidoune [18] (see also [2, Thm. 3.7 (a)]).

Theorem 3.1 (Hamidoune [18]**).** *If G is a strongly connected vertex-transitive graph of out-degree d, then G is strongly $\left(\lfloor \frac{d}{2} \rfloor + 1\right)$-connected.*

The *automaton graph* of $\mathcal{A} = (Q, \Sigma, \delta)$ is the graph $G(\mathcal{A}) = (Q, E)$ with $E = \{(s, t) \mid \exists x \in \Sigma : \delta(s, x) = t\}$. In the context where the automaton graph is used in Sect. 4 we only use the property that there exists at least one letter inducing each edge in $G(\mathcal{A})$. Hence, the possibility that multiple transitions $\delta(s, x) = \delta(s, y)$ in $\mathcal{A}$ with different letters $x, y \in \Sigma$ correspond to the same edge in $G(\mathcal{A})$ does not need special attention.

Rystsov [33,34] introduced a graph-theoretical construction which he used to investigate shortest synchronizing words for SI-automata and proved the following statement (see for example [34, Prop. 1 & Cor. 1]). The statement incorporates the definition of the graph construction.

Theorem 3.2 (Rystsov [33,34]**).** *Let $\mathcal{A} = (Q, \Sigma, \delta)$ be an SI-automaton. The automaton $\mathcal{A}$ is strongly connected and synchronizing if and only if the graph $I(\mathcal{A}) = (Q, E)$ with*

$$E = \{(\delta(p, u), \delta(q, u)) \mid \exists a \in \Sigma_I : \delta(p, a) = q \land p \neq q \land u \in \Sigma_G^*\} \tag{3}$$

is strongly connected.

We call $I(\mathcal{A})$ the *Rystsov graph* of $\mathcal{A}$. From the definition of the edge set of $I(\mathcal{A})$, it directly follows that for every $u \in \Sigma_G^*$ we have

$$(s,t) \in E \Rightarrow (\delta(s,u), \delta(t,u)) \in E. \tag{4}$$

Bondar & Volkov [4,6] introduced a related construction to give a sufficient criterion for complete reachability.

Theorem 3.3 (Bondar & Volkov [4,6]). *Let $\mathcal{A} = (\Sigma, Q, \delta)$. If the graph $\Gamma_1(\mathcal{A}) = (Q, E)$ with edge set*

$$E = \{(p,q) \mid \exists w \in \Sigma^* : p \notin \delta(Q,w), |\delta^{-1}(q,w)| = 2, \ w \text{ has defect one}\}.$$

is strongly connected, then $\mathcal{A}$ is completely reachable.

The construction was extended in [5,6] to give a sufficient and necessary criterion for complete reachability. The graph $\Gamma_1(\mathcal{A})$ is computable in polynomial time [17].

In [8] $\Gamma_1(\mathcal{A})$ was called the Rystsov graph, but we preferred to stick with the original construction due to Rystsov to call it by that name.

Every edge in the Rystsov graph corresponds to a simple idempotent word.

Proposition 3.4. *Let $\mathcal{A} = (Q, \Sigma, \delta)$ be an SI-automaton with Rystsov graph $I(\mathcal{A}) = (Q, E)$. Then for every edge $(s,t) \in E$ there exists a simple idempotent word $u \in \Sigma^*$ such that $\delta(s,u) = t$.*

This implies that every edge in $I(\mathcal{A})$ is also an edge in $\Gamma_1(\mathcal{A})$ (but this also follows as words of the form au with $a \in \Sigma_I$ and $u \in \Sigma_G^*$ inducing an edge in $I(\mathcal{A})$ by Eq. (3) have defect one). We note the following corollary from this observation.

Corollary 3.5. *Let $\mathcal{A} = (Q, \Sigma, \delta)$ be an SI-automaton. If $I(\mathcal{A})$ is strongly connected, then $\mathcal{A}$ is completely reachable.*

3.2 Application to SI-Automata

The first result is implied by Theorem 3.2 and Cor. 3.5.

Theorem 3.6. *Let $\mathcal{A} = (Q, \Sigma, \delta)$ be an SI-automaton. Then $\mathcal{A}$ is completely reachable if and only if it is strongly connected and synchronizing.*

Next, we show that for a subclass of SI-automata all our notions characterize synchronizability.

Theorem 3.7. *Let $\mathcal{A} = (Q, \Sigma, \delta)$ be an SI-automaton with $|Q| \geq 3$. If $\mathcal{T}_M(\mathcal{A})$ contains a transitive permutation group G (or equivalently $\mathcal{A}_{|Q,\Sigma_G}$ is strongly connected), then the following are equivalent:*

1. *$\mathcal{A}$ is synchronizing.*
2. *$\mathcal{A}$ is completely reachable.*

3. *$\mathcal{A}$ is completely distinguishable.*
4. *$\mathcal{A}$ is completely distinguishable and completely reachable.*
5. *$\mathcal{A}$ is sync-maximal.*

Proof (sketch). We only sketch the implication (2) ⇒ (3). Let $A \subseteq Q$ with distinct $p, q \in A$ and $r \in Q \setminus A$. As $\mathcal{A}$ is completely reachable, and so synchronizing, with Thm. 3.2 the graph $I(\mathcal{A})$ is strongly connected. By Eqn. (4) and the assumption the graph $I(\mathcal{A})$ is vertex-transitive. Suppose the outdegree is greater than one (the other case is left out due to space). With Thm. 3.1, $I(\mathcal{A})$ is strongly 2-connected. Using a path from p to q that does not contain r, we can construct $u \in \Sigma^*$ (using Prop. 3.4) such that $|\delta(A,u)| < |A|$ and $\delta(r,u) \notin \delta(A,u)$. With Thm. 2.2 $\mathcal{A}$ is completely distinguishable. □

In [4,6,8] (see also [21] for a more general statement) it was noted that a binary, i.e., where $|\Sigma| = 2$, completely reachable automaton contains a letter cyclically permuting the states. With this, Theorem 3.7 and 2.3 imply the following.

Corollary 3.8 *For a binary SI-automaton $\mathcal{A}$ the following are equivalent:*

1. *$\mathcal{A}$ is complete reachabilty.*
2. *$\mathcal{A}$ is sync-maximal.*
3. *$\mathcal{A}$ is synchronizing and strongly connected.*

Our result generalizes a characterization of primitive permutation groups from [19,22] (the paper [22] used the faulty argument, see Remark 1). A permutation group G on X is primitive if the only equivalence relations $\sim$ on X such that

$$x \sim y \Rightarrow g(x) \sim g(y) \tag{5}$$

holds are the equality and the universal relation where all elements are equivalent. Primitive permutation groups go back to Évariste Galois [30].

We give a proof using Theorem 3.7 (and a theorem due to Rystsov [1,34]).

Theorem 3.9 *Let G be a permutation group on the finite set Q with $|Q| \geq 3$. Then the following are equivalent.*

1. *G is primitive,*
2. *Every $\mathcal{A} = (Q, \Sigma, \delta)$ having a letter of defect one with $G \subseteq T_M(\mathcal{A})$ is synchronizing.*
3. *Every $\mathcal{A} = (Q, \Sigma, \delta)$ having a letter of defect one with $G \subseteq T_M(\mathcal{A})$ is completely reachable.*
4. *Every $\mathcal{A} = (Q, \Sigma, \delta)$ having a letter of defect one with $G \subseteq T_M(\mathcal{A})$ is completely distinguishable.*
5. *Every $\mathcal{A} = (Q, \Sigma, \delta)$ having a letter of defect one with $G \subseteq T_M(\mathcal{A})$ is completely distinguishable and completely reachable.*
6. *Every $\mathcal{A} = (Q, \Sigma, \delta)$ having a letter of defect one with $G \subseteq T_M(\mathcal{A})$ is sync-maximal.*

Proof (sketch). A primitive group with a permutation domain containing at least three elements is transitive. Rystsov's theorem [33], explicitly stated in [1], states that a transitive permutation group on a domain with at least three elements is primitive if and only if for any transformation t of defect one there exists a constant function written as a product of elements from the group and t. This yields that any of the conditions (2) to (6) implies that G must be primitive. Now, suppose G is primitive. If we find a simple idempotent w for $\mathcal{A}$, then consider the automaton $\mathcal{B} = (Q, \Gamma, \mu)$ with $\Gamma = \{x \in \Sigma \mid x \text{ is a permutational letter}\} \cup \{w\}$ (the word w viewed as a letter) and transition function $\mu(q, w) = \delta(q, w)$ and $\mu(q, x) = \delta(q, x)$ for all $q \in Q$ and permutational letters $x \in \Sigma$. Then $\mathcal{B}$ is an SI-automaton. We have $G \subseteq \mathcal{T}_M(\mathcal{B})$ and by Rystsov's theorem, $\mathcal{T}_M(\mathcal{B})$ contains a constant map, i.e., $\mathcal{B}$ is synchronzing. Then Theorem 3.7 gives all the equivalences.

The proof of the existence of a simple idempotent together with a proof that G is transitive is left out due to space. □

4 Decision Problems

We use the following two NL-complete problems from [26] in our hardness proofs: STRONG-CONN and GAP. The former asks whether a given graph is strongly connected, in the latter we have given an acyclic graph, i.e., a graph where there exists no path starting and ending at the same vertex, two vertices s (start) and t (target) and the problem asks whether there exists a path from s to t.

We show that deciding complete reachability, complete distinguishability and sync-maximality are NL-hard (Theorem 4.1, 4.2 and 4.3) for SI-automata. The constructions rely on the results from the previous section (Sect. 3). After acceptance of the present work, Marek Szykuła & Adam Zyzik (personal communication) informed me that they proved that the two problems to decide complete distinguishability and sync-maximality are both PSPACE-complete.

To ease the definition of simple idempotents, we use the notation $[a \mapsto b] \colon X \to X$ to denote a simple idempotent mapping $a \in X$ to $b \in X$ with $a \neq b$. A similar notation was used in [7].

Theorem 4.1. *Deciding whether a given automaton is completely reachable is* NL*-hard, even for SI-automata as input.*

Proof. We give a reduction from STRONG-CONN. Let $G = (V, E)$ be a graph. Set $\mathcal{A} = (V, \Sigma, \delta)$ with $\Sigma = \{a_e \mid e \in E\}$ where for each edge $e = (s, t)$ we introduce a simple idempotent letter a_e mapping s to t, i.e., $\delta_{a_e} = [s \mapsto t]$. As $\mathcal{A}$ has no permutational letters, for the automaton graph we have $G(\mathcal{A}) = I(\mathcal{A})$. By construction $G(\mathcal{A}) = G$.

If $\mathcal{A}$ is completely reachable, then by Lemma 1.3 it is strongly connected. So G is strongly connected. If G is strongly connected, then as $G = I(\mathcal{A})$ by Cor. 3.5 the automaton $\mathcal{A}$ is completely reachable. □

Theorem 4.2. *Deciding whether a given automaton is completely distinguishable is* NL*-hard, even for SI-automata as input.*

Proof. We give a reduction from GAP. We can assume s has no incoming transitions and t no outgoing transitions. Let $\hat{s} \notin V$ be a new element. Set $\mathcal{A} = (Q, \Sigma, \delta)$ with $Q = V \cup \{\hat{s}\}$, $\Sigma = \Sigma_1 \cup \Sigma_2$ where $\Sigma_1 = \{a_e \mid e \in E\}$, $\Sigma_2 = \{c_v, d_v \mid v \in Q \setminus \{t\}\}$ and, for $q \in Q$, $v \in Q \setminus \{t\}$ and $e = (v_1, v_2)$, set $\delta_{a_e} = [v_1 \mapsto v_2]$, $\delta_{c_v} = [v \mapsto s]$, $\delta_{d_v} = [v \mapsto \hat{s}]$.

First, suppose there exists a path $p_0, p_1, \ldots, p_n$ ($n \geq 1$) in G from s to t with edges $e_i = (p_{i-1}, p_i)$ for $i \in \{1, \ldots, n\}$. Set $u = a_{e_1} \cdots a_{e_n}$. Let $A \subseteq Q$ with $|A| \geq 2$ and $p \in Q \setminus A$. If $t \notin A$, let $q_1, q_2 \in A$ be two distinct states. Then set $w = c_{q_1} c_{q_2}$ if $p \neq s$ and $w = d_{q_1} d_{q_2}$ if $p = s$. We have $\delta(A, w) = (A \setminus \{q_1, q_2\}) \cup \{s\}$ if $p \neq s$ and $\delta(A, w) = (A \setminus \{q_1, q_2\}) \cup \{\hat{s}\}$ if $p = s$. If $t \in A$, then $p \neq t$. Let $A = \{t, q_1, \ldots, q_m\}$ with $|A| = m + 1$. Set $w = c_{q_1} \cdots c_{q_m} d_p u$. Then $\delta(p, w) = \hat{s}$ and $\delta(A, w) = \delta(\{s, t\}, u) = \{t\}$. With Theorem 2.2 it follows that $\mathcal{A}$ is completely distinguishable.

Now, suppose there exists no path from s to t. Then, for every $u \in \Sigma^*$, we have $\delta(s, u) \neq t$. Also $\delta(t, u) = t$. This implies that $|\delta(\{s, t\}, u)| = 2$ for all $u \in \Sigma^*$. Considering $A = \{s, t\}$ and $p = \hat{s}$, with Theorem 2.2 $\mathcal{A}$ is not completely distinguishable. □

Theorem 4.3. *Deciding whether a given automaton is sync-maximal is* NL*-hard, even for SI-automata as input.*

Proof. Let (G, s, t) with $G = (V, E)$ be an instance of GAP. Set $\mathcal{A} = (Q, \Sigma, \delta)$ with $Q = V \cup \{\hat{s}\}$ ($\hat{s} \notin V$ being a new element), $\Sigma = \Sigma_1 \cup \Sigma_2$ where $\Sigma_1 = \{a_e \mid e \in E\}$, $\Sigma_2 = \{b_v, c_v, d_v \mid v \in V\}$ and, for each $v \in Q$ and $e = (v_1, v_2)$, set $\delta_{a_e} = [v_1 \mapsto v_2]$, $\delta_{b_v} = [t \mapsto v]$, $\delta_{c_v} = [v \mapsto s]$, $\delta_{d_v} = [v \mapsto \hat{s}]$. We can assume $|V| \geq 3$. As $\mathcal{A}$ has no letter of defect one, by Theorem 2.3 $\mathcal{A}$ is sync-maximal if and only if it is completely rechable and completely distinguishable. Also, note that, contrary to the reduction in the proof of Theorem 4.2, we have added transitions leaving t.

If $A \subseteq Q$ is non-empty with two distinct $q_1, q_2 \in A$ and $p \in Q \setminus A$, then for $p \neq s$, we have $\delta(A, c_{q_1} c_{q_2}) = (A \setminus \{q_1, q_2\}) \cup \{s\}$ and $\delta(p, c_{q_1} c_{q_2}) = p$, and for $p = s$, we have $\delta(A, d_{q_1} d_{q_2}) = (A \setminus \{q_1, q_2\}) \cup \{\hat{s}\}$ and $\delta(p, d_{q_1} d_{q_2}) = p$. So, with Theorem 2.2 $\mathcal{A}$ is completely distinguishable.

If there exists a path $p_0, p_1, \ldots p_n$ from s to t in G with $e_i = (p_{i-1}, p_i) \in E$ for $i \in \{1, \ldots, n\}$, then $\mathcal{A}$ is strongly connected, as $\delta(p, c_p a_{e_1} \cdots a_{e_n} b_q) = q$ for all $p, q \in Q$. As $\mathcal{A}$ is an SI-automaton without permutational letters, its automaton graph equals $I(\mathcal{A})$, and so $\mathcal{A}$ is completely reachable with Cor. 3.5. If there does not exist a path from s to t in G, then t is not reachable from s in $\mathcal{A}$, so $\mathcal{A}$ is not completely reachable by Lemma 1.3. □

With Theorem 3.6 we can decide in NL whether an SI-automaton is completely reachable, as deciding synchronizability [24, Lem. 14], [20, Thm. 5] and deciding strongly connectedness [26] are both in NL. As the reductions used SI-automata, deciding complete reachability is NL-complete for SI-automata.

Corollary 4.4. *Deciding complete reachability is* NL*-complete for SI-automata.*

5 Conclusion

After the review process Marek Szykuła & Adam Zyzik (personal communication) informed me that they have found a proof showing that deciding complete distinguishability and deciding sync-maximality are both PSPACE-complete. Ferens & Szykuła [12] have shown that complete reachability can be decided in P. It remains an *open problem* whether complete reachability can be decided in NL.

Acknowledgement. I am thankful to an anonymous referee of a previous version for providing one example from Remark 1 and for having some good suggestions leading to Theorem 2.2 (Item 6). I also thank the referees of the present version for careful reading and identifying some typos. Furthermore, I thank Marek Szykuła & Adam Zyzik for contacting me and telling me about their PSPACE-completeness result.

References

1. Araújo, J., Cameron, P.J., Steinberg, B.: Between primitive and 2-transitive: synchronization and its friends. EMS Surv. Math. Sci. **4**(2), 101–184 (2017). http://www.ems-ph.org/doi/10.4171/EMSS
2. Babai, L.: Automorphism Groups, Isomorphism, pp. 1447–1540. Reconstruction. MIT Press, Cambridge, MA, USA (1996)
3. Berstel, J., Perrin, D., Reutenauer, C.: Codes and Automata, Encyclopedia of mathematics and its applications, vol. 129. Cambridge University Press, Cambridge (2010)
4. Bondar, E.A., Volkov, M.V.: Completely reachable automata. In: Câmpeanu, C., Manea, F., Shallit, J. (eds.) DCFS 2016. LNCS, vol. 9777, pp. 1–17. Springer, Cham (2016). https://doi.org/10.1007/978-3-319-41114-9_1
5. Bondar, E.A., Volkov, M.V.: A characterization of completely reachable automata. In: Hoshi, M., Seki, S. (eds.) DLT 2018. LNCS, vol. 11088, pp. 145–155. Springer, Cham (2018). https://doi.org/10.1007/978-3-319-98654-8_12
6. Bondar, E.A., David, Volkov, M.V.: Completely reachable automata: an interplay between automata, graphs, and trees. CoRR abs/2201.05075 (2022). https://arxiv.org/abs/2201.05075. (accepted for publication in Int. J. Found. Comput. Sci.)
7. Cameron, P.J., Castillo-Ramirez, A., Gadouleau, M., Mitchell, J.D.: Lengths of words in transformation semigroups generated by digraphs. J. Algebraic Combin. **45**(1), 149–170 (2016). https://doi.org/10.1007/s10801-016-0703-9
8. Casas, D., Volkov, M.V.: Binary completely reachable automata. In: Castañeda, A., Rodríguez-Henríquez, F. (eds.) LATIN 2022: Theoretical Informatics - 15th Latin American Symposium, Guanajuato, Mexico, 7–11 November 2022, Proceedings. Lecture Notes in Computer Science, vol. 13568, pp. 345–358. Springer, Cham (2022). https://doi.org/10.1007/978-3-031-20624-5_21
9. Černý, J.: Poznámka k homogénnym experimentom s konečnými automatmi. Matematicko-fyzikálny časopis **14**(3), 208–216 (1964). (Translation: A Note on Homogeneous Experiments with Finite Automata. Journal of Automata, Languages and Combinatorics **24** (2019) 2–4, 123–132)
10. Cerný, J., Pirická, A., Rosenauerová, B.: On directable automata. Kybernetika **7**(4), 289–298 (1971). http://www.kybernetika.cz/content/1971/4/289
11. Eppstein, D.: Reset sequences for monotonic automata. SIAM J. Comput. **19**(3), 500–510 (1990). https://doi.org/10.1137/0219033

12. Ferens, R., Szykuła, M.: Completely reachable automata: A polynomial solution and quadratic bounds for the subset reachability problem. CoRR abs/2208.05956 (2022). 10.48550/arXiv. 2208.05956, https://doi.org/10.48550/arXiv.2208.05956
13. Fernau, H., Gusev, V.V., Hoffmann, S., Holzer, M., Volkov, M.V., Wolf, P.: Computational complexity of synchronization under regular constraints. In: Rossmanith, P., Heggernes, P., Katoen, J. (eds.) 44th International Symposium on Mathematical Foundations of Computer Science, MFCS 2019, 26–30 August 2019, Aachen, Germany. LIPIcs, vol. 138, pp. 1–14. Schloss Dagstuhl - Leibniz-Zentrum für Informatik (2019). https://doi.org/10.4230/LIPIcs.MFCS.2019.63
14. Gao, Y., Moreira, N., Reis, R., Yu, S.: A survey on operational state complexity. J. Automata, Lang. Comb. **21**(4), 251–310 (2017). https://doi.org/10.25596/jalc-2016-251
15. Gawrychowski, P., Straszak, D.: Strong inapproximability of the shortest reset word. In: Italiano, G.F., Pighizzini, G., Sannella, D.T. (eds.) MFCS 2015. LNCS, vol. 9234, pp. 243–255. Springer, Heidelberg (2015). https://doi.org/10.1007/978-3-662-48057-1_19
16. Goldberg, K.Y.: Orienting polygonal parts without sensors. Algorithmica **10**(2–4), 210–225 (1993). https://doi.org/10.1007/BF01891840
17. Gonze, F., Jungers, R.M.: Hardly reachable subsets and completely reachable automata with 1-deficient words. J. Automata, Lang. Comb. **24**(2–4), 321–342 (2019). https://doi.org/10.25596/jalc-2019-321
18. Hamidoune, Y.O.: Quelques problèmes de connexité dans les graphes orientés. J. Comb. Theory, Ser. B **30**(1), 1–10 (1981). https://doi.org/10.1016/0095-8956(81)90085-X
19. Hoffmann, S.: Completely reachable automata, primitive groups and the state complexity of the set of synchronizing words. In: Leporati, A., Martín-Vide, C., Shapira, D., Zandron, C. (eds.) LATA 2021. LNCS, vol. 12638, pp. 305–317. Springer, Cham (2021). https://doi.org/10.1007/978-3-030-68195-1_24
20. Hoffmann, S.: Constrained synchronization and commutativity. Theor. Comput. Sci. **890**, 147–170 (2021). https://doi.org/10.1016/j.tcs.2021.08.030
21. Hoffmann, S.: State complexity of the set of synchronizing words for circular automata and automata over binary alphabets. In: Leporati, A., Martín-Vide, C., Shapira, D., Zandron, C. (eds.) LATA 2021. LNCS, vol. 12638, pp. 318–330. Springer, Cham (2021). https://doi.org/10.1007/978-3-030-68195-1_25
22. Hoffmann, S.: Sync-maximal permutation groups equal primitive permutation groups. In: Han, Y., Ko, S. (eds.) Descriptional Complexity of Formal Systems - 23rd IFIP WG 1.02 International Conference, DCFS 2021, Virtual Event, 5 September 2021, Proceedings. Lecture Notes in Computer Science, vol. 13037, pp. 38–50. Springer, Cham (2021). https://doi.org/10.1007/978-3-030-93489-7_4
23. Hoffmann, S.: Reset complexity and completely reachable automata with simple idempotents. In: Han, Y., Vaszil, G. (eds.) Descriptional Complexity of Formal Systems - 24rd IFIP WG 1.02 International Conference, DCFS 2022, 29–31 August 2022, Debrecen, Hungary, Proceedings. Lecture Notes in Computer Science, vol. 13439, pp. 85–99. Springer, Cham (2022). https://doi.org/10.1007/978-3-031-13257-5_7
24. Holzer, M., Jakobi, S.: On the computational complexity of problems related to distinguishability sets. Inf. Comput. **259**(2), 225–236 (2018). https://doi.org/10.1016/j.ic.2017.09.003
25. Hopcroft, J.E., Ullman, J.D.: Introduction to Automata Theory, Languages, and Computation. Addison-Wesley Publishing Company, Boston (1979)

26. Jones, N.D.: Space-bounded reducibility among combinatorial problems. J. Comput. Syst. Sci. **11**(1), 68–85 (1975). https://doi.org/10.1016/S0022-0000(75)80050-X
27. Jürgensen, H.: Synchronization. Inf. Comput. **206**(9–10), 1033–1044 (2008). https://doi.org/10.1016/j.ic.2008.03.005
28. Maslennikova, M.I.: Reset complexity of ideal languages over a binary alphabet. Int. J. Found. Comput. Sci. **30**(6–7), 1177–1196 (2019). https://doi.org/10.1142/S0129054119400343
29. Natarajan, B.K.: An algorithmic approach to the automated design of parts Orienters. In: 27th Annual Symposium on Foundations of Computer Science, Toronto, Canada, 27–29 October 1986. pp. 132–142. IEEE Computer Society (1986). https://doi.org/10.1109/SFCS.1986.5
30. Neumann, P.M.: The Mathematical Writings of Évariste Galois. European Mathematical Society, Helsinki, Heritage of European Mathematics (2011). https://doi.org/10.4171/104
31. Olschewski, J., Ummels, M.: The complexity of finding reset words in finite automata. In: Hliněný, P., Kučera, A. (eds.) MFCS 2010. LNCS, vol. 6281, pp. 568–579. Springer, Heidelberg (2010). https://doi.org/10.1007/978-3-642-15155-2_50
32. Rystsov, I.K.: On minimizing the length of synchronizing words for finite automata. In: Theory of Designing of Computing Systems, pp. 75–82. Institute of Cybernetics of Ukrainian Acad. Sci. (1980). (in Russian)
33. Rystsov, I.K.: Estimation of the length of reset words for automata with simple idempotents. Cybern. Syst. Anal. **36**(3), 339–344 (2000). https://doi.org/10.1007/BF02732984
34. Rystsov, I.K.: Cerny's conjecture for automata with simple idempotents. Cybern. Syst. Anal. **58**(1), 1–7 (2022). https://doi.org/10.1007/s10559-022-00428-3
35. Sandberg, S.: 1 homing and synchronizing sequences. In: Broy, M., Jonsson, B., Katoen, J.-P., Leucker, M., Pretschner, A. (eds.) Model-Based Testing of Reactive Systems. LNCS, vol. 3472, pp. 5–33. Springer, Heidelberg (2005). https://doi.org/10.1007/11498490_2
36. Starke, P.H.: Eine Bemerkung über homogene Experimente. Elektronische Informationverarbeitung und Kybernetik (later Journal of Information Processing and Cybernetics) **2**(2), 61–82 (1966), (Translation: A remark about homogeneous experiments. Journal of Automata, Languages and Combinatorics 24 (2019) 2–4, 133–237)
37. Trahtman, A.N.: The road coloring problem. Israel J. Math. **172**(1), 51–60 (2009). https://doi.org/10.1007/s11856-009-0062-5
38. Volkov, M.V.: Synchronizing automata and the Černý conjecture. In: Martín-Vide, C., Otto, F., Fernau, H. (eds.) LATA 2008. LNCS, vol. 5196, pp. 11–27. Springer, Heidelberg (2008). https://doi.org/10.1007/978-3-540-88282-4_4
39. Volkov, M.V.: Preface. J. Automata, Lang. Comb. **24**(2–4), 119–121 (2019). https://doi.org/10.25596/jalc-2019-119
40. Volkov, M.V., Kari, J.: Černý's conjecture and the road colouring problem. In: Pin, J.É. (ed.) Handbook of Automata Theory, Volume I, pp. 525–565. European Mathematical Society Publishing House (2021)
41. Yu, S., Zhuang, Q., Salomaa, K.: The state complexities of some basic operations on regular languages. Theor. Comput. Sci. **125**(2), 315–328 (1994). https://doi.org/10.1016/0304-3975(92)00011-F

Zielonka DAG Acceptance and Regular Languages over Infinite Words

Christopher Hugenroth(✉)

TU Ilmenau, Ilmenau, Germany
christopher.hugenroth@tu-ilmenau.de

Abstract. We study an acceptance type for regular ω-languages called Zielonka DAG acceptance. We focus on deterministic automata with Zielonka DAG acceptance (DZA) and show that they are the first known automaton type with all of the following properties:

1. Emptiness can be decided in non-deterministic logspace.
2. Boolean operations on DZA cause at most polynomial blowup.
3. DZA capture exactly the ω-regular languages.

We compare Zielonka DAG acceptance to many other known acceptance types and give a complete description of their relative succinctness. Further, we show that non-deterministic Zielonka DAG automata turn out to be polynomial time equivalent to non-deterministic Büchi automata. We introduce extension acceptance as a helpful tool to establish these results. Extension acceptance also leads to new acceptance types, namely existential and universal extension acceptance.

Keywords: ω-Automata · Zielonka Trees · Extension Acceptance

1 Introduction

Deterministic finite automata on finite words are fundamental in theoretical computer science. This is, at least partly, because they have many good properties. First, decision problems, like the emptiness problem, can be decided efficiently, even in non-deterministic logarithmic space (NL). Second, Boolean operations cause at most quadratic blowup. Third, they capture exactly the class of regular languages over finite words.

In contrast, there is no standard automaton type for regular languages over *infinite* words. The key difference between deterministic finite automata on finite and on infinite words is the way that acceptance is defined. For automata on finite words a run ends in some state, so acceptance can be defined based on states. For infinite words, however, a run does not end. Instead, the set of states visited infinitely often by a run, called a loop, is used to define acceptance. So, acceptance is defined by a set of loops. Different kinds of representations of such sets yield different acceptance types.

The classic acceptance types are loop, Büchi, Co-Büchi, parity, Rabin, Streett and Muller acceptance and they form the foundation for research on infinite

F. Drewes and M. Volkov (Eds.): DLT 2023, LNCS 13911, pp. 143–155, 2023.
https://doi.org/10.1007/978-3-031-33264-7_12

words. The corresponding automaton types differ significantly with respect to the complexity of Boolean operations and decision problems, see Table 1.

For none of the aforementioned acceptance types the corresponding deterministic automaton type has all the good properties of DFAs. And to the best of the author's knowledge this hasn't been shown for any acceptance type. So, none of these deterministic automaton types has been known to satisfy all of the following conditions:

1. Decidability: The emptiness problem of the automaton type is in NL.
2. Closure: The automaton type is closed under Boolean operations and Boolean operations cause at most polynomial blowup.
3. Completeness: The automaton type is complete, i.e. for every regular ω-language there is an automaton of the type that recognizes the language.

1.1 Zielonka-DAG Acceptance and Extension Acceptance

Zielonka trees were introduced to analyze infinite duration games over a finite arena with a regular acceptance condition, also called winning condition in the context of games [16]. A Zielonka tree is inductively defined for a given set of loops (a given acceptance condition). This set of loops can be recovered from the Zielonka tree alone. In such a tree each node is labeled by a set of states and by a bit that alternates along each path.

Hunter and Dawar introduced Zielonka DAGs which are Zielonka trees where nodes with the same labels are merged [7]. For some sets of loops, Zielonka DAGs are exponentially more succinct than Zielonka trees but they still allow to compute the original set of loops which makes them an interesting data structure. In Zielonka DAG games the acceptance condition is represented as a Zielonka DAG and it was shown that these games are PSpace-complete [7].

In this paper, we show that Zielonka-DAG acceptance is also interesting for automata. In particular, we show that Zielonka DAG automata satisfy all of the conditions 1.-3.. To prove this, and to compare Zielonka DAG acceptance to other acceptance types, we introduce extension acceptance which is PTime-equivalent to Zielonka DAG acceptance. So, the conditions 1.-3. hold for Zielonka DAG acceptance iff they hold for extension acceptance.

An extension condition consists of two sets of sets of states, $\mathcal{F}, \mathcal{G}$. Roughly speaking, $\mathcal{F}$ corresponds to the sets labeled 0 in a Zielonka DAG and $\mathcal{G}$ to those labeled 1. Based on $\mathcal{F}$ and $\mathcal{G}$ we define an extension relation that relates each set of states to its minimal supersets in $\mathcal{F} \cup \mathcal{G}$. This relation can be computed efficiently and does not have to be part of the extension condition. An extension condition is then given by two sets $\mathcal{F}, \mathcal{G}$ such that each set of states has only extensions in $\mathcal{F}$ or only extension in $\mathcal{G}$ but at least one extension. For example, the Explicit-Muller condition $\mathcal{F}_e = \{\{2,3,4\},\{2,3\},\{2,4\},\{3,4\},\{2\},\{3\}\}$ is equivalent to the extension condition $(\mathcal{F}, \mathcal{G})$ where $\mathcal{F} = \{\{2,3,4\}\}$ and $\mathcal{G} = \{\{1,2,3,4\},\{4\}\}$. The sets in $\mathcal{F}_e$ are exactly the sets that can be extended to a set in $\mathcal{F}$ with respect to $(\mathcal{F}, \mathcal{G})$.

Even though PTime-equivalent, extension acceptance is more general than Zielonka DAG acceptance. In fact, extension conditions can be minimized efficiently for a fixed automaton structure and Zielonka DAG conditions correspond exactly to these minimal extension conditions, see Theorem 1. So, for a fixed automaton structure there are more extension conditions than Zielonka DAGs. Further, for Zielonka DAGs only an inductive definition based on a given acceptance condition is known while extension acceptance has a descriptive definition. Simple modifications of this definition yield existential and universal extension acceptance, for Zielonka DAG acceptance there do not seem to be such variants. Working with extension acceptance allows us to easily prove results for these more general variants and not just for Zielonka DAG acceptance.

Further, constructions are simpler for extension acceptance than for Zielonka DAG acceptance. Since the constructed object is allowed to be more general it has to satisfy fewer conditions. In fact, for union and intersection of deterministic extension automata standard product constructions suffice, yielding at most quadratic blowup. For complementation it suffices to swap $\mathcal{F}$ and $\mathcal{G}$, causing no blowup. For these reasons we work with extension acceptance instead of Zielonka DAG acceptance.

After some further investigations, we conclude that deterministic extension automata, and therefore deterministic Zielonka DAG automata, are the only known automaton types with the good properties 1.-3. which is our first contribution.

1.2 Other Acceptance Types

As a second contribution, we compare Zielonka DAG acceptance to several other acceptance types. We say that an acceptance type T is PTime-reducible to another acceptance type T' if for every T-automaton an equivalent T'-automaton *over the same automaton structure* can be computed in polynomial time, this was done similarly in [7]. Using this concept we establish the reductions shown in Fig. 1. We also establish for several pairs of acceptance types that no PTime-reduction is possible by specifying a family of T-automata such that equivalent T'-automata must be exponentially larger. So, if an acceptance type T is PTime-reducible to an acceptance type T' and not the other way around then we say that T' is more succinct than T. We give a complete description of the most common acceptance types and their relative succinctness, see theorem 2.

Next, we consider acceptance types whose corresponding automaton types are close to satisfying conditions 1.-3.. This is the case for hyper-dual automata, only the non-emptiness problem is P-hard [2]. However, it is co-NP-hard to decide whether a given tuple is a proper hyper-dual automaton [2]. This also means that the emptiness problem for hyper-dual automata is co-NP-hard if the given automaton is not known to be a hyper-dual automaton. In contrast, it can be checked in polynomial time whether an extension automaton is proper, see Lemma 1.

For color-Muller automata it is not known whether Boolean operations cause an exponential blow up. If it turns out that the blowup is small then they satisfy the conditions 1.-3.. Extension acceptance is, however, more succinct than color-Muller acceptance. Further, we show that not even parity acceptance is PTime-reducible to color-Muller acceptance. Also, minimizing the number of colors used in a color-Muller condition for a fixed automaton structure is NP-hard, while extension conditions can be minimized efficiently.

Next, we consider some variants of extension automata. For deterministic existential and deterministic universal extension automata we investigate the non-emptiness problem and the blowup of Boolean operations, see table 1. Further, we determine the succinctness of the corresponding acceptance types compared to other acceptance types. Finally, we show that non-deterministic existential extension automata are PTime-equivalent to non-deterministic Büchi automata which is an important and well-studied automaton type [1] [12].

In the appendix, we give formal proofs for our results.

Table 1. Overview of properties 1., 2., 3. for some deterministic automaton types. The results of the three bottom rows are established in this paper, all other results can be found in [1] [2].

Acceptance Type	1. Non-emptiness is complete for	2. Blowup: Union	2. Blowup: Intersection	2. Blowup: Complement	3. Complete
Loop	NL	Quadratic		No blowup	No
Büchi				No blowup (if possible)	
Co-Büchi					
Parity		Exponential		No blowup	Yes
Rabin		Quad.	Exp.	Exponential	
Streett	P	Exp.	Quad.		
Explicit-Muller	NL	Exponential			
Generalized Rabin	NL	Quadratic		Exponential	Yes
Generalized Streett	NP				
Hyper Rabin	P				
Hyper Streett	NP				
Emmerson-Lei	NP			No blowup	
Hyper-Dual	P				
∀-Extension	NP	Quadratic		Exponential	Yes
∃-Extension	NL				
Zielonka DAG, Extension	NL			No blowup	

2 Preliminaries

We denote the *natural numbers* by $\mathbb{N} = \{0, 1, 2, \dots\}$. The strict linear order on the natural numbers is denoted by $<$. All numbers in this paper are natural, we do not state that for every number explicitly.

An *alphabet* Σ is a finite, non-empty set of symbols. With Σ^* we denote the set of finite words over Σ and with Σ^ω the set of infinite words. A symbol $a \in \Sigma$

occurs infinitely often in a word $\alpha \in \Sigma^\omega$ if there are infinitely many $i \in \mathbb{N}$ with $\alpha(i) = a$. For an infinite word $\alpha \in \Sigma^\omega$ let $\text{Inf}(\alpha)$ be the *set of symbols that occur infinitely often* in α.

2.1 Automaton Structures

An *automaton structure* is a tuple $A = (Q, \Sigma, \Delta, q_0)$ where Q is a finite set of states, Σ is a finite alphabet, $\Delta \subseteq Q \times \Sigma \times Q$ is a set of transitions and $q_0 \in Q$ is the initial state.

We say that $q \in Q$ is a *successor* state of $p \in Q$ if there is some $a \in \Sigma$ such that $(p, a, q) \in \Delta$. In the following we assume that all states have a successor. A state $q \in Q$ is *reachable* from a state $p \in Q$ if there are $p_0, \ldots, p_n$ such that p_i is a successor state of p_{i-1} for $1 \leq i \leq n$, $p_0 = p$ and $p_n = q$. A *run* of an automaton structure $A = (Q, \Sigma, \Delta, q_0)$ on a word $\alpha \in \Sigma^\omega$ is a word $\rho \in Q^\omega$ such that $\rho(0) = q_0$ and $(\rho(i), \alpha(i), \rho(i+1)) \in \Delta$ for all $i \in \mathbb{N}$.

A *loop* in an automaton structure $A = (Q, \Sigma, \Delta, q_0)$ is a set of states $P \subseteq Q$ such that there is a run ρ of A on some word $\alpha \in \Sigma^\omega$ with $\text{Inf}(\rho) = P$. A *strongly connected component* (SCC) of A is a loop that is maximal with respect to inclusion. An *automaton* $\mathcal{A} = (A, Acc)$ consists of an automaton structure $A = (Q, \Sigma, \Delta, q_0)$ and a set of loops Acc. A word α is accepted by $\mathcal{A}$ if there is a run ρ of A on α such that $\text{Inf}(\rho) \in Acc$. The *accepted language* of an automaton $\mathcal{A}$ is the set $L(\mathcal{A})$ of words that are accepted by $\mathcal{A}$. An automaton structure A is *deterministic* if for all $p \in Q$, $a \in \Sigma$ there is exactly one transition $(p, a, q) \in \Delta$, we then refer to Δ as a function $\delta : Q \times \Sigma \to Q$.

2.2 Acceptance Conditions

We define acceptance conditions of automata and a notion of reducibility between them, this was done similarly in [7]. An *acceptance type* $\mathcal{T}$ is a set of triples (A, C, Acc) where (A, Acc) is an automaton and C some finite object (a set, a formula, etc.) such that $(A, C, Acc_1), (A, C, Acc_2) \in \mathcal{T}$ implies $Acc_1 = Acc_2$. Given A we can understand C as a representation of the set of loops Acc and we denote this set by $Acc\mathcal{T}(A, C)$. Further, we call the pair (A, C) a $\mathcal{T}$-automaton and C a $\mathcal{T}$-acceptance condition on A. If the acceptance type is clear from the context, we identify the $\mathcal{T}$-automaton (A, C) with the automaton $(A, Acc\mathcal{T}(A, C))$.

Next, we define several acceptance types. For each automaton structure we state the acceptance conditions and the corresponding set of loops. For an automaton structure $A = (Q, \Sigma, \Delta, q_0)$ a

- Loop condition is a set of states $S \subseteq Q$ that is a union of some SCCs of A and a loop $P \subseteq Q$ is in $AccLoop(S)$ iff $P \subseteq S$.
- Büchi (Co-Büchi) condition is a set of accepting states $F \subseteq Q$ and a loop $P \subseteq Q$ is in $AccBüchi(F)$ iff $P \cap F \neq \emptyset$ (in $AccCoBüchi(F)$ iff $P \cap F = \emptyset$, respectively).
- Parity condition is a function $\Omega : Q \to \{0, \ldots, |Q|\}$ and a loop $P \subseteq Q$ is in $AccParity(\Omega)$ iff $\max(\Omega(P)) = \max(\{\Omega(p) \mid p \in P\})$ is even.

- Rabin condition is a set $\Omega = \{(I_1, L_1), \ldots, (I_k, L_k)\}$ with $I_i, L_i \subseteq Q$ and a loop $P \subseteq Q$ is in $AccRabin(\Omega)$ iff there is an $1 \leq i \leq k$ such that $P \cap I_i \neq \emptyset$ and $P \cap L_i = \emptyset$.
- Streett condition is a set $\Omega = \{(I_1, L_1), \ldots, (I_k, L_k)\}$ with $I_i, L_i \subseteq Q$ and a loop $P \subseteq Q$ is in $AccStrett(\Omega)$ iff for all $1 \leq i \leq k$ it holds $P \cap I_i \neq \emptyset$ or $P \cap L_i = \emptyset$.
- Explicit-Muller condition is a set $\mathcal{F} \subseteq 2^Q$ and a loop $P \subseteq Q$ is $AccExplicitMuller(\mathcal{F})$ iff it is in $\mathcal{F}$.
- win-set condition is a pair $(W, \mathcal{W})$ where $W \subseteq Q$ and $\mathcal{W} \subseteq 2^Q$. A loop $P \subseteq Q$ is in $AccWinset(W, \mathcal{W})$ if $P \cap W \in \mathcal{W}$.
- color-Muller condition is a pair $(c, \mathcal{C})$ with $c : Q \to C$ and $\mathcal{C} \subseteq 2^C$ for some set of colors C. A loop $P \subseteq Q$ is in $AccColor(c, \mathcal{C})$ if $c(P) \in \mathcal{C}$.
- Zielonka DAG condition is a Zielonka DAG G that is defined by some subset $\mathcal{F} \subseteq 2^Q$, see Sect. 4.1 for a definition of Zielonka DAGs. A loop $P \subseteq Q$ is in $AccZielonkaDAG(G)$ if $P \in \mathcal{F}$.
- Emerson-Lei condition is an arbitrary Boolean formula φ where $\mathrm{Fin}(S)$ and $\mathrm{Inf}(S)$ are propositional variables for all $S \subseteq Q$. A loop $P \subseteq Q$ is in $AccEL(\varphi)$ if φ evaluates to true under the evaluation that maps $\mathrm{Inf}(S)$ to 1 iff $S \cap P \neq \emptyset$ and that maps $\mathrm{Fin}(S)$ to 1 iff $S \cap P = \emptyset$.
- generalized Rabin condition is an Emerson-Lei condition where the formula has the form $\bigvee_{i=1}^{k} \mathrm{Fin}(L_i) \wedge \mathrm{Inf}(I_{i_1}) \wedge \cdots \wedge \mathrm{Inf}(I_{i_{k_i}})$.
- generalized Streett condition is an Emerson-Lei condition where the formula has the form $\bigwedge_{i=1}^{k} \mathrm{Inf}(I_i) \vee \mathrm{Fin}(L_{i_1}) \vee \cdots \vee \mathrm{Fin}(L_{i_{k_i}})$.
- hyper Rabin condition is an Emerson-Lei condition where the formula has the form $\bigvee_{i=1}^{k} \bigwedge_{j=1}^{l} \mathrm{Fin}(L_{i,j}) \vee \mathrm{Inf}(I_{i,j})$.
- hyper Streett condition is an Emerson-Lei condition where the formula has the form $\bigwedge_{i=1}^{k} \bigvee_{j=1}^{l} \mathrm{Inf}(I_{i,j}) \wedge \mathrm{Fin}(L_{i,j})$.
- hyper Dual condition is a pair (H_R, H_S) where H_R is a hyper Rabin condition and H_S an equivalent hyper Streett condition. A loop $P \subseteq Q$ is in $AccHyperDual(H_R, H_S)$ if it is in $AccEL(H_R)$ and in $AccEL(H_S)$.

The size $|C|$ of an acceptance condition C is the size of the finite object that defines it. The size of an automaton $\mathcal{A} = (Q, \Sigma, \Delta, q_0, Acc)$ where Acc is represented by an acceptance condition C is $|\mathcal{A}| = |Q| + |C|$. We ignore the size of Δ and Σ as they are not relevant for the results of this paper.

An acceptance type $\mathcal{T}_1$ is *PTime-reducible* to an acceptance type $\mathcal{T}_2$, denoted $\mathcal{T}_1 \leq_p \mathcal{T}_2$, if there is an algorithm that given (A, C_1) with $(A, C_1, Acc) \in \mathcal{T}_1$ computes (A, C_2) with $(A, C_2, Acc) \in \mathcal{T}_2$ in polynomial time. So, we can translate every acceptance conditions of type $\mathcal{T}_1$ into an equivalent acceptance conditions of type $\mathcal{T}_2$ without changing the automaton structure and this is possible in polynomial time. Similarly, $\mathcal{T}_1$ is *ExpTime-reducible* to $\mathcal{T}_2$ if C_2 can be computed from C_1 in exponential time. Two acceptance types $\mathcal{T}_1, \mathcal{T}_2$ are *PTime-equivalent* if $\mathcal{T}_1 \leq_p \mathcal{T}_2$ and $\mathcal{T}_2 \leq_p \mathcal{T}_1$. Some known reductions are shown in Fig. 1.

We will frequently use the following remark.

Remark 1. Let A be a deterministic automaton structure in which every state is reachable and C_1, C_2 two acceptance conditions. Then $L(A, C_1) = L(A, C_2)$ iff C_1 and C_2 define the same set of accepting loops.

3 Extension Acceptance

In this section we introduce extension acceptance. For this, we define an extension relation that is used throughout the paper. Let Q be a set, $\mathcal{H} \subseteq 2^Q$ a set of sets and $P, P' \subseteq Q$. The set P *can be extended* to P' with respect to $\mathcal{H}$, denoted by $P \lhd_{\mathcal{H}} P'$, if $P \subseteq P'$, $P' \in \mathcal{H}$ and there is no $P'' \in \mathcal{H}$ with $P \subseteq P'' \subsetneq P'$. So, P' is a minimal superset of P in $\mathcal{H}$ with respect to inclusion. Notice that P is not required to be in $\mathcal{H}$.

Before we define extension conditions (ECs) we first define existential and universal extension conditions. Each of these conditions defines a set of sets. Let $A = (Q, \Sigma, \Delta, q_0)$ be an automaton structure. An *existential extension condition* is given by a pair $(\mathcal{F}, \mathcal{G})$ where $\mathcal{F}, \mathcal{G} \subseteq 2^Q$ and it defines the set of sets $Acc^{\exists}_{(\mathcal{F},\mathcal{G})} = \{P \subseteq Q \mid \exists P' \subseteq Q : P \lhd_{\mathcal{F}\cup\mathcal{G}} P' \wedge P' \in \mathcal{F}\}$. Analogously, a universal extension condition is a tuple $(\mathcal{F}, \mathcal{G})$ where $\mathcal{F}, \mathcal{G} \subseteq 2^Q$. The defined set of sets is $Acc^{\forall}_{(\mathcal{F},\mathcal{G})} = \{P \subseteq Q \mid \forall P' \subseteq Q : P \lhd_{\mathcal{F}\cup\mathcal{G}} P' \rightarrow P' \notin \mathcal{G}\}$. So, a set is in $Acc^{\forall}_{(\mathcal{F},\mathcal{G})}$ if none of its extensions with respect to $(\mathcal{F}, \mathcal{G})$ are in $\mathcal{G}$.

An *extension condition* (EC) is a pair $(\mathcal{F}, \mathcal{G})$ where $\mathcal{F}, \mathcal{G} \subseteq 2^Q$ such that $Acc^{\exists}_{(\mathcal{F},\mathcal{G})} = Acc^{\forall}_{(\mathcal{F},\mathcal{G})}$. That is, every set $P \subseteq Q$ can be extended to a set in $\mathcal{F} \cup \mathcal{G}$ but not both to a set in $\mathcal{F}$ and to a set in $\mathcal{G}$. Equivalently, $(\mathcal{F}, \mathcal{G})$ is an EC if $\mathcal{F}, \mathcal{G} \subseteq 2^Q$ and for all $P \subseteq Q$:

$$\exists P' \subseteq Q : (P \lhd_{\mathcal{F}\cup\mathcal{G}} P' \wedge P' \in \mathcal{F}) \Leftrightarrow \forall P' \subseteq Q : (P \lhd_{\mathcal{F}\cup\mathcal{G}} P' \rightarrow P' \notin \mathcal{G}) . \quad (1)$$

Notice that $\mathcal{F}$ and $\mathcal{G}$ are disjoint for every extension condition $(\mathcal{F}, \mathcal{G})$. For existential and universal ECs this does not have to be the case. We can decide efficiently whether a given tuple $(\mathcal{F}, \mathcal{G})$ is an EC using the following lemma.

Lemma 1. *Let (Q, Σ, Δ, q_0) be an automaton structure and $\mathcal{F}, \mathcal{G} \subseteq 2^Q$. Then $(\mathcal{F}, \mathcal{G})$ is an extension condition iff all of the following conditions hold:*

1. *The set of all states, Q, is in $\mathcal{F} \cup \mathcal{G}$.*
2. *For all $F \in \mathcal{F}$, $G \in \mathcal{G}$: $F \cap G \not\lhd_{\mathcal{F}\cup\mathcal{G}} F$ or $F \cap G \not\lhd_{\mathcal{F}\cup\mathcal{G}} G$.*

Proof. "$\Rightarrow$" First, if there is a set $F \in \mathcal{F}$ with $Q \lhd_{\mathcal{F}\cup\mathcal{G}} F$ then $Q = F$ and therefore $Q \in \mathcal{F}$. Otherwise, there is no such set F and because of equivalence (1) there must be a set G with $Q \lhd_{\mathcal{F}\cup\mathcal{G}} G$ and $G \in \mathcal{G}$. Again, $G = Q$ and therefore $Q \in \mathcal{F} \cup \mathcal{G}$. Second, if there are $F \in \mathcal{F}$, $G \in \mathcal{G}$ such that $F \cap G \lhd_{\mathcal{F}\cup\mathcal{G}} F$ and $F \cap G \lhd_{\mathcal{F}\cup\mathcal{G}} G$ then equivalence (1) is clearly not satisfied.

"$\Leftarrow$" To establish equivalence (1) we show two implications.

Let $P \subseteq Q$ such that $P \lhd_{\mathcal{F}\cup\mathcal{G}} F$ for some $F \in \mathcal{F}$. Assume there is a set $G \in \mathcal{G}$ with $P \lhd_{\mathcal{F}\cup\mathcal{G}} G$. Then, $F \cap G \lhd_{\mathcal{F}\cup\mathcal{G}} F, G$, contradicting property 2..

Let $P \subseteq Q$ such that for all $P' \subseteq Q$ with $P \lhd_{\mathcal{F}\cup\mathcal{G}} P'$ we have $P' \notin \mathcal{G}$. It holds $P \subseteq Q$ and $Q \in \mathcal{F} \cup \mathcal{G}$ by property 1.. So, there is a set $P' \in \mathcal{F} \cup \mathcal{G}$ with $P \lhd_{\mathcal{F}\cup\mathcal{G}} P'$. Further, $P' \notin \mathcal{G}$, so $P' \in \mathcal{F}$ because $P' \in \mathcal{F} \cup \mathcal{G}$.

An extension condition $(\mathcal{F}, \mathcal{G})$ defines the set of sets $Acc^{\exists}_{(\mathcal{F},\mathcal{G})} = Acc^{\forall}_{(\mathcal{F},\mathcal{G})}$. The set of loops defined by (existential/universal) extension conditions is the set of loops that are in the corresponding set of sets. Extension acceptance (EA) is the corresponding acceptance type for EC. An *extension automaton* is a tuple $\mathcal{A} = (A, \mathcal{F}, \mathcal{G})$ where A is an automaton structure and $(\mathcal{F}, \mathcal{G})$ is an extension condition. The size of $\mathcal{A}$ is $|\mathcal{A}| = |Q| + |\mathcal{F}| + |\mathcal{G}|$. For existential and universal ECs the definitions are analogous.

Reduced Extension Conditions. Next, we show that extension conditions can be minimized efficiently for a fixed automaton structure. Let $A = (Q, \Sigma, \Delta, q_0)$ be an automaton structure and $(\mathcal{F}, \mathcal{G})$ an EC. There might be sets in $\mathcal{F} \cup \mathcal{G}$ that can be removed without changing the set of sets defined by $\mathcal{F} \cup \mathcal{G}$. First, we show that removing one of these sets leads to an equivalent EC.

Lemma 2. *Let $(\mathcal{F}, \mathcal{G})$ be an EC with $\mathcal{F}, \mathcal{G} \subseteq 2^Q$. Further, let $P \subsetneq Q$ be a set such that for all P' with $P \lhd_{\mathcal{F}\cup\mathcal{G}\setminus\{P\}} P'$ we have $P' \in \mathcal{F}$ iff $P \in \mathcal{F}$.*
Then the tuple $(\mathcal{F} \setminus \{P\}, \mathcal{G} \setminus \{P\})$ is an EC and $Acc_{(\mathcal{F}\setminus\{P\},\mathcal{G}\setminus\{P\})} = Acc_{(\mathcal{F},\mathcal{G})}$.

For an EC $(\mathcal{F}, \mathcal{G})$ we can repeatedly remove sets using the above lemma until some minimal equivalent EC $(\mathcal{F}', \mathcal{G}')$ is reached that does not contain any set $P \subsetneq Q$ such that for all P' with $P \lhd_{\mathcal{F}'\cup\mathcal{G}'\setminus\{P\}} P'$ we have $P \in \mathcal{F}'$ iff $P' \in \mathcal{F}'$. We call $(\mathcal{F}', \mathcal{G}')$ the reduced EC of $(\mathcal{F}, \mathcal{G})$. We show that the reduced EC is uniquely determined, independent of the initial EC and the order in which the sets were removed.

Proposition 1. *Let $(\mathcal{F}', \mathcal{G}')$ be a reduced EC and $(\mathcal{F}'', \mathcal{G}'')$ an EC with $Acc_{(\mathcal{F}',\mathcal{G}')}$*
$= Acc_{(\mathcal{F}'',\mathcal{G}'')}$. Then $\mathcal{F}' \subseteq \mathcal{F}''$ and $\mathcal{G}' \subseteq \mathcal{G}''$.

Thus, the reduced EC is indeed minimal and unique for the fixed automaton structure. Further, it can be computed in polynomial time by repeatedly removing sets as in Lemma 2 until no set can be removed anymore.

4 Comparison to Other Acceptance Types

In this section, we relate extension acceptance to several other acceptance types. First, we show that extension acceptance and Zielonka DAG acceptance are PTime-equivalent. Next, we compare Zielonka DAG and extension acceptance to several other acceptance types as shown in Fig. 1. Finally, we take a closer look at color-Muller acceptance.

4.1 Zielonka DAG Acceptance

Zielonka trees were introduced in [16], we restate the definition here. Let Q be a set, $\mathcal{F}_0 \subseteq 2^Q$ and $\mathcal{F}_1 = 2^Q \setminus \mathcal{F}_0$. A *Zielonka tree* associated with a tuple $(\mathcal{F}_0, \mathcal{F}_1)$ is a finite tree T with vertices labeled by tuples (i, P), where $i \in \{0, 1\}$ and $P \in \mathcal{F}_i$. The root of T is labeled by (i, Q) where $Q \in \mathcal{F}_i$. Suppose that a node v of T is labeled (i, P).

- If there are subsets of P that belong to $\mathcal{F}_{1-i}$ then for the $\subseteq$-maximal ones $P_1, \ldots, P_\ell$ of these the node v has ℓ children labeled $(1-i, P_1), \ldots, (1-i, P_\ell)$.
- Otherwise, if all subsets of P belong to $\mathcal{F}_i$ then the node v is a leaf of T.

A *Zielonka DAG* is a Zielonka tree where we identify nodes with the same label. This allows us to uniquely describe a node by its label (i, P). A well-named Zielonka DAG is one where the nodes are named by their labels. The set of sets defined by a (well-named) Zielonka DAG is $\mathcal{F}_0$.

Theorem 1. *Let $A = (Q, \Sigma, \Delta, q_0)$ be an automaton structure. There is a bijection from the set of reduced ECs for A to the set of well-named Zielonka DAGs for Q that preserves the set of sets defined by the condition.*

The idea here is to define $\mathcal{F}$ as the set of sets that occur in the Zielonka DAG and are labeled 0 and $\mathcal{G}$ as the set of sets labeled 1. A *Zielonka DAG condition* is given by a Zielonka DAG and such a condition defines the set of loops that are contained in the set of sets defined by the corresponding DAG. *Zielonka DAG acceptance* is the corresponding acceptance type.

Corollary 1. *Zielonka DAG acceptance and extension acceptance are* PTime*-equivalent.*

Proof. We can translate between the two acceptance types using Theorem 1. Starting with an extension condition it might be necessary to reduce it first but this can be done in polynomial time using proposition 1.

To give an analogy, extension automata are to Zielonka DAG automata what DFAs are to minimal DFAs: they are PTime-equivalent and the latter are minimal instances of the former.

4.2 Relative Succinctness

The combination of several results yields the following theorem.

Theorem 2. *The reductions shown in Fig. 1 hold and no other reductions hold between these acceptance types (besides those contained in the transitive closure).*

We give a short analysis of Fig. 1. An arrow from one acceptance type A to an acceptance type B indicates that B is more succinct than A as a condition of type A can be translated in one of type B in polynomial time but not the other way around, this was similarly done in [7]. So, among the considered acceptance types, extension acceptance is the least succinct general purpose acceptance type to which parity acceptance is polynomial time reducible. Further, extension acceptance is PTime -reducible to hyper-dual acceptance, so extension acceptance is in-between parity and Hyper-dual acceptance.

Rabin and Streett acceptance can not be translated to extension acceptance in polynomial time. But they can be translated to $\exists$-extension acceptance and $\forall$-extension acceptance, respectively. Further, these are the largest

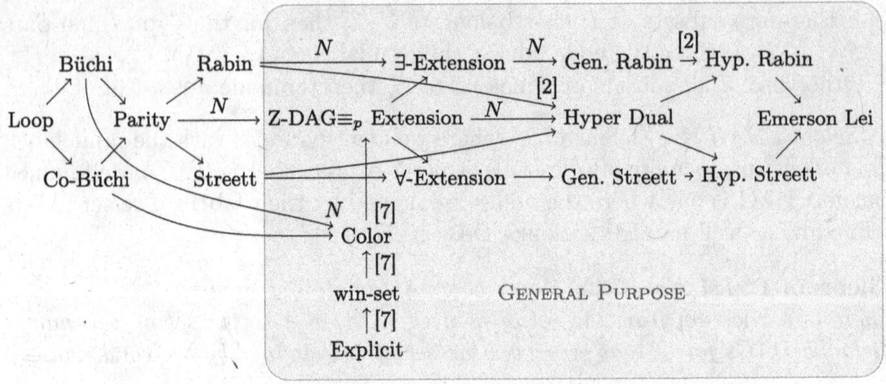

Fig. 1. The figure shows reductions between acceptance types. An arrow from acceptance type A to acceptance type B indicates that A can be reduced to B in polynomial time without changing the automaton structure. Reductions that are established in this paper are marked with an N. For each reduction a reference to the proof is given, the reduction is the dual of a referenced reduction or the reduction directly follows from the definition. The rectangle labeled General Purpose contains all of the depicted acceptance types that can define every set of loops in every automaton.

general purpose acceptance types with this property. ∃-extension acceptance and ∀-extension acceptance can in turn be translated to generalized Rabin and Streett acceptance, respectively. Thus ∃-extension is in-between Rabin and generalized Rabin acceptance and similarly ∀-extension is in-between Streett and generalized Streett acceptance.

4.3 Color-Muller Acceptance

Every acceptance condition can be modified such that a coloring of the states is added. Then the acceptance condition is defined on the set of colors and translated back to states using the coloring. This can lead to a smaller acceptance condition for automata with a special structure. This idea is often applied to explicit Muller acceptance which leads to color-Muller acceptance. In the following, we investigate color-Muller acceptance and show that there are disadvantages of color-Muller acceptance compared to Zielonka DAG and extension acceptance.

Consider the problem color-minimization which asks for a given automaton structure A, a number $k \in \mathbb{N}$ and an explicit-Muller condition $\mathcal{F}$ to determine whether there is a color-Muller condition $c, \mathcal{C}$ with at most k colors such that $L(A, c, \mathcal{C}) = L(A, \mathcal{F})$. Here, the explicit-Muller condition can be thought of as a color-Muller condition where each state is mapped to a unique color.

Lemma 3. *The problem color-minimization is* NP*-hard.*

Next, we consider the minimal number of colors needed if the automaton structure is not fixed but only the language. For this we use the Wagner hierarchy [12,15]. The following lemma shows that in general the number of colors needed by a color-Muller automaton over some *arbitrary* automaton structure to recognize a regular ω-language is unbounded. Further, we use this lemma to establish some reductions shown in Fig. 1.

Lemma 4. *Let $L \subseteq \Sigma^\omega$ be a regular language and $m \in \mathbb{N}$. There is a set C with $|C| = m$ such that there is a color-Muller automaton $(A, c, \mathcal{C})$ with $\mathcal{C} \subseteq 2^C$ and $L(A, c, \mathcal{C}) = L$ iff $L \in E_m$, where E_m is a class of the infinite Wagner hierarchy.*

Thus, the number of colors required might be close to the number of states and cannot be minimized efficiently, so introducing colors does not help in general.

5 Zielonka DAG and Extension Automata

In this section we consider non-deterministic and deterministic extension automata. First, we show that intersection and union are well-behaved for all kinds of extension automata considered here. For deterministic extension automata complementation is possible without blowup. Further, non-emptiness is NL-complete for existential extension automata and NP-complete for universal extension automata. With the results of the previous section it follows that Zielonka DAG automata and DEA posses all the good properties 1–3. Finally, we show that for a given non-deterministic existential extension an equivalent non-deterministic Büchi automata can be computed in polynomial time.

5.1 Boolean Closure

As already mentioned, the complementation of DEAs is very simple.

Proposition 2. *Let $A = (Q, \Sigma, \delta, q_0)$ be a deterministic automaton structure and $(A, \mathcal{F}, \mathcal{G})$ a DEA. Then $\Sigma^\omega \setminus L(A, \mathcal{F}, \mathcal{G}) = L(A, \mathcal{G}, \mathcal{F})$. So, DEAs can be complemented without blowup and in polynomial time.*

For intersection and union there is a simple product construction which also works for existential and universal extension automata.

Proposition 3. *Let $\mathcal{A}_1$, $\mathcal{A}_2$ be two extension automata over the same alphabet. An extension automaton $\mathcal{A}$ with $L(\mathcal{A}) = L(\mathcal{A}_1) \cap L(\mathcal{A}_2)$ and $|\mathcal{A}| \leq |\mathcal{A}_1| \cdot |\mathcal{A}_2|$ can be computed in polynomial time. If $\mathcal{A}_1$, $\mathcal{A}_2$ are DEAs then $\mathcal{A}$ is a DEA, too.*

Proof. The idea for the construction is to use the product automaton $\mathcal{A}_1 \times \mathcal{A}_2$ where the transition functions are applied componentwise with $\mathcal{F} = \{P_1 \times P_2 \mid P_1 \in \mathcal{F}_1 \text{ and } P_2 \in \mathcal{F}_2\}$ and $\mathcal{G} = \{P_1 \times P_2 \mid P_1 \in \mathcal{G}_1 \text{ or } P_2 \in \mathcal{G}_2, P_i \in \mathcal{F}_i \cup \mathcal{G}_i\}$.

However, complementation causes an exponential blowup for the variants.

Lemma 5. *Complementation of deterministic existential and universal extension automata causes an exponential blowup.*

5.2 Decision Problems

The non-emptiness problem for deterministic extension automata only requires to find a reachable and accepting loop. In the appendix, we show that the problem is NL-complete, just like graph-reachability.

Proposition 4. *The non-emptiness problem for DEA is* NL*-complete.*

For DEA Boolean operations can be performed in polynomial time, so problems like equivalence and universality can be reduced to emptiness in polynomial time and are therefore in P. For existential and universal extension acceptance this is not the case.

Proposition 5. *The non-emptiness problem for deterministic universal extension automata is* NP*-complete.*

5.3 Non-deterministic Extension Automata

Non-deterministic parity automata, and therefore non-deterministic Büchi automata (NBAs), can be translated into non-deterministic existential extension automata (NEEAs) in polynomial time by Theorem 2. The opposite direction holds as well as Theorem 3 shows. So, these automata types are polynomial time equivalent. Thus, all closure properties and decision problems for NEEAs are in PTime iff the corresponding problems for NBAs are in PTime.

Theorem 3. *For an NEEA $\mathcal{A}$ an equivalent NBA $\mathcal{B}$ can be computed in polynomial time.*

Proof. Let $\mathcal{A} = (Q, \Sigma, \Delta, q_0, \mathcal{F}, \mathcal{G})$ be an NEEA. We construct an NBA $\mathcal{B}$ that guesses for an input word α a run ρ of $\mathcal{A}$ on α as well as a set $F \in \mathcal{F}$. The construction ensures that only states from F are visited infinitely often in ρ. In addition, we give multiple Büchi conditions on the guessed run, one for each proper subset H of F that belongs to $\mathcal{F} \cup \mathcal{G}$, namely the condition that ρ visits states outside of H infinitely often. Then Choueka's flag construction [3] allows to join these polynomially many Büchi conditions (at most $|\mathcal{F} \cup \mathcal{G}| \leq |\mathcal{A}|$ many) into a single one using only polynomial many states. The complete proof can be found in the appendix.

References

1. Boker, U.: Why these automata types? In: LPAR. vol. 18, pp. 143–163 (2018)
2. Boker, U.: Inherent size blowup in ω-automata. In: Hofman, P., Skrzypczak, M. (eds.) DLT 2019. LNCS, vol. 11647, pp. 3–17. Springer, Cham (2019). https://doi.org/10.1007/978-3-030-24886-4_1
3. Choueka, Y.: Theories of automata on ω-tapes: A simplified approach. J. Comput. Syst. Sci. **8**(2), 117–141 (1974)

4. Clarke, E.M., Browne, I., Kurshan, R.P.: A unified approach for showing language containment and equivalence between various types of ω-automata. In: Colloquium on Trees in Algebra and Programming, pp. 103–116. Springer (1990)
5. Clarke, E.M., Emerson, E.A., Sistla, A.P.: Automatic verification of finite-state concurrent systems using temporal logic specifications. ACM Trans. Program. Lang. Syst. (TOPLAS) **8**(2), 244–263 (1986)
6. Emerson, E.A., Lei, C.L.: Modalities for model checking: branching time logic strikes back. Sci. Comput. Program. **8**(3), 275–306 (1987)
7. Hunter, P., Dawar, A.: Complexity bounds for regular games. In: Jędrzejowicz, J., Szepietowski, A. (eds.) MFCS 2005. LNCS, vol. 3618, pp. 495–506. Springer, Heidelberg (2005). https://doi.org/10.1007/11549345_43
8. King, V., Kupferman, O., Vardi, M.Y.: On the complexity of parity word automata. In: Honsell, F., Miculan, M. (eds.) FoSSaCS 2001. LNCS, vol. 2030, pp. 276–286. Springer, Heidelberg (2001). https://doi.org/10.1007/3-540-45315-6_18
9. Kupferman, O., Morgenstern, G., Murano, A.: Typeness for ω-regular automata. In: Wang, F. (ed.) ATVA 2004. LNCS, vol. 3299, pp. 324–338. Springer, Heidelberg (2004). https://doi.org/10.1007/978-3-540-30476-0_27
10. Löding, C.: Optimal bounds for transformations of ω-automata. In: Rangan, C.P., Raman, V., Ramanujam, R. (eds.) FSTTCS 1999. LNCS, vol. 1738, pp. 97–109. Springer, Heidelberg (1999). https://doi.org/10.1007/3-540-46691-6_8
11. Löding, C., Yue, H.: Memory bounds for winning strategies in infinite games (2008)
12. Perrin, D., Pin, J.É.: Infinite Words: Automata, Semigroups, Logic and Games. Academic Press, Cambridge (2004)
13. Safra, S.: Complexity of automata on infinite objects. Ph.D. thesis (1989)
14. Vardi, M.Y., Wolper, P.: Reasoning about infinite computations. Inf. Comput. **115**(1), 1–37 (1994)
15. Wagner, K.: On ω-regular sets. Inf. Control **43**(2), 123–177 (1979)
16. Zielonka, W.: Infinite games on finitely coloured graphs with applications to automata on infinite trees. Theor. Comput. Sci. **200**(1–2), 135–183 (1998)

On Word-Representable and Multi-word-Representable Graphs

Benny George Kenkireth(✉) and Ahaan Sameer Malhotra(✉)

Indian Institute of Technology Guwahati, Guwahati 781039, India
{bgkiitg,ahaanmalhotra001}@gmail.com

Abstract. The notion of word-representable graphs has been extensively studied. It is well known that the set of word-representable graphs are exactly the graphs whose edges can be ordered in a semi-transitive manner. Thus the set of word-representable graphs is decidable. This paper gives an alternative and simpler proof of decidability of word-representable graphs. The second part of the paper introduces a notion called multi-word-representability. Many classes of graphs - planar graphs, interval graphs, split graphs, co-bipartite graphs and line graphs - are shown to be two word-representable. An upper bound on the number of words needed to represent $k-$colourable graphs has also been calculated.

Keywords: Combinatorics on words · word-representable graphs · Semi-transitive orientation · Alternation graphs · $k-$Colourable graphs · Planar graphs · Split graphs · Co-bipartite graphs · Line graphs · Interval graphs

1 Introduction

Graphs and words are two central objects in computer science. The two most common representations of (finite) graphs are the adjacency matrix and the adjacency list representations. One can view these representations as words over a suitable alphabet. There have been many attempts to develop alternative representations of graphs [8].

An interesting connection between words and graphs was developed in [6]. The notion of alternating graph of a word was introduced in that paper. The notion of word representable graphs was more systematically studied in [5] Interested reader can know more about these graphs in [4]. Halldórsson et al. [3] characterised word-representable graphs as graphs admitting semi-transitive orientations. This characterisation shows that the class of word-representable graphs is decidable.

The organisation of the paper is as follows; Sect. 2 gives a brief introduction to the notion of word-representable graphs. The main content of Sect. 3 is Theorem 6, which gives a novel proof for the decidability of word-representable graphs. In Sect. 4, we present a generalisation of the notion of word-representability.

F. Drewes and M. Volkov (Eds.): DLT 2023, LNCS 13911, pp. 156–167, 2023.
https://doi.org/10.1007/978-3-031-33264-7_13

Theorem 14 from this section is one of the main contributions of this paper. It shows that many classes of graphs are representable using two words.

2 Mathematical Preliminaries

Let Σ be a finite set. Any finite sequence w of elements from Σ will be called a *word* over the *alphabet* Σ. Any element of the alphabet will be called a *letter*. The set of all words over Σ will be denoted by Σ^*. Any contiguous subsequence of w is called a *factor* of w. The concatenation of two words, w_1 and w_2 is represented as $w_1 \cdot w_2$ or simply as $w_1 w_2$. If a word w can be written as uv then the word u (respectively v) is called a *prefix* (respectively *suffix*) of w.

Consider a word w on Σ. Let a, b be letters of Σ. The word $w_{\{a,b\}}$ is defined as the word obtained by deleting all letters other than a and b from w. For any word w, Σ_w denotes the set of letters occurring in w. The length of a word w is the number of letters, counting multiplicities, appearing in w and is denoted by $\mid w \mid$.

Let w be a word on a two-letter alphabet, say $\{a, b\}$. We say that w is an alternating word if it doesn't contain either aa or bb as a factor. Note that this definition makes the empty word alternating.

Example 1. Let $\Sigma = \{a, b, c, d\}$. Consider the word $w = abcdadbaacd$. a is a letter and the word $abcd$ is a factor of w as well as a prefix of w. The word cd is a suffix of w. The length of w is 11. The word $w_{\{a,b\}} = ababaa$ is not an alternating word, while the word $w_{\{b,c\}} = bcbc$ is an alternating word.

2.1 Word-Representable Graphs

Definition 1. *[6] The* ***alternating graph*** *of word w is $G_w(V, E)$ such that*

$$V \triangleq \Sigma_w$$

$$E \triangleq \{(a, b) \mid w_{\{a,b\}} \text{ is alternating in } w.\}$$

A graph G is said to be ***word-representable****, if there is a word w such that $G_w = G$. The word w will be called a* ***representant*** *word of G.*

Example 2. Consider the word $w = aebacbdced$. Note that the alternating graph corresponding to w, i.e., G_w is the graph shown in Figure 1a. Thus the graph shown in Fig. 1a is a word-representable graph. On the other hand, it can be shown that the graph in Fig. 1b is not the alternating graph of any word. Thus it is a non-word-representable graph. In fact, this graph known as W_5 is the smallest non-word-representable graph; all other 6-vertex graphs are word-representable.

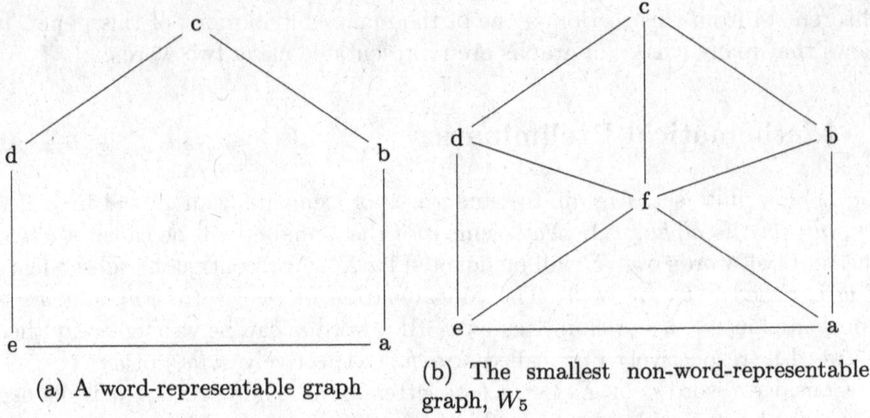

(a) A word-representable graph

(b) The smallest non-word-representable graph, W_5

Fig. 1. Word-representable and non-word-representable Graphs

Consider a graph $G = (V, E)$. If it is word-representable, then there exists a word w in V^* such that $G_w = G$. Thus we could search through all the elements of V^* to check if G is word-representable. But clearly, the search will not terminate in case of non-word-representable graphs. In other words, exhaustive search shows that the set of word-representable graphs are semi-decidable.

The natural question is whether the class of word-representable graphs is decidable. Halldórsson et. al [3] showed that it is the case by giving a characterisation of word-representable graphs, that also acts as a decidability proof. The next section gives a brief description of this.

2.2 Semi-transitive Orientation

Definition 2. *A directed graph G is* ***semi-transitive*** *if, for any directed path $v_1, v_2, \ldots, v_k$, one of the following conditions holds:*

(i) (v_1, v_k) is not an edge in G.
(ii) (v_i, v_j) is an edge in G for all $1 \leq i < j \leq k$.

Theorem 1. *[3] A graph is word-representable iff it admits a semi-transitive orientation.*

By going through all possible orientations of a graph and checking if any one of them is semi-transitive, we now have a finite time algorithm to check if a graph is word-representable. In fact, the proof of Theorem 1 shows that the representant word of the graph is of length less that $2n(n - \kappa)$ where κ is the size of the maximum clique in G.

Corollary 2. *The class of word-representable graphs is* ***decidable****.*

In the next section we give a more natural proof of decidability of word-representable graphs. This proof doesn't require the notion of semi-transitive orientation.

3 A Novel Decidability Proof of Word-Representable Graphs

We show that the set of word-representable graphs is decidable by converting the problem into a reachability problem on a suitably defined graph $\mathcal{G}$. The path in $\mathcal{G}$ will correspond to a representant word. We will introduce some basic notions with respect to words before giving a formal definition of $\mathcal{G}$.

Definition 3. *Let w be a word. The* ***suffix permutation*** *of w, denoted by $SP(w)$, is the subsequence of w that consists of the last occurrence of each $b \in \Sigma_w$ in w. The notion of* ***prefix permutation*** *can be defined similarly.*

Example 3. Let w_1 be *abcdbdad* and w_2 be *abccba*. Then $SP(w_1) = cbad$, $SP(w_2) = cba$, $PP(w_1) = abcd$ and $PP(w_2) = abc$.

When the alphabet is of size $n \in \mathbb{N}$, we will assume that Σ is the set $\{1, 2, \ldots, n\}$ and we shall denote the set of all suffix permutations of length n by S_n.

Note that the graph G_u for any word $u = wv$ (i.e. words which have w as a prefix) can be entirely determined by G_w, v and the suffix permutation of w. This observation will lead to a decidability algorithm for word-representability.

For $n \in \mathbb{N}$, let $\mathcal{V}_n$ be defined as below:

$$\mathcal{V}_n \triangleq \{(G, \sigma) \mid G \text{ is a graph with vertex set } [n] \text{ and } \sigma \in S_n\}$$

Let $V_1 = (G_1, \sigma_1)$ and $V_2 = (G_2, \sigma_2) \in \mathcal{V}_n$. Let $i \in [n]$. We say that V_2 is *adjacent* to V_1 *by* i, and denote it by $V_1 \xrightarrow{i} V_2$, if the following conditions are true:

(i) $G_2 = G_1 \setminus \{(i, j) \mid j \text{ comes before } i \text{ in } \sigma_1\}$
(ii) $\sigma_2 = SP(\sigma_1 i)$

The above definition helps in identifying the changes to the alternation graph of a word on appending letter i.

Definition 4. *The* ***word-representability graph*** *of length n is $\mathcal{G}_n = (V, E)$ such that*

$$V \triangleq \mathcal{V}_n$$

$$E \triangleq \{(V_1, V_2) \mid \text{There is an } i \in [n] \text{ such that } V_1 \xrightarrow{i} V_2\}$$

Note that the word-representability graph can be viewed as an edge labelled directed graph. For any pair of adjacent vertices, the direction as well as the labels can be deduced from the vertices.

The edge relation given captures the effect of adding letter i to word w s.t. $G_w = G_1$ and $SP(w) = \sigma_1$, as shown in the following theorem.

Theorem 3. *Let* (G_1, σ_1) *and* $(G_2, \sigma_2) \in \mathcal{V}_n$ *and let* $i \in [n]$ *be such that* $(G_1, \sigma_1) \xrightarrow{i} (G_2, \sigma_2)$. *If* w *is a word such that* $G_w = G_1$ *and* $SP(w) = \sigma_1$, *then* $G_{wi} = G_2$ *and* $SP(wi) = \sigma_2$.

Proof. The word w is a representant word for the graph G_1. The theorem asserts a relationship between G_1 and the graph corresponding to the word wi. Note that for any word w and any letter i, the graph G_{wi} is a subgraph of G_w. This is because if the letters x and y are not alternating in w, then they cannot be alternating in any word containing w as a factor. Of course, there could be alternating pairs in w that are not alternating in wi. If we remove these edges from G_w, we get the graph corresponding to G_{wi}.

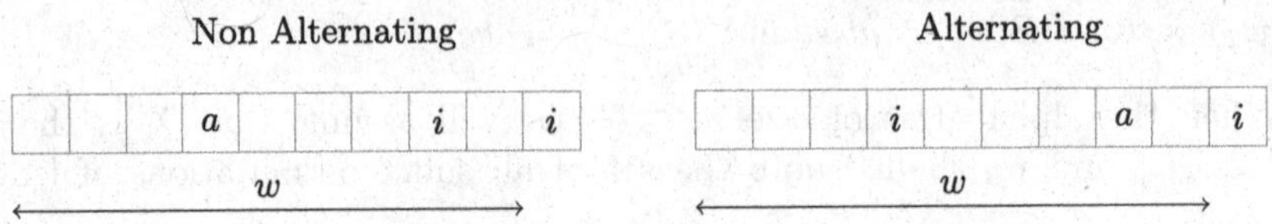

Fig. 2. Alternation on adding a letter i to a word w

Note that if the alternation of a pair of letters is affected by the addition of letter i, then either one of those letters has to be i. Let us call the other letter a. Furthermore, an alternating pair of letters (a, i) in w becomes non alternating in wi only if last occurence of a comes before the last occurence of i in w. (see Fig 2)

(G_2, σ_2) is adjacent to (G_1, σ_1) by i. By the definition of adjacency in $\mathcal{G}_n$, G_2 is exactly G_{wi}. Again from definition of adjacency, it is easy to see that $SP(wi) = \sigma_2$. □

For any path in $\mathcal{G}_\backslash$, we can concatenate the edge labels to obtain a label for the path. The following corollary connects path labels to representant words.

Corollary 4. *Let* G *be a word-representable graph on* n *vertices and let* w *be a representant word of* G. *Let* H *be any graph on* n *vertices and let* $\tau \in S_n$. *If* (H, τ) *is reachable from* $(G, SP(w))$ *in* $\mathcal{G}_n$, *then* H *is word-representable. Also, if the path* $(G, SP(w)) \rightarrow (H, \tau)$ *has label* w', *then* ww' *represents* H.

The complete graph K_n is word-representable. Clearly, any permutation σ of $[n]$ represents K_n. In fact, if we start searching from nodes of the form (K_n, σ), then all word-representable graphs will be found. We shall prove this in the following lemma.

Lemma 5. *If a* G *is a word-representable graph on* n *vertices, then* $\exists \sigma, \sigma'$ *s.t.* (G, σ) *will be reachable from* (K_n, σ').

Proof. As G is word-representable, there exists a word w s.t. $G = G_w$. We may choose σ to be $SP(w)$. Let the prefix permutation of w be $PP(w) = \sigma_0$. Note that the word $\sigma_0 w$ also represents G. For each x, y, if $w_{\{x,y\}}$ starts with, say x, then $(\sigma_0 w)_{\{x,y\}}$ is equal to $xyw_{\{x,y\}}$. Therefore, $(\sigma_0 w)_{\{x,y\}}$ will alternate iff $w_{\{x,y\}}$ alternates.

We can choose σ' to be σ_0. In $\mathcal{G}_n$, start from (K_n, σ'). Follow the edges in $\mathcal{G}_n$, as per the labels in w. By Corollary 4, the final vertex will be $(G, SP(\sigma_0 w))$, and $SP(\sigma_0 w)$ is σ. □

By corollary 4 and lemma 5, we have the following result.

Theorem 6. *The set* $\{G | (G, \sigma)$ *is reachable from* $(K_n, \sigma')\}$ *is precisely the set of word-representable graphs on* n *vertices.*

Thus, finding word-representable graphs, can be seen as a graph traversal problem where the set of reachable vertices gives us all word-representable graphs.

The state space is of size $O(2^{\frac{n^2}{2}} n!) = O(2^{\frac{n^2}{2} + n \log n})$. For each state, there are n transitions. Given any vertex $v = (G, \sigma)$, computing all the outgoing edges from v takes $O(n^2)$ time. Hence, the entire graph traversal will take $O(2^{\frac{n^2}{2} + n \log n + 2 \log n})$ time and $O(2^{\frac{n^2}{2} + n \log n})$ space. The earlier decidability proof requires checking for all words of length $O(n^2)$ and thus has a time complexity $O(2^{n^2 \log n + c \log n})$.

4 Multi-Word-Representability

Not all graphs are word-representable. In this section, we explore the possibility of representing a given graph as a union of word-representable graphs. Note that the union of two graphs G_1 and G_2 will be the graph whose vertex set is the union of the vertex sets of G_1 and G_2 and the edge set will be the union of their edge sets. If G is the union of k word-representable graphs, then we can represent G by the set of the k representant words, one corresponding to each subgraph.

Given a set $W \subseteq \Sigma^*$, the graph G_W will be defined as the $\bigcup_{w \in W} G_w$. Note that when W is a set of size one, the notion we get is exactly that of word-representability.

We are interested only in finite subsets W. Amongst all the sets W such that $G_W = G$, the minimum cardinality of W will be defined as the **multi-word-representation number** of G. Note that multi-word-representation number of all word-representable graphs is 1. Further since any graph with a single edge is word-representable, the multi-word-representation number of every graph is finite. We now present the exact values as well as some upper bounds on the multi-word-representation number for different graph classes.

Let G_1 and G_2 be word-representable graphs with representant words w_1 and w_2 respectively. If G_1 and G_2 are vertex disjoint, then the word $w_1SP(w_1)w_2SP(w_2)$ clearly represents $G_1 \cup G_2$. Thus word-representable graphs are closed under disjoint unions.

4.1 Colourability and Multi-word-Representability

In this section, we bound the multi-word-representation number of a graph based on its chromatic number. Note that three colourable graphs are word-representable [2,3]. The next theorem is a generalisation of this result.

Theorem 7. *The multi-word-representation number of* 3^k-*colourable graphs is at most* k.

Proof. The proof will be by induction. The base case is true at $k = 1$ as every 3-colourable graph is word-representable. Let G be a $3^{k+1}-$colourable graph. We will show that G can be split as a union of disjoint 3^k-colourable graphs and a single tripartite graph. Let A_1 denote the set of vertices coloured in $1, \ldots, 3^k$. Let A_2 denote the set of vertices coloured in $3^k+1, \ldots, 2 \cdot 3^k$. Let A_3 denote the set of vertices coloured in $2 \cdot 3^k + 1, \ldots, 3^{k+1}$. Let G_i, $1 \leq i \leq 3$ denote the induced subgraphs on A_i. By induction, each G_i, being 3^k-colourable, is representable by at most k words. Further $G \setminus \bigcup_{i=1}^{3} G_i$ is a tripartite graph T, with the parts being A_1, A_2 and A_3. T contains all the edges not included in $\bigcup_{i=1}^{3} G_i$. As G_1, G_2 and G_3 are vertex disjoint, the graph $\bigcup_{i=1}^{3} G_i$ can be represented using at most k words. As T is 3-colourable, we can represent T by a single word thus proving that G can be represented using $k+1$ words. □

Corollary 8. *The complement of a* 3^k-*colourable graph has multi-word-representation number at most* $k+1$.

Proof. Let G be the complement of a 3^k-colourable graph, H, with colouring $\chi(\cdot)$. Each colour class in H is an independent set. Thus, the vertices from these classes form disjoint cliques, in G. The subgraph formed by the remaining edges in G are properly 3^k-coloured using $\chi(\cdot)$. By Theorem 7, this subgraph can be represented using at most k words. Since the cliques mentioned above are disjoint, a single word can represent their union. Hence, the graph G has multi-word-representation number at most $k+1$. □

Corollary 9. *Let* G *be a graph on* n *vertices such that the chromatic number* $\chi(G) \geq n + 1 - 3^k$. *Then,* G *has multi-word-representation number* $\leq k+1$.

Proof. For any graph G on n vertices and its complement G', by a result of Nordhaus and Gaddum [7], we have $\chi(G) + \chi(G') \leq n+1$. As $\chi(G) \geq n+1-3^k$ by assumption, $\chi(G') \leq 3^k$. By the above corollary, G can be represented using $k+1$ words. □

From the above results, we can show that a general graph on 3^k vertices, since it can certainly be coloured with 3^k colours, has multi-word-representation number at most k. We can slightly strengthen this result by noting that all 5 vertex graphs are word-representable.

Theorem 10. *All graphs with up to* $5 \cdot 3^{k-1}$ *vertices have multi-word-representation number at most* k.

Proof. We follow a similar proof as given in Theorem 7. Every graph on 5 vertices is word-representable. We can split a $5 \cdot 3^{k-1}$ vertex graph into disjoint $5 \cdot 3^{k-2}$ vertex graphs and a tripartite graph. Inductively, we can prove that $5 \cdot 3^{k-1}$ vertex graphs can be represented using at most k words. □

4.2 Graphs Representable by 2 Words

In this section, we show that many classes of graphs have multi-word-representation number two. Kitaev and Lozin [4] had given a list of graph classes which contain non-word-representable graphs. We show that most of the graph classes mentioned there have multi-word-representation number two.

We begin by deriving some useful results on "small" graphs. From Theorem 10, any graph with at most 15 vertices has multi-word-representation number less than or equal to 2. We now extend this result by showing that every graphs with at most 18 vertices has multi-word-representation number less than or equal to 2.

The next lemma gives a handle on splitting a graph into smaller word-representable graphs.

Lemma 11. *Let* G *be any graph on seven vertices. Then, there is at least one subset* V' *of six vertices such that the induced subgraph on* V' *is word-representable.*

Proof. Let $V = \{V_1, \ldots, V_7\}$. Let us say that the subgraph induced by vertices $\{V_1, \ldots, V_6\}$ is not word-representable. As the wheel graph W_5 is the only non-word-representable graph on six vertices, we can without loss of generality, assume that the subgraph induced by these vertices must be W_5 (ref. Figure 3a). For the sake of contradiction, let us assume that G does not have a word-representable induced subgraph on six vertices. Given the structure of the induced subgraph on $\{V_1, \ldots, V_6\}$, the subgraph induced by $\{V_2, \ldots, V_7\}$, must similarly take the structure shown in Fig. 3b.

Now, looking at the subgraph induced by $\{V_1, V_3, \ldots, V_7\}$, we have 2 vertices ($V_1$ and V_7) with *degree* ≥ 5. It is thus a 6-vertex, non-wheel graph. Hence, it must be word-representable. □

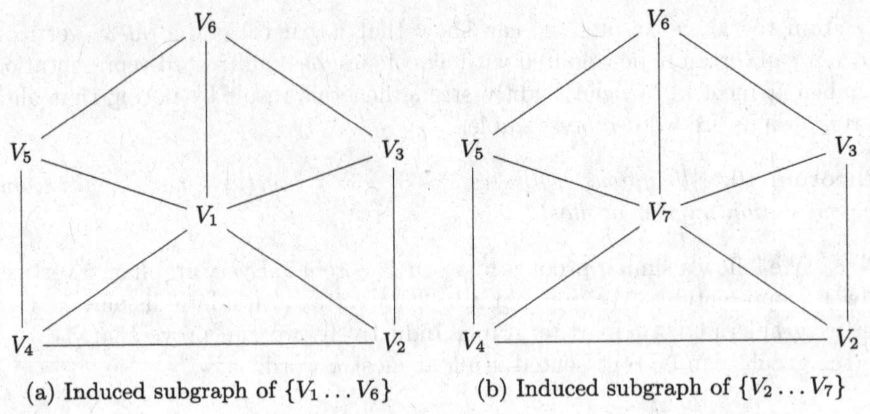

(a) Induced subgraph of $\{V_1 \ldots V_6\}$ (b) Induced subgraph of $\{V_2 \ldots V_7\}$

Fig. 3. Induced subgraphs of $V_1 \ldots V_7$ if each is not word-representable

Lemma 12. *Let $G = (V, E)$ be a graph on 12 vertices. Then, we can partition V into V_1, V_2 such that the subgraph induced by both V_1 and V_2 is word-representable.*

Proof. Let $V = \{A_1, \ldots, A_{12}\}$. Consider $V_1 = \{A_1, \ldots, A_6\}$ and $V_2 = \{A_7, \ldots, A_{12}\}$. Let the graphs induced by V_1 and V_2 be G_1, G_2 respectively. If neither G_1, nor G_2 are W_5, we get a valid partition.

In the event that V_1 and V_2 chosen as above does not give us a valid partition, we try to swap vertices of V_1 and V_2. Swap the vertex A_6 in V_1 with an arbitrary vertex in V_2. There are six possible partitions. For the swap of vertex i with A_6, let us refer to the partitions as $(G_{i,1}, G_{i,2})$ where $G_{i,1}$ is the induced graph on $V_1 \setminus \{A_6\} \cup \{i\}$ and $G_{i,2}$ the induced subgraph on the remaining vertices. Clearly, $G_{i,2}$ can be W_5 for at most one value of i. Assume that $G_{i,2}$ is word-representable for five out of the six possible choices of i. If $G_{i,1}$, too, is word-representable for any of these five choices, we get a valid partition. Otherwise, we can conclude that all the vertices in $\{A_7, \ldots, A_{11}\}$ have identical edges w.r.t $\{A_1, \ldots, A_5\}$. Furthermore, as they are part of a W_5, each of them is connected to at least 3 vertices from $\{A_1, \ldots, A_5\}$, say $\{A_1, A_2, A_3\}$.

We choose V_2 to consist of any 6 vertices from $\{A_4, A_5, A_6, A_9, A_{10}, A_{11}, A_{12}\}$, such that the induced subgraph is word-representable, using Lemma 11. From the remaining vertices, A_7 and A_8 are connected to 3 vertices A_1, A_2, A_3. This is not possible in a W_5 as there are no pairs of vertices which are connected to the same three vertices. Hence, the subgraph induced by V_1 is word-representable as well. □

Theorem 13. *Let $G = (V, E)$ be a graph $G(V, E)$ with at most 18 vertices. Then, G has multi-word-representation number at most 2.*

Proof. Looking at 18 vertex graphs would suffice, as the class of word-representable graphs is hereditary [4]. Say $V = \{V_1, \dots, V_{18}\}$. Using Lemma 12, we partition $\{V_1, \dots, V_{12}\}$ into sets A, X containing 6 vertices each, s.t. the subgraph induced by each set - G_A, G_X - is word-representable. Similarly, we partition $V \setminus A$ into B, C where G_B, G_C are word-representable. Take $G_1 = G_A \cup G_B \cup G_C$, and $G_2 = G \setminus G_1$.

By construction, G_2 is tripartite. On the other hand G_1, being the union of disjoint word-representable graphs, is itself word-representable. Hence, the given graph is representable using two words. □

The next theorem shows that many of the classes of graphs that contain both word-representable and non-word-representable graphs, listed by Kitaev and Lozin [4], are representable by two words.

Theorem 14. *Let G be a graph in one of the following classes. The multi-word-representation number of G is at most 2.*

a) Planar Graphs
b) Co-bipartite Graphs
c) Split Graphs
d) Interval Graphs
e) Line Graphs

Proof. Note that each of the classes mentioned above contain non-word-representable graphs. We shall now show that each of them is representable using two words.

a) Planar Graphs: As each planar graph is known to be 4–colourable [1], it can be represented using two words using Theorem 7.
b) Co-bipartite Graphs: Any co-bipartite graph is the union of two disjoint cliques and a bipartite graph. Clearly, each of these is word-representable. Hence, co-bipartite graphs can be represented using two words.
c) Split Graphs: Any split graph is the union of a clique and a bipartite graph. Clearly, each of these is word-representable. Hence, split graphs can be represented using two words.
d) Interval Graphs:
Let $G(V, E)$ be an interval graph. Every vertex $v \in V$ can be associated with a interval of $\mathbb{R}$ such that two vertices share an edge iff their corresponding intervals overlap. We shall denote the interval associated with a vertex i by (s_i, e_i) where s_i denotes the start of the interval and e_i denotes the end.
The overlaps between intervals can be of two types. The first type is when one of the interval is contained in the other. The second type of overlap is when none of the intervals is contained within the other. We spilt the graph G based on these two types of overlap.

Let i and j be two connected vertices in G. Based on the above discussion the edge (i, j) can be one of the two types given below:

- Containment edges: $(s_i - s_j)(e_i - e_j) \leq 0$
- Non containment edges: $(s_i - s_j)(e_i - e_j) > 0$

Let G_1 be the subgraph containing all the containment edges and let G_2 be the subgraph containing all the non containment edges. Clearly, the union gives the graph G. We will now show that both G_1 and G_2 have semi-transitive orientations. In G_1, we orient the edge from i to j if interval i is contained inside interval j. Since containment is a transitive relation of intervals, it is semi-transitive as well. Thus G_1 is word-representable.
In G_2, we orient the edge from i to j if $s_i < s_j$. Let us assume that the said orientation of G_2 is not semi-transitive. There must exist a path $I_1 I_2, \ldots, I_m$ s.t. there exists an edge from I_1 to I_m, but $\exists\ j < k$ s.t. there is no edge from I_j to I_k. As there is an edge from I_p to $I_{p+1} \forall\ 1 \leq p < m$, we can say that $s_1 < s_2 \cdots < s_m$ and $e_1 < e_2 \cdots < e_m$. As (I_1, I_m) is an edge, $e_1 > s_m$. Combining the three inequalities, $e_j > s_k$ for all j, k, i.e. there is an edge from I_j to I_k. This is a contradiction. Hence, G_2 is word-representable.
Clearly, the words representing G_1 and G_2 together represent G. Hence, any interval graph is representable using at most two words.

e) Line Graphs:
A line graph G can be represented as the union of n cliques, where n is the number of vertices of the base graph. This can be achieved by making one clique for all edges passing through a given vertex in the base graph. Each vertex of the line graph is part of exactly two such cliques. Hence, two vertices of the line graph share an edge only if one of these two cliques is common.
Let $C_1, C_2, \ldots, C_n$ be an arbitrary ordering of these cliques. For every clique C_i, let L_i consist of all the vertices $v \in C_i$ such that v has appeared previously in C_j, i.e. $v \in C_j, j < i$. Let R_i be $C_i \setminus L_i$, i.e. R_i consists of those vertices in C_i whose second clique appears later.
The edges of G can be of three types as below:

- LL edges: These are the edges (a, b) such that $a, b \in L_i$
- RR edges: These are the edges (a, b) such that $a, b \in R_i$
- LR edges: These are the edges (a, b) such that $a \in L_i$ and $b \in R_i$

Let G_1 be the subgraph of G containing all the RR edges and let G_2 contain all the LL and LR edges. Clearly, G_1 is a union of disjoint cliques and thus word-representable. We will show that G_2 is also semi-transitive.
For a vertex a such that $a \in L_i$, let $L(a)$ be defined as i. Similarly, if $a \in R_i$, let $R(a)$ be defined as i. Note that for any vertex v, we have $L(v) > R(v)$. We orient the edges of G_2 in the increasing order of their $R(\cdot)$ values, i.e. we direct the edge from a to b iff $R(a) < R(b)$. We show that this is a semi-transitive ordering.
Consider a path P given by $v_1, v_2, \ldots, v_k$ in G_2. Suppose this path violates semi-transitivity property, then (v_1, v_k) is an edge. Suppose this is an LL edge. We then have $L(v_1) = L(v_k) > R(v_k)$. If this is an LR edge, then we have $L(v_1) = R(v_k)$. Thus in either case we have

$$R(v_k) \leq L(v_1) \tag{1}$$

We claim that all edges except for the last edge on the path P must be of LL type. Note that this would imply that all vertices are present in the clique $C_{L(i)}$. For the sake of contradiction, let $(v_j, v_{j+1}), j < k-1$, be the first LR edge. Since all the previous edges along the path are LL edges we have

$$L(v_j) = R(v_{j+1}) = L(v_1) \quad (2)$$

Since the edges were oriented in terms of the increasing order of their $R(\cdot)$ value, we have $R(v_{j+1}) < R(v_k)$. Putting this together with equations (1) and (2), we have

$$L(v_1) = R(v_{j+1}) < R(v_k) \leq L(v_1)$$

This is a contradiction. □

5 Conclusion and Open Problems

We have shown that the multi-word-representation number of any graph G on n vertex is at most $O(\log n)$. From Theorem 5.3.1 in Kitaev and Lozin [4], we know that the number of n vertex word-representable graphs is $2^{\frac{2n^2}{3}+o(n^2)}$. We could thus generate $O(2^{\frac{2n^2}{3}})$ pairs, which is greater than $O(2^{\frac{n^2}{2}})$, the total number of n vertex graphs. Hence, it is likely that two words can generate almost all graphs. However, we are yet to find a graph that is provably not representable using two words. We conjecture a constant bound on the multi-word-representation number of an arbitrary graph.

We could also try to optimise the sum of lengths of the set of words used to represent a graph. This would give us the minimum size of the representation of a particular graph. We have not explored this direction that is in some sense analogous to representation number in the case of word-representable graphs.

References

1. Appel, K., Haken, W.: Every Planar Map is Four Colorable. American Mathematical Society, Contemporary mathematics (1989)
2. Halldórsson, M.M., Kitaev, S., Pyatkin, A.: Alternation graphs. In: Kolman, P., Kratochvíl, J. (eds.) WG 2011. LNCS, vol. 6986, pp. 191–202. Springer, Heidelberg (2011). https://doi.org/10.1007/978-3-642-25870-1_18
3. Halldórsson, M.M., Kitaev, S., Pyatkin, A.V.: Semi-transitive orientations and word-representable graphs. Discret. Appl. Math. **201**, 164–171 (2016)
4. Kitaev, S., Lozin, V.: Words and Graphs. Springer, Cham (2016)
5. Kitaev, S., Pyatkin, A.: On representable graphs. J. Automata Lang Comb. **13**(1), 45–54 (2008)
6. Kitaev, S., Seif, S.: Word problem of the Perkins semigroup via directed acyclic graphs. Order **25**(3), 177–194 (2008)
7. Nordhaus, E.A., Gaddum, J.W.: On complementary graphs. Am. Math. Monthly **63**(3), 175–177 (1956)
8. Spinrad, J.: Efficient Graph Representations.: The Fields Institute for Research in Mathematical Sciences. Fields Institute monographs, American Mathematical Society (2003)

On the Simon's Congruence Neighborhood of Languages

Sungmin Kim[1], Yo-Sub Han[1], Sang-Ki Ko[2(✉)], and Kai Salomaa[3]

[1] Department of Computer Science, Yonsei University, 50 Yonsei-Ro, Seodaemun-Gu, Seoul 03722, Republic of Korea
{rena_rio,emmous}@yonsei.ac.kr

[2] Department of Computer Science and Engineering, Kangwon National University, 1 Kangwondaehak-gil, Chuncheon-si, Gangwon-do 24341, Republic of Korea
sangkiko@kangwon.ac.kr

[3] School of Computing, Queen's University, Kingston, ON K7L 3N6, Canada
salomaa@queensu.ca

Abstract. Given an integer k, Simon's congruence relation says that two strings u and v are $\sim_k$-congruent if they have the same set of subsequences of length at most k. We extend Simon's congruence to languages. First, we define the Simon's congruence neighborhood of a language L to be a set of strings that have a $\sim_k$-congruent string in L. Next, we define two languages L_1 and L_2 to be $\equiv_k$-congruent if both have the same Simon's congruence neighborhood. We prove that it is PSPACE-complete to check $\equiv_k$-congruence of two regular languages and decidable up to recursive languages. Moreover, we tackle the problem of computing the maximum k that makes two given languages $\equiv_k$-congruent. This problem is PSPACE-complete for two regular languages, and undecidable for context-free languages.

Keywords: Simon's congruence · k-piecewise testable languages · congruence relation · computational complexity

1 Introduction

Simon's congruence is a binary relation over the set of strings, where, for an integer k, two strings w_1 and w_2 are $\sim_k$-congruent if the set of subsequences of w_1 and w_2 with length at most k is equal. For example, strings *ababb* and *babaa* are $\sim_2$-congruent but not $\sim_3$-congruent, because the subsequence *bbb* of the second string is not a subsequence of the first string. After Simon [20] first defined the congruence relation in the 1970s, the search for an optimal algorithm that decides whether or not two strings are $\sim_k$-congruent continued for decades [3,6]. Fleischer and Kufleitner [2] devised algorithms for computing normal form string that represents each congruence class. Barker et al. [1] optimally solved $\sim_k$ in linear time. Furthermore, Gawrychowski et al. [4] presented an algorithm using tree data structures that computes the maximum k that makes two given

F. Drewes and M. Volkov (Eds.): DLT 2023, LNCS 13911, pp. 168–181, 2023.
https://doi.org/10.1007/978-3-031-33264-7_14

strings $\sim_k$-congruent. They also extended the Simon's congruence problem and proposed a new problem called LANGSIMK [4], which asks to decide if there exists a $\sim_k$-congruent string of a given string in a given language. Recently, Kim et al. [11] solved LANGSIMK for a single string in polynomial time.

This motivates us to extend LANGSIMK for a set of strings instead of a single string. In other words, we consider LANGSIMK between two languages; given two languages L_1 and L_2 and an integer k, we want to decide whether or not every string from L_1 has a $\sim_k$-congruent string in L_2. The problem is formalized using the *Simon's congruence neighborhood* $\mathcal{N}_k(L)$ of a language L, which is a set of all strings $\sim_k$-congruent to some string in L. Then, LANGSIMK between two languages is equal to testing $\mathcal{N}_k(L_1) \subseteq \mathcal{N}_k(L_2)$. Furthermore, we study three more problems related to the equivalence of the Simon's congruence neighborhoods of two languages. We show that the considered problems are PSPACE-complete if L_1 and L_2 are given as NFAs.

The concept of the neighborhood of a language is useful in pattern matching problems. For example, the approximate pattern matching problem (or the fuzzy string searching problem) requires finding all substrings of an input text that can be converted into a given pattern within a certain number of edits [5,22]. We can solve the approximate pattern matching problem by first computing the edit distance neighborhood of the set of subsequences of the text, and then testing membership of the pattern in the edit distance neighborhood. Another approach is to compute the edit distance neighborhood of the pattern first, and then intersect the resulting neighborhood with the set of substrings of the text.

The pattern matching problem has been generalized to many contexts, such as matching between two languages [14] or matching that allows inversions [10,19]. In the context of Simon's congruence, the Simon's congruence pattern matching problem requires finding all substrings of a given text that is $\sim_k$-congruent to a given pattern. Kim et al. [12] recently solved the Simon's congruence pattern matching problem in linear time under a fixed-sized alphabet. If we have a finite (or an infinite) set P of patterns for the approximate matching, then the Simon's congruence neighborhood of P can be useful in designing efficient algorithms.

The Simon's congruence neighborhood is deeply related to the k-piecewise testable languages. A language is *piecewise testable* if it can be expressed as a finite Boolean combination of languages of the form $\Sigma^* a_1 \Sigma^* a_2 \Sigma^* \cdots \Sigma^* a_n \Sigma^*$ [20]. If $n \leq k$ for all languages used in the Boolean formula, the language is k-piecewise testable [15,17]. Indeed, we find that the class of k-piecewise testable languages is equal to the class of the Simon's congruence neighborhoods. However, piecewise testable languages were mainly studied in contexts where the piecewise testable language itself is given as input. For example, one of the most studied questions related to k-piecewise testable languages is the problem of deciding if a given regular language is piecewise testable [21] or if a given piecewise testable language is k-piecewise testable for a given integer k [13]. Note that we take an arbitrary language as input for which the Simon's congruence neighborhood is a k-piecewise testable language (Table 1).

Table 1. Summarization of the complexity results for problems related to $\equiv_k$.

problem\input	regular	context-free	recursive	recursively enumerable
$\sup\{k \mid L_1 \equiv_k L_2\} = \infty$?	PSPACE-c Theorem 1	undecidable Corollary 1		
$\mathcal{N}_k(L_1) \subseteq \mathcal{N}_k(L_2)$?	PSPACE-c. Theorem 3	decidable Theorem 5		undecidable, incomputable
$L_1 \equiv_k L_2$?				
$\max\{k \mid L_1 \equiv_k L_2\}$	PSPACE-c Theorem 4	incomputable Corollary 1		Theorem 5

Moreover, when measuring the complexity of a language, piecewise testable languages use the concept of finite-state automaton depth, rather than the traditional state complexity. This is based on the characterization by Simon [20] that the minimal DFA for a piecewise testable language must be acyclic. However, a direct construction of a finite-state automaton that recognizes a Simon's congruence neighborhood of a recursively enumerable language is impossible, which is proved by the end of Sect. 3.2. Therefore, we emphasize the difference in the initial representation by using the term "Simon's congruence neighborhood".

In Sect. 2, we define the Simon's congruence neighborhood and introduce core notations related to Simon's congruence. In Sect. 3, we propose four problems based on the equivalence of the Simon's congruence neighborhood. Then, in Sect. 3.1, we tackle the finiteness problem first to obtain a hardness lower bound on the other three problems. We provide NPSPACE algorithms for the remaining problems in Sect. 3.2 and conclude the paper in Sect. 4.

2 Preliminaries

Let Σ denote an alphabet. We define the size $|N|$ of an NFA $N = (Q, \Sigma, \delta, S, F)$ to be the cardinality of the transition function $\delta \subseteq Q \times \Sigma \times Q$. We denote the ith character of w as $w[i-1]$, $i = 1, 2, \ldots, |w|$. For example, if $w = abcdef$, then the 1st character a is $w[0]$. We say that a string u is a *subsequence* of string w if u can be obtained by deleting some characters from w, and the relation is written as $u \preceq w$. For example, string $u = aaba$ is a subsequence of string $v = cabaaccbab$, because $u = v[1]v[3]v[7]v[8]$. We say that a subsequence v of w is a *substring* of w if the characters of v occur consecutively in w. We denote a substring $v = w[i]w[i+1] \cdots w[j-1]$ as $w[i:j]$. We use $\texttt{alph}(w)$ to denote the set of characters that appears in string w.

Let M be a set and $*$ be a totally defined binary operation on M. A *monoid* is a tuple $(M, *)$ if, for all elements a, b, and c in M, associativity holds, namely, $(a * b) * c = a * (b * c)$, and there exists an identity element $\lambda \in M$, where $a * \lambda = \lambda * a = a$. A binary relation R over a monoid $(M, *)$ is a *congruence relation* when R is an equivalence relation and for all elements $m_1, m_2, n_1, n_2 \in M$ that satisfy $m_1 R m_2$ and $n_1 R n_2$, we have $(m_1 * n_1) R (m_2 * n_2)$.

For a given string w and an integer k, let $\mathbb{S}_k(w)$ denote the set of subsequences of w that have length at most k. In other words, $\mathbb{S}_k(w) = \{u \in \Sigma^{\leq k} \mid u \preceq w\}$. We say that two strings w_1 and w_2 are $\sim_k$-congruent if $\mathbb{S}_k(w_1) = \mathbb{S}_k(w_2)$. The relation is called *Simon's congruence.* The *congruence closure* or *congruence class* of a string with respect to the Simon's congruence is $\texttt{Closure}_k(w) = \{x \in \Sigma^* \mid x \sim_k w\}$. Note that $\texttt{Closure}_k(w_1) = \texttt{Closure}_k(w_2)$ if and only if $w_1 \sim_k w_2$.

It is easy to verify that the relation $\sim_k$ is a congruence by the following:

- $w \sim_k w$ because $\mathbb{S}_k(w) = \mathbb{S}_k(w)$
- $w_1 \sim_k w_2$ iff $\mathbb{S}_k(w_1) = \mathbb{S}_k(w_2)$ iff $w_2 \sim_k w_1$
- $w_1 \sim_k w_2 \wedge w_2 \sim_k w_3 \Rightarrow w_1 \sim_k w_3$ because $\mathbb{S}_k(w_1) = \mathbb{S}_k(w_2) = \mathbb{S}_k(w_3)$
- $w_1 \sim_k w_2 \wedge w_3 \sim_k w_4 \Rightarrow w_1w_3 \sim_k w_2w_4$ because $\mathbb{S}_k(w_1w_3) = \mathbb{S}_k(w_2w_4)$

Since Simon's congruence is a relation over the free monoid $(\Sigma^*, \cdot)$ where $\cdot$ is the concatenation operation, our goal is to extend the Simon's congruence relation to work on languages. It is known that the power set of a given monoid equipped with the operation $\cdot : L_1 \cdot L_2 = \{u \cdot v \mid u \in L_1, v \in L_2\}$ is a monoid as well. Now, we extend Simon's congruence to the set of all languages, i.e. the monoid $(2^{\Sigma^*}, \cdot)$. The *Simon's congruence between languages* $\equiv_k$ extends the definition of the congruence closure so that a string u is in the congruence closure of language L if and only if there exists a string in L that is $\sim_k$-congruent to u. Then, two languages L_1 and L_2 are $\equiv_k$-congruent, denoted as $L_1 \equiv_k L_2$, if their congruence closures are equivalent. Also, we call the congruence closure of a language L the *Simon's congruence neighborhood* of L, although k is not a distance metric.

Definition 1. *Let L_1 and L_2 be languages over an alphabet Σ and k be a positive integer. We define the Simon's congruence neighborhood $\mathcal{N}_k(L_1)$ of L_1 to be $\bigcup_{w \in L_1} \texttt{Closure}_k(w)$. Moreover, we denote the equivalence of Simon's congruence neighborhood of two languages $\mathcal{N}_k(L_1) = \mathcal{N}_k(L_2)$ by $L_1 \equiv_k L_2$.*

Without loss of generality, assume a fixed ordered alphabet from now on. The *shortlex normal form (SNF) string* of a congruence closure is the lexicographically least string in the set of shortest strings in the congruence closure. With slight abuse of jargon, a *saturated block* of an SNF string w is a maximal substring of w where rearranging characters in the substring results in a string $\sim_k$-congruent to w. For a given string w and an integer i between 0 and $|w|$, an *X-vector* (or a *Y-vector*) is an array with $|\Sigma|$ cells indexed by characters in Σ, where the cell indexed σ contains the minimum between the length of the shortest unique subsequence of $w[0:i]\sigma$ (or $\sigma w[i:|w|]$) that ends (or starts) with σ, or $k+1$, respectively. The X-vector for string w and integer i is denoted as $\overrightarrow{X}(w,i)$, and the Y-vector for string w and integer i is denoted as $\overrightarrow{Y}(w,i)$. For example, for string $w = ababcab$, we have $\overrightarrow{X}(w,4) = [3,3,1]$ and $\overrightarrow{Y}(w,4) = [2,2,2]$, where each cell corresponds to character a, b, and c in order. The value of the cell $\overrightarrow{X}(w,i)[w[i]]$ and the value of the cell $\overrightarrow{Y}(w,i+1)[w[i]]$ are called the *X- and Y-coordinates*, and denoted as $\mathbb{C}_X(w,i)$ and $\mathbb{C}_Y(w,i)$, respectively. If $\mathbb{C}_X(w,i) + \mathbb{C}_Y(w,i) > k+1$, then $w \sim_k w[0:i]w[i+1:|w|]$ [2]. Let function $\texttt{Iter}(\vec{v_1}, \sigma)$ return a vector $\vec{v_2}$ such that $\vec{v_2}[\sigma] = \min(\vec{v_1}[\sigma] + 1, k + 1)$ and $\vec{v_2}[\sigma'] = \min(\vec{v_2}[\sigma], \vec{v_1}[\sigma'])$. The X-vector $\vec{X}(w, i)$ can be computed from $\vec{X}(w, i-1)$ by the relation $\vec{X}(w, i) = \texttt{Iter}(\vec{X}(w, i-1), w[i-1])$. Likewise, the Y-vector $\vec{Y}(w, i)$ can be computed from $\vec{Y}(w, i+1)$ by the relation $\vec{Y}(w, i) = \texttt{Iter}(\vec{Y}(w, i+1), w[i])$ [11]. We define the *inverse* of the function Iter as $\texttt{Iter}^{-1}(\vec{v_1}, \sigma) = \{\vec{v_2} \in \{1, 2, \ldots, k+1\}^{|\Sigma|} \mid \texttt{Iter}(\vec{v_2}, \sigma) = \vec{v_1}\}$.

A regular language is *k-piecewise testable* if the language is equivalent to a finite Boolean combination of languages in the form $\Sigma^* a_1 \Sigma^* a_2 \Sigma^* \cdots \Sigma^* a_n \Sigma^*$, $n \leq k$. The class of *piecewise testable languages* is the set of all k-piecewise testable languages for all positive integers k. Then the following statement is immediate from the definitions of piecewise testable languages and Simon's congruence neighborhood:

Proposition 1 ([16]). *The class of piecewise testable languages is equal to the class of the Simon's congruence neighborhoods.*

3 Simon's Congruences and Their Neighborhoods

Proposition 2. *For an integer k, the binary relation $\equiv_k$ over the monoid $(2^{\Sigma^*}, \cdot)$ is a congruence relation.*

We tackle three problems. The first problem is to determine the inclusion of a Simon's congruence neighborhood of one language in the Simon's congruence neighborhood of another language. The second problem is to determine the equivalence of the Simon's congruence neighborhood of two languages. Finally, the third problem is to compute the maximum k value that makes two languages $\equiv_k$-congruent. The following problem serves as a stepping stone for the next three main problems.

Problem 1. Given two languages L_1 and L_2, decide whether or not $\sup\{k \in \mathbb{N} \mid L_1 \equiv_k L_2\}$ is finite.

Problem 2 (multi-string LangSimK*).* Given two languages L_1, L_2, and an integer k, decide whether or not $\mathcal{N}_k(L_1) \subseteq \mathcal{N}_k(L_2)$.

Problem 3. Given two languages L_1, L_2, and an integer k, decide whether or not $L_1 \equiv_k L_2$.

Problem 4. Given two languages L_1 and L_2, compute $\max\{k \in \mathbb{N} \mid L_1 \equiv_k L_2\}$.

3.1 Finding Maximal $\equiv_k$-Congruence of Two Languages

Lemma 1 characterizes the case when k is unbounded for Problem 1. The if and only if condition implies that the length of the shortest string that belongs to only one of the two languages is an upper bound of the value of k for $L_1 \equiv_k L_2$, which is formalized in Lemma 2. Note that the length of such a string is polynomial for two DFAs and can be exponential for two NFAs.

Lemma 1. *For two languages* L_1 *and* L_2, $\sup\{k \in \mathbb{N} \mid L_1 \equiv_k L_2\}$ *is infinite if and only if* $L_1 = L_2$.

Lemma 2. *Let* $L_1 \equiv_k L_2$ *for two different languages* L_1 *and* L_2. *If we define* $D = (L_1 \backslash L_2) \cup (L_2 \backslash L_1)$, *then* $k \leq \min_{w \in D} |w|$.

Now we are ready to solve Problem 1 when two languages are given as NFAs. Our approach tests the equivalence of the two given languages, and thus, the complexity of the problem follows from the equivalence test between each language representation.

Theorem 1 (Problem 1). *Given two regular languages* L_1 *and* L_2, *the problem of deciding whether or not the supremum of the values of* k *such that* $L_1 \equiv_k L_2$ *is finite is* PSPACE*-complete.*

If regular languages L_1 and L_2 are given as DFAs, then we can decide if $\sup\{k \in \mathbb{N} \mid L(N_1) \equiv_k L(N_2)\} = \infty$ in $O(n \log n)$ time following the DFA equivalence checking algorithm by Hopcroft [7]. Moreover, Corollary 1 proves that the problem is undecidable when either language is at least context-free. This is because the equivalence between a regular language and a context-free language as well as the equivalence between two context-free languages is undecidable [8]. Since Lemma 1 necessitates checking $L_1 = L_2$, the problem is undecidable.

Corollary 1. *Given a context-free language* L_1 *and a regular or context-free language* L_2, *the problem of deciding if* $\sup\{k \in \mathbb{N} \mid L_1 \equiv_k L_2\} = \infty$ *is undecidable.*

3.2 Determining the Equivalence and Containment of Two Simon's Congruence Neighborhoods

Recall the Simon's congruence neighborhood for any language L is regular by Proposition 1. Therefore, one way to solve Problem 3 is to construct a DFA that recognizes each Simon's congruence neighborhood of L_1 and L_2, and test their equivalence. However, the proof is non-constructive and we do not know how to build such a DFA for an arbitrary language. Indeed, we shall see that Problem 3 is only decidable for languages up to recursive languages, and is undecidable for recursively enumerable languages. Therefore, we first examine the complexity of Problem 3 when the languages are regular. Unfortunately, given a regular language L and an integer k, the deterministic state complexity of $\mathcal{N}_k(L)$ is at most $2^{O(k^{|\Sigma|-1} \log k)}$ [9], which follows from the number of congruence classes with respect to $\sim_k$, and $\Omega\left(\binom{k+|\Sigma|}{k}\right)$ [15], obtained from the length of the longest string w for which each distinct prefix pair of w does not satisfy the Nerode congruence of a k-piecewise testable language. Recall that two strings u and v are Nerode congruent with respect to language L if, for every string x, the membership of ux and vx are equal [8]. The problem is EXPTIME at best and 2-EXPTIME in the worst case if we use the automata construction method. Indeed, the following lemma implies that even the best-case EXPTIME bound is a loose complexity bound of Problem 3.

Lemma 3. *Problem 3 is* PSPACE*-hard when* L_1 *and* L_2 *are regular.*

Now, our goal is to find a PSPACE solution for Problem 3. Since constructing a finite-state automaton for the Simon's congruence neighborhood of a language takes at least exponential time, we shift our attention towards the congruence classes with respect to $\sim_k$. If a congruence class has a nonempty intersection with only one of the two languages, then it means that the two languages are not $\equiv_k$-congruent. The most compact representation of a congruence class must take at most $O(k^{|\Sigma|-1} \log k)$ bits because of the number of congruence classes is $2^{\Theta(k^{|\Sigma|-1} \log k)}$ [9]. If we do find such an encoding, then the encoding itself would take polynomial space only if k is given in unary form. Therefore, we cannot remember the full representation of such a congruence class in memory if we are to design a PSPACE algorithm for k in binary representation. The ideal representation of a congruence class therefore must be able to convey enough information about the congruence class while each bit of the input is read through exactly once.

A good choice for such a representation is the SNF string of a congruence class. The string representation gives us an intuitive idea of how the input should be read, and also enables us to exploit various characterizations of the SNF string. Then, the most important technicality of the algorithm that solves Problem 2 in PSPACE is that the SNF string cannot be fully loaded on memory at any given time. Indeed, the length of the SNF string of a congruence class with respect to Simon's congruence is at most $\binom{k+|\Sigma|}{k}$ [11,15]. Therefore, we use a generator that outputs each saturated block of an SNF string instead of the entire SNF string. We first make two assumptions to give an idea of how the main algorithms for Problems 2 and 3 work.

- There exists a nondeterministic SNF string generator G equipped with two methods that run in NPSPACE with respect to $\log k$ and $|\Sigma|$:
  `next()` returns the next saturated block of the target SNF string, and `hasNext()` returns whether or not there are more saturated blocks to be returned by `next()`.
- For a language L and an SNF generator G that generates w, there exists a PSPACE algorithm (with respect to $\log k$, $|\Sigma|$, and the size of the representation of L) that decides whether or not there exists a string in L which is $\sim_k$-congruent to w. Denote the algorithm that decides the above in PSPACE as LangSimK(G, L).

Assuming that the above assumptions are true, we can nondeterministically choose an SNF string w and run LangSimK(G, L_1) and LangSimK(G, L_2) using a deterministic SNF string generator G that generates w. Here, we can use a nondeterministic generator instead of a deterministic one in order to burden the task of nondeterministically choosing the SNF string to the generator. Then, we must additionally verify that the strings generated in the two calls of LangSimK are identical. Thus, the key idea is to run LangSimK(G, L_1) and LangSimK(G, L_2) in parallel, where the two algorithms synchronize calling

Algorithm 1. Simon's congruence neighborhood inclusion $\mathcal{N}_k(L_1) \subseteq \mathcal{N}_k(L_2)$

Given: an integer k, two languages L_1 and L_2
construct a nondeterministic SNF string generator G that outputs a saturated block each time
initialize and execute processes LANGSIMK(G, L_1) and LANGSIMK(G, L_2) until the first call of G`.next()` in each process
repeat
    $u \leftarrow G$`.next()`
    plug in u as the return value for G`.next()` in the two suspended processes LANGSIMK(G, L_1) and LANGSIMK(G, L_2)
    resume execution of LANGSIMK(G, L_1) and LANGSIMK(G, L_2) until the next call of G`.next()` in each process
until G`.hasNext()` returns `False`
$(r_1, r_2) \leftarrow$ results of the processes LANGSIMK(G, L_1) and LANGSIMK(G, L_2)
return $r_1 =$ `True` and $r_2 =$ `False`

G`.next()`. Then, LANGSIMK(G, L_1) and LANGSIMK(G, L_2) will be tested for the same SNF string. Algorithm 1 describes such a process in detail.

Note that, if deciding $\mathcal{N}_k(L_1) \subseteq \mathcal{N}_k(L_2)$ is possible in NPSPACE through Algorithm 1, then Savitch's theorem [18] entails the existance of a PSPACE algorithm that decides $\mathcal{N}_k(L_1) \subseteq \mathcal{N}_k(L_2)$. Moreover, we can also test $L_1 \equiv_k L_2$ in PSPACE through two tests of $\mathcal{N}_k(L_1) \subseteq \mathcal{N}_k(L_2)$. Therefore, we design two NPSPACE algorithms that prove the two assumptions made for Algorithm 1. The first algorithm is a generator that outputs a saturated block of an SNF string w at a time. The target string w is nondeterministically chosen and then verified that it is in the required form just before the generator outputs the last saturated block. If the algorithm runs in NPSPACE, then the first assumption is true. The second algorithm decides whether there exists a string in a given language that is $\sim_k$-congruent to an SNF string w, which is provided by a generator one saturated block at a time. Note that the length of a saturated block is at most $|\Sigma|$ [11]. If the algorithm runs in NPSPACE with respect to $\log k, |\Sigma|$, and the size of the representation of the given language, then the second assumption is true.

Proof of the First Assumption. Algorithm 2 describes the generator function `next`(k, Σ), which returns a saturated block of an SNF string over $\sim_k$ each call.

Recall that we need an SNF string generator that outputs characters of w from front to back. Lemma 5 proves that the concatenation of all strings returned by `next`(k, Σ) in order is a string in SNF. Moreover, Lemma 6 proves that every congruence class under $\sim_k$ can be represented by some string generated through repeated calls of `next`(k, Σ), except for the singleton class containing the empty string. Therefore, the nondeterministic Algorithm 2 can generate any SNF string except λ, and does not generate strings that are not in SNF.

Algorithm 2. SNF string generator function $\texttt{next}(k, \Sigma)$

1: Given: an integer k and an alphabet Σ
2: nondeterministically choose any character σ from Σ
3: set X-vector $\overrightarrow{X} \leftarrow \overrightarrow{X}(\sigma, 1)$
4: nondeterministically choose any Y-vector $\overrightarrow{Y} \in \{1, 2, \ldots, k+1\}^{|\Sigma|}$
5: $(x, y) \leftarrow (1, \overrightarrow{Y}[\sigma])$
6: $s \leftarrow \sigma$
7: **repeat**
8:   nondeterministically choose to end generation or choose c in Σ
9:   **if** did not choose to end string **then**
10:     nondeterministically choose next Y-vector $\overrightarrow{Y}$ from $\texttt{Iter}^{-1}(\overrightarrow{Y}, c)$ so that $\overrightarrow{X}[c'] + \overrightarrow{Y}[c'] \leq k+1$ for all $c' \in \Sigma$
11:     **if** $x = \overrightarrow{X}[c]$, $y = \overrightarrow{Y}[c]$, and $\overrightarrow{X}[c] + \overrightarrow{Y}[c] = k+1$ **then**
12:       **assert** $\sigma < c$
13:       $\sigma \leftarrow c$, append c to s
14:     **else**
15:       **yield** s
16:       $(s, x, y) \leftarrow (c, \overrightarrow{X}[c], \overrightarrow{Y}[c])$
17:     **end if**
18:     $\overrightarrow{X} \leftarrow \texttt{Iter}(\overrightarrow{X}, c)$
19:   **end if**
20: **until** chose to end string or a total of $\binom{k+|\Sigma|}{k}$ characters are generated
21: **assert** $\overrightarrow{Y} = [1, 1, \ldots, 1]$
22: **return** s

Assume the algorithm terminates by the **return** statement at the ith call. Let us denote the jth output of $\texttt{next}(k, \Sigma)$ as u_j for $1 \leq j \leq i$, and the concatenated string $u_1 u_2 \cdots u_i$ as w. The intermediate Lemma 4 shows that the vectors $\overrightarrow{X}$ and $\overrightarrow{Y}$ are X- and Y-vectors of the generated string w. This characterization is useful when we prove that the generated string is in SNF.

Lemma 4. *On the overall jth execution of line 8, we have $\overrightarrow{X} = \overrightarrow{X}(w, j)$ and $\overrightarrow{Y} = \overrightarrow{Y}(w, j)$.*

Lemma 5. *$w = u_1 u_2 \cdots u_i$ is a string in SNF with respect to $\sim_k$.*

Lemma 6. *Any SNF string except λ can be generated using Algorithm 2 by concatenating the string output of calls of $\texttt{next}(k, \Sigma)$ until $\texttt{False}$ is output.*

Now that the algorithm is correct, we show that the algorithm only uses space polynomial in $\log k$ and $|\Sigma|$. Then, we can imagine an SNF string generator that is capable of nondeterministically generating any SNF string by either outputting the empty string or concatenating all outputs of $\texttt{next}(k, \Sigma)$. Finally, Lemma 8 characterizes the returned string of each call of $\texttt{next}(k, \Sigma)$. Based on these characterizations, we are able to construct a nondeterministic SNF string generator G able to generate any SNF string w from k and Σ, which is equipped with two methods. The first method G.`next()` returns the next saturated block of w returned by `next`(k, Σ), or the empty string if $w = \lambda$. The second method G.`hasNext()` returns a Boolean value, where it returns `True` if and only if the generated string is not an empty string and Algorithm 2 has not terminated through the **return** statement on line 22. This proves the first assumption for Algorithm 1.

Lemma 7. *Algorithm 2 uses $O(|\Sigma| \log k)$ space.*

Corollary 2. *Given an integer k and an alphabet Σ, there exists an SNF string generator capable of nondeterministically generating any SNF string with respect to $\sim_k$ using $O(|\Sigma| \log k)$ space.*

Lemma 8. *If the SNF string generator generates string $w = u_1 u_2 \cdots u_i$, then each u_j for $j = 1, 2, \ldots, i$ returned by Algorithm 2 is a saturated block of w.*

Proof of the Second Assumption. Now that we have an SNF string w, we can check whether there exist strings in L_1 and L_2 that are $\sim_k$-congruent to w. The problem is exactly the same with the single-string LangSimK [4], but the known solution uses a DFA construction that is polynomial in k, exponential in $|\Sigma|$, and linear in the length of the SNF string w. This is problematic because the size of the DFA is already exponential in $\log k$ and $|\Sigma|$ [11]. Therefore, we describe an alternative, nondeterministic algorithm that decides LangSimK using space polynomial in $|\Sigma|$, $\log k$, and the size of the given NFA N.

Let the SNF string generator G output strings $u_1, u_2, \ldots, u_m$ in order, where $w = u_1 u_2 \cdots u_m$ is in SNF. Assume for now that G is deterministic. Then, w is fixed by the start of Algorithm 3. Let c_n be the character chosen on the overall nth execution of line 8 of Algorithm 3, for $n = 1, 2, \ldots m \leq (\binom{k+|\Sigma|}{k} \times |Q|)$, where m is the total number of chosen characters. Moreover, denote $x = c_1 c_2 \cdots c_m$ for a non-failing computation of Algorithm 3. Lemma 9 proves that x is $\sim_k$-congruent to w because Algorithm 3 is an on-the-fly simulation of a DFA that recognizes `Closure`$_k(w)$. Furthermore, Lemma 10 proves that Algorithm 3 solves LangSimK$(w, L(N))$.

Lemma 9. *The string x is $\sim_k$-congruent to w.*

Lemma 10. *There exists a string in $L(N)$ that is $\sim_k$-congruent to w if and only if there exists a non-failing computation of* `LangSimK`(k, N, G) *that outputs* `True`.

Since the algorithm is correct, we now prove that the algorithm runs in PSPACE with respect to $\log k$, $|\Sigma|$, and the size of the NFA. This proves the second assumption for Algorithm 1.

Lemma 11. *Algorithm 2 uses $O(|Q| + |\Sigma| \log k)$ space, where Q is the state set of the NFA N.*

Algorithm 3. Simon's congruence membership testing $\texttt{LangSimK}(k, N, G)$

1: Given: an integer k, an NFA $N = (Q, \Sigma, \delta, S, F)$, and an SNF string generator G that returns a saturated block at a time
2: nondeterministically choose any Y-vector $\overrightarrow{Y}$ from $\{1, 2, \ldots, k+1\}^{|\Sigma|}$
3: set X-vector $\overrightarrow{X} \leftarrow [1, 1, \ldots, 1]$
4: set N's current state set $C \leftarrow S$
5: **while** $G.\texttt{hasNext()}$ **do**
6:     $u \leftarrow G.\texttt{next()}$
7:     **repeat**
8:         nondeterministically choose to end NFA simulation or choose c in Σ
9:         **if** chose to end NFA simulation **then**
10:             **break while** loop at line 5
11:         **end if**
12:         **if** c is in $\texttt{alph}(u)$ **then**
13:             nondeterministically choose next Y-vector $\overrightarrow{Y}$ from $\texttt{Iter}^{-1}(\overrightarrow{Y}, c)$
14:             remove c from u
15:         **else**
16:             **assert** $\overrightarrow{X}[c] + \overrightarrow{Y}[c] > k+1$
17:         **end if**
18:         $\overrightarrow{X} \leftarrow \texttt{Iter}(\overrightarrow{X}, c)$
19:         update N's current state set $C \leftarrow \delta(C, c)$
20:     **until** (u is empty and $G.\texttt{hasNext()}$) or a total of $\binom{k+|\Sigma|}{k} \times |Q|$ characters are chosen in line 8
21: **end while**
22: **assert** $\overrightarrow{Y} = [1, 1, \ldots, 1]$, $u = \lambda$, G finished string generation without failing
23: **return** $C \cap F \neq \emptyset$

Corollary 3. *Given an integer k, an NFA N, and the SNF string generator G generating w, we can decide whether there exists a string in $L(N)$ that is $\sim_k$-congruent to w in* PSPACE.

Complexity Results. Since we have proven that the two assumptions are true for Algorithm 1, testing LANGSIMK between all strings in a language and another language given in the form of NFAs is PSPACE-complete. Since testing $\equiv_k$ for two regular languages given as NFAs are easily done through two tests of Problem 2, Problem 3 is also PSPACE-complete. The hardness bound follows from Lemma 3.

Theorem 2 (Problem 2 regular case). *Given two NFAs N_1 and N_2, the problem of testing whether every string in $L(N_1)$ has a $\sim_k$-congruent string in $L(N_2)$ is* PSPACE*-complete.*

Theorem 3 (Problem 3 NFA case). *Given an integer k and two NFAs N_1 and N_2, the problem of deciding whether or not $L(N_1) \equiv_k L(N_2)$ is* PSPACE*-complete.*

Theorem 4 (Problem 4 NFA case). *Given two regular languages in the form of NFAs N_1 and N_2, the problem of computing the maximum k such that $L(N_1) \equiv_k L(N_2)$ is* PSPACE*-complete.*

Note that Problems 2, 3, and 4 are decidable up to recursive languages whereas they are undecidable for recursively enumerable languages. Now, if it is possible to construct an FA that recognizes the Simon's congruence neighborhood of a recursively enumerable language, then a simple equivalence test between two FAs would solve Problem 3. This implies that it is impossible to construct an FA that recognizes the Simon's congruence neighborhood of a recursively enumerable language, even though the Simon's congruence neighborhood for any given language is regular.

Theorem 5 (Problem 3 general case). *Given an integer k and two languages L_1 and L_2, the problem of deciding whether or not $L_1 \equiv_k L_2$ is decidable when L_1 and L_2 are recursive, but undecidable when L_1 and L_2 are recursively enumerable.*

Corollary 4. *Given a recursively enumerable language L, it is impossible to construct a finite state automaton that recognizes $\mathcal{N}_k(L)$.*

Remark 1. We can lift the assumption that the alphabet is a constant-sized alphabet.

4 Conclusions

Given two languages L_1 and L_2, and an integer k, we have studied four problems related to the congruence relation $\equiv_k$. We have focused on cases where L_1 and L_2 are regular, and have pinpointed the language classes that make each problem undecidable or incomputable. From the viewpoint of extending Simon's congruence, the subsequence set of the language L, defined as $\{u \in \Sigma^{\leq k} \mid \exists w \in L : u \preceq w\}$ is an interesting future research topic. Just as Simon's congruence, we can define a congruence relation between languages based on the equivalence of the subsequence set of languages. The equivalence and containment problems of the subsequence sets as well as finding the maximum k value that makes the two sets equivalent stands as open problems. Finally, the tight complexity bound of Problem 3 when the languages are context-free remains open.

Acknowledgments. We thank the reviewers that pointed us to the book "Varieties of Formal Languages". Kim, Han and Ko were supported by the NRF grant (RS-2023-00208094) and Salomaa was supported by the Natural Sciences and Engineering Research Council of Canada (NSERC).

References

1. Barker, L., Fleischmann, P., Harwardt, K., Manea, F., Nowotka, D.: Scattered factor-universality of words. In: Jonoska, N., Savchuk, D. (eds.) DLT 2020. LNCS, vol. 12086, pp. 14–28. Springer, Cham (2020). https://doi.org/10.1007/978-3-030-48516-0_2
2. Fleischer, L., Kufleitner, M.: Testing Simon's congruence. In: 43rd International Symposium on Mathematical Foundations of Computer Science, pp. 62:1–62:13 (2018)
3. Garel, E.: Minimal separators of two words. In: Apostolico, A., Crochemore, M., Galil, Z., Manber, U. (eds.) CPM 1993. LNCS, vol. 684, pp. 35–53. Springer, Heidelberg (1993). https://doi.org/10.1007/BFb0029795
4. Gawrychowski, P., Kosche, M., Koß, T., Manea, F., Siemer, S.: Efficiently testing Simon's congruence. In: 38th International Symposium on Theoretical Aspects of Computer Science, vol. 187, pp. 34:1–34:18 (2021)
5. Hall, P.A.V., Dowling, G.R.: Approximate string matching. ACM Comput. Surv. **12**(4), 381–402 (1980)
6. Hébrard, J.: An algorithm for distinguishing efficiently bit-strings by their subsequences. Theor. Comput. Sci. **82**(1), 35–49 (1991)
7. Hopcroft, J.E.: An $n \log n$ algorithm for minimizing states in a finite automaton. In: Theory of Machines and Computations, pp. 189–196. Academic Press (1971)
8. Hopcroft, J.E., Motwani, R., Ullman, J.D.: Introduction to Automata Theory, Languages, and Computation, 3rd edn. Pearson International Edition. Addison-Wesley (2007)
9. Karandikar, P., Kufleitner, M., Schnoebelen, P.: On the index of Simon's congruence for piecewise testability. Inf. Process. Lett. **115**(4), 515–519 (2015)
10. Kim, H., Han, Y.-S.: Space-efficient approximate string matching allowing inversions in fast average time. In: Chen, J., Hopcroft, J.E., Wang, J. (eds.) FAW 2014. LNCS, vol. 8497, pp. 141–150. Springer, Cham (2014). https://doi.org/10.1007/978-3-319-08016-1_13
11. Kim, S., Han, Y.-S., Ko, S.-K., Salomaa, K.: On Simon's congruence closure of a string. In: Han, Y.S., Vaszil, G. (eds.) DCFS 2022. LNCS, vol. 13439, pp. 127–141. Springer, Cham (2022). https://doi.org/10.1007/978-3-031-13257-5_10
12. Kim, S., Ko, S.-K., Han, Y.-S.: Simon's congruence pattern matching. In: 33rd International Symposium on Algorithms and Computation, ISAAC 2022, Seoul, Korea, 19–21 December 2022, vol. 248, pp. 60:1–60:17 (2022)
13. Klíma, O., Polák, L.: Alternative automata characterization of piecewise testable languages. In: Béal, M.-P., Carton, O. (eds.) DLT 2013. LNCS, vol. 7907, pp. 289–300. Springer, Heidelberg (2013). https://doi.org/10.1007/978-3-642-38771-5_26
14. Ko, S.-K., Han, Y.-S., Salomaa, K.: Approximate matching between a context-free grammar and a finite-state automaton. Inf. Comput. **247**, 278–289 (2016)
15. Masopust, T., Thomazo, M.: On the complexity of k-piecewise testability and the depth of automata. In: Potapov, I. (ed.) DLT 2015. LNCS, vol. 9168, pp. 364–376. Springer, Cham (2015). https://doi.org/10.1007/978-3-319-21500-6_29
16. Pin, J.E.: Varieties of Formal Languages. North Oxford Academic (1989). Trans. by A. Howie
17. Ruiz, J., García, P.: Learning k-piecewise testable languages from positive data. In: Miclet, L., de la Higuera, C. (eds.) ICGI 1996. LNCS, vol. 1147, pp. 203–210. Springer, Heidelberg (1996). https://doi.org/10.1007/BFb0033355

18. Savitch, W.J.: Relationships between nondeterministic and deterministic tape complexities. J. Comput. Syst. Sci. **4**(2), 177–192 (1970)
19. Schöniger, M., Waterman, M.S.: A local algorithm for DNA subsequence alignment with inversions. Bull. Math. Biol. **54**, 521–536 (1992)
20. Simon, I.: Piecewise testable events. In: Brakhage, H. (ed.) GI-Fachtagung 1975. LNCS, vol. 33, pp. 214–222. Springer, Heidelberg (1975). https://doi.org/10.1007/3-540-07407-4_23
21. Stern, J.: Complexity of some problems from the theory of automata. Inf. Control **66**(3), 163–176 (1985)
22. Ukkonen, E.: Algorithms for approximate string matching. Inf. Control **64**(1–3), 100–118 (1985)

Tree-Walking-Storage Automata

Martin Kutrib[1(✉)] and Uwe Meyer[2]

[1] Institut für Informatik, Universität Giessen, Arndtstr. 2, 35392 Giessen, Germany
kutrib@informatik.uni-giessen.de
[2] Technische Hochschule Mittelhessen, Wiesenstr. 14, 35390 Giessen, Germany
uwe.meyer@mni.thm.de

Abstract. We introduce and investigate tree-walking-storage automata, which are finite-state devices equipped with a tree-like storage. The automata are generalized stack automata, where the linear stack storage is replaced by a non-linear tree-like stack. Therefore, tree-walking-storage automata have the ability to explore the interior of the tree storage without altering the contents, where the possible moves of the tree pointer correspond to those of tree walking automata. In addition, a tree-walking-storage automaton can append (push) non-existent descendants to a tree node and remove (pop) leaves from the tree. As for classical stack automata, we also consider non-erasing and checking variants. As first steps to investigate these models we consider the computational capacities of deterministic one-way variants. In particular, a main focus is on the comparisons of the different variants of tree-walking-storage automata as well as on the comparisons with classical stack automata, and we can draw a complete picture.

1 Introduction

Stack automata were introduced in [6] as a theoretical model motivated by compiler theory, and the implementation of recursive procedures with parameters. Their computational power lies between that of pushdown automata and Turing machines. Basically, a stack automaton is a finite-state device equipped with a generalization of a pushdown store. In addition to be able to push or pop at the top of the pushdown store, a stack automaton can move its storage head (stack pointer) *inside* the stack to read stack symbols, but without altering the contents. In this way, it is possible to read but not to change the stored information. Over the years, stack automata have aroused great interest and have been studied in different variants. Apart from distinguishing deterministic and nondeterministic computations, the original two-way input reading variant has been restricted to one-way [7]. One-way nondeterministic stack automata can be simulated by deterministic linear-bounded automata, so the accepted languages are deterministic context sensitive [14]. Further investigated restrictions concern the usage of the stack storage. A stack automaton is said to be *non-erasing* if no symbol may be popped from the stack [12], and it is *checking* if it cannot push any symbols once the stack pointer has moved into the stack [9]. While the early studies of stack

F. Drewes and M. Volkov (Eds.): DLT 2023, LNCS 13911, pp. 182–194, 2023.
https://doi.org/10.1007/978-3-031-33264-7_15

automata have extensively been done in relation with AFL theory as well as time and space complexity [11,13,15,21,23], more recent papers consider the computational power gained in generalizations by allowing the input head to jump [19], allowing multiple input heads, multiple stacks [18], and multiple reversal-bounded counters [17]. The stack size required to accept a language by stack automata has been considered as well [16]. In [20] the property of working input-driven has been imposed to stack automata, and their capacities as transducer are studied in [2].

All these models have in common that their storage structures are linear. Therefore, it is a natural idea to generalize stack automata by replacing the stack storage by some non-linear data structure. Here we introduce the tree-walking-storage automata that are essentially stack automata with a tree-like stack. As for classical stack automata, tree-walking-storage automata have the additional ability to move the storage head (here tree pointer) inside the tree without altering the contents. The possible moves of the tree pointer correspond to those of tree walking automata. In this way, it is possible to read but not to change the stored information. In addition, a tree-walking-storage automaton can append (push) a non-existent descendant to a tree node and remove (pop) a leaf from the tree.

Before we turn to our main results and the organization of the paper, we briefly mention different approaches to introduce tree-like stacks. So-called pushdown tree automata [10] extend the usual string pushdown automata by allowing trees instead of strings in both the input and the stack. So, these machines accept trees and may not explore the interior of the stack. Another big difference is the way the tree-storage can be altered. In particular, the machine can access only the root of the stack tree and alters the stack by substituting a tree for the root (pushing) or deleting the root and selecting one of its subtrees as new stack tree content (popping). Essentially, this model has been adapted to string inputs and tree-stacks where the so-called *tree-stack automaton* can explore the interior of the tree-stack in read-only mode [8]. However, in the writing-mode a new tree can be pushed on the stack employing the subtrees of the old tree-stack, that is, subtrees can be permuted, deleted, or copied. If the root of the tree-stack is popped, exactly one subtree is left in the store. Another model also introduced under the name *tree-stack automaton* gave up the bulky way of pushing and popping at the root of the tree-stack [5]. However, this model may alter the interior nodes of the tree-stack. Therefore, the tree-stack is actually a non-linear Turing tape. Therefore, we have chosen the name *tree-walking-storage automaton*, so as not to have one more model under the name of tree-stack automaton.

The idea of a tree-walking process originates from [1]. A tree-walking automaton is a sequential model that processes input trees. For example, it is known that deterministic tree-walking automata are strictly weaker than nondeterministic ones [3] and that even nondeterministic tree-walking automata cannot accept all regular tree languages [4].

Here we introduce checking, non-erasing and general tree-walking-storage automata. As first steps to investigate these models we consider the computational capacities of deterministic one-way variants. In particular, a main focus is on the comparisons of the different variants of tree-walking-storage automata as well as on the comparisons with classical stack automata. The paper is organized as follows. The definition of the models and an illustrating example are given in Sect. 2. In Sect. 3 we study the computational capacity of general deterministic tree-walking-storage automata, where *general* means that we do not impose a time limit on the machines a priori. It turns out that their checking variant is no more powerful than classical checking stack automata. Moreover, we show that in the case of unlimited time deterministic tree-walking-storage automata are as powerful as Turing machines. This is in strong contrast to the classic models since it is known that all languages accepted by one-way stack automata are context sensitive [14]. This result suggests considering time constraints for deterministic tree-walking-storage automata. It is known that any classical DSA (deterministic stack automaton) can effectively be transformed into an equivalent DSA that accepts in quadratic time (see [6,14] and [24] for a direct proof). Therefore, it is natural to consider deterministic tree-walking-storage automata that accept in polynomial time as well. Clearly, these devices are no longer computationally universal. In Sect. 4 the computational capacities of polynomial-time non-erasing tree-walking-storage automata and non-erasing stack automata are separated. Section 5 is devoted to compare non-erasing tree-walking-storage and tree-walking-storage automata. For classical stack automata it is known that the former are strictly weaker than the latter. However, the situation for tree-walking-storage automata is quite different, both models are equally powerful.

2 Definitions and Preliminaries

Let Σ^* denote the *set of all words* over the finite alphabet Σ. The *empty word* is denoted by λ. The reversal of a word w is denoted by w^R and for the length of w we write $|w|$. Set inclusion is denoted by $\subseteq$, and strict set inclusion by $\subset$.

A tree-walking-storage automaton is an extension of a classical stack automaton to a tree storage. As for classical stack automata, tree-walking-storage automata have the additional ability to move the storage head (here tree pointer) inside the tree without altering the contents. The possible moves of the tree pointer correspond to those of tree walking automata. In this way, it is possible to read but not to change the stored information. However, a classical stack automaton can push and pop at the top of the stack. Accordingly, a tree-walking-storage automaton can append (push) a non-existent descendant to a tree node and remove (pop) a leaf from the tree.

Here we consider mainly deterministic one-way devices. The trees in this paper are finite, binary trees whose nodes are labeled by a finite alphabet Γ. A Γ-tree T is represented by a mapping from a finite, non-empty, prefix-closed subset of $\{l, r\}^*$ to $\Gamma \cup \{\bot\}$, such that $T(w) = \bot$ if and only if $w = \lambda$. The elements of the domain of T are called *nodes of the tree*. Each node of the tree has a *type* from $\mathsf{TYPE} = \{-, l, r\} \times \{-, +\}^2$, where the first component expresses whether the node *is* the root ($-$), a left descendant (l), or a right descendant (r), and the second and third components tell whether the node *has* a left and right descendant (+), or not (−). A *direction* is an element from $\mathsf{DIRECT} = \{u, s, d_l, d_r\}$, where u stands for 'up', s stands for 'stay', d_l stands for 'left descendant' and d_r for 'right descendant'.

A *deterministic tree-walking-storage automaton (abbreviated* twsDA*)* is a system $M = \langle Q, \Sigma, \Gamma, \delta, q_0, \triangleleft, \perp, F\rangle$, where Q is the finite set of *internal states*, Σ is the finite set of *input symbols* not containing the *endmarker* $\triangleleft$, Γ is the finite set of *tree symbols*, $q_0 \in Q$ is the *initial state*, $\perp \notin \Gamma$ is the *root symbol*, $F \subseteq Q$ is the set of *accepting states*, and

$$\begin{aligned}\delta\colon Q \times (\Sigma \cup \{\lambda, \triangleleft\}) \times \mathsf{TYPE} \times (\Gamma \cup \{\perp\}) \to\\ Q \times (\mathsf{DIRECT} \cup \{\mathtt{pop}\} \cup \{\, \mathtt{push}(x, d) \mid x \in \Gamma, d \in \{l, r\}\,\})\end{aligned}$$

is the *transition function*. There must never be a choice of using an input symbol or of using λ input. So, it is required that for all q in Q, $(t_1, t_2, t_3) \in \mathsf{TYPE}$, and x in $\Gamma \cup \{\perp\}$: if $\delta(q, \lambda, (t_1, t_2, t_3), x)$ is defined, then $\delta(q, a, (t_1, t_2, t_3), x)$ is undefined for all a in $\Sigma \cup \{\triangleleft\}$.

A *configuration* of a twsDA M is a quadruple (q, v, T, P), where $q \in Q$ is the current state, $v \in \Sigma^*\{\triangleleft, \lambda\}$ is the unread part of the input, T is the current Γ-tree, and P is an element of the domain of T, called the *tree pointer*, that is the current node of T. The *initial configuration* for input w is set to $(q_0, w\triangleleft, T_0, \lambda)$, where $T_0(\lambda) = \perp$ and T_0 is undefined otherwise.

During the course of its computation, M runs through a sequence of configurations. In a given configuration (q, v, T, P), M is in state q, reads the first symbol of v or λ, knows the type of the current node P, and sees the label $T(P)$ of the current node. Then it applies δ and, thus, enters a new state and either moves the tree pointer along a direction, removes the current node (if it is a leaf) by pop, or appends a new descendant to the current node (if this descendant does not exist) by push. Here and in the sequel, it is understood that δ is well defined in the sense that it will never move the tree pointer to a non-existing node, will never pop a non-leaf node and will never push an existing descendant. This normal form is always available through effective constructions.

One step from a configuration to its successor configuration is denoted by $\vdash$, and the reflexive and transitive (resp., transitive) closure of $\vdash$ is denoted by $\vdash^*$ (respectively $\vdash^+$). Let $q \in Q$, $av \in \Sigma^*\triangleleft$ with $a \in \Sigma \cup \{\lambda, \triangleleft\}$, T be a Γ-tree, P be a tree pointer of T, and $(t_1, t_2, t_3) \in \mathsf{TYPE}$ be the type of the current node P. We set

1. $(q, av, T, P) \vdash (q', v, T, P')$ with $P = P'l$ or $P = P'r$, if $t_1 \neq -$ and $\delta(q, a, (t_1, t_2, t_3), T(P)) = (q', u)$, (move the tree pointer up),
2. $(q, av, T, P) \vdash (q', v, T, P)$, if $\delta(q, a, (t_1, t_2, t_3), T(P)) = (q', s)$, (do not move the tree pointer),
3. $(q, av, T, P) \vdash (q', v, T, P')$ with $P' = Pl$, if $t_2 = +$ and $\delta(q, a, (t_1, t_2, t_3), T(P)) = (q', d_l)$, (move the tree pointer to the left descendant),

4. $(q, av, T, P) \vdash (q', v, T, P')$ with $P' = Pr$,
   if $t_3 = +$ and $\delta(q, a, (t_1, t_2, t_3), T(P)) = (q', d_r)$, (move the tree pointer to the right descendant),
5. $(q, av, T, P) \vdash (q', v, T', P')$ with $P = P'l$ or $P = P'r$, $T'(P)$ is undefined and $T'(w) = T(w)$ for $w \neq P$,
   if $t_2 = t_3 = -$ and $\delta(q, a, (t_1, t_2, t_3), T(P)) = (q', \texttt{pop})$, (remove the current leaf node, whereby the tree pointer is moved up),
6. $(q, av, T, P) \vdash (q', v, T', P')$ with $P' = Pl$, $T'(Pl) = x$ and $T'(w) = T(w)$ for $w \neq Pl$,
   if $t_2 = -$ and $\delta(q, a, (t_1, t_2, t_3), T(P)) = (q', \texttt{push}(x, l))$, (append a left descendant to the current node, whereby the tree pointer is moved to the descendant),
7. $(q, av, T, P) \vdash (q', v, T', P')$ with $P' = Pr$, $T'(Pr) = x$ and $T'(w) = T(w)$ for $w \neq Pr$, $t_3 = -$ and $\delta(q, a, (t_1, t_2, t_3), T(P)) = (q', \texttt{push}(x, r))$, (append a right descendant to the current node, whereby the tree pointer is moved to the descendant).

Figure 1 illustrates the transitions that move the tree pointer up, respectively to the left descendant. Figure 2 illustrates the push, respectively the pop transitions. All remaining transitions are analogous.

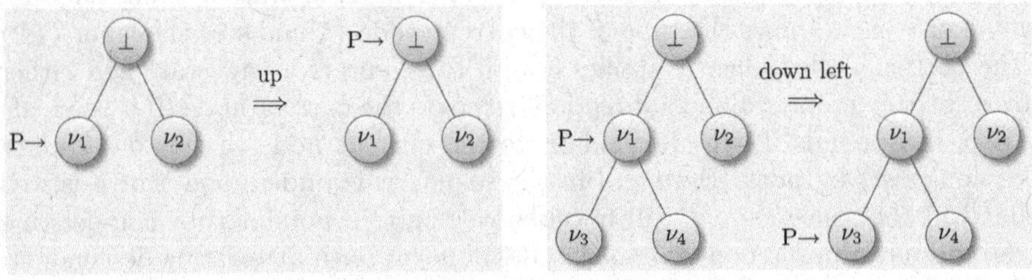

Fig. 1. Up and left transitions.

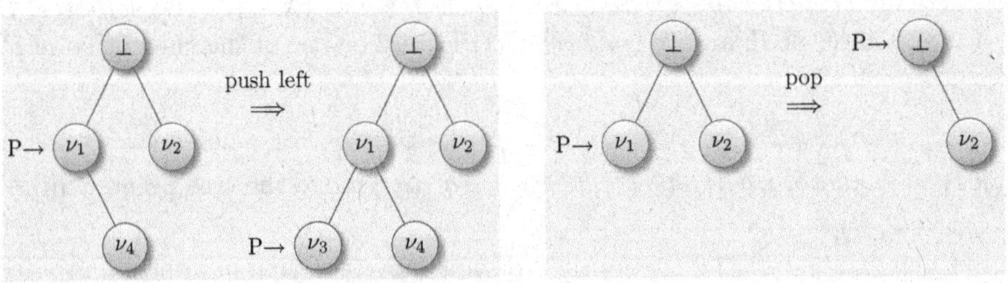

Fig. 2. Push left and pop operations.

So, a classical stack automaton can be seen as a tree-walking-storage automaton all of whose right descendants of the tree storage are not present. In accordance with stack automata, a twsDA is said to be *non-erasing* (twsDNEA) if it is not allowed to pop from the tree. A twsDNEA is *checking* (twsDCA) if it cannot push any further nodes once the treepointer has moved into the tree by moving upwards from a leaf.

A twsDCA M *halts* if the transition function is not defined for the current configuration. A word w is *accepted* if the machine halts in an accepting state after having read the input $w\lhd$ entirely, otherwise it is *rejected.* The *language accepted* by M is $L(M) = \{ w \in \Sigma^* \mid w \text{ is accepted by } M \}$.

We write DSA for deterministic one-way stack automata, DNESA for the non-erasing, and DCSA for the checking variant. The family of languages accepted by a device of type X is denoted by $\mathscr{L}(\mathsf{X})$.

In order to clarify our notion, we continue with an example.

Example 1. The language $L = \{ a^n b^{n^2} \mid n \geq 1 \} \cup \{ a^n b^{2n^2} c \mid n \geq 1 \}$ is accepted by the twsDA $M = \langle Q, \{a, b, c\}, \{\bullet, x\}, \delta, q_0, \lhd, \perp, \{q_+\} \rangle$ with state set $Q = \{q_+, q_0, q_1, q_2, p_1, p_2, p_3, p_4, s_1, r_1, r_2, \bar{p}_1\} \cup Q_1$ in linear time as follows.

In a first phase, M reads the a's from the input entirely and stores them into its storage, where the tree-walking-storage is a degenerated tree with only left descendants (Transitions 1 and 2). In this phase, the condition $n \geq 1$ is verified as well.

Let $(t_1, t_2, t_3) \in \mathsf{TYPE}$ be the type of the current node in the tree-walking-storage of M. We set:

1. $\delta(q_0, a, (-, -, -), \perp) = (q_1, \mathbf{push}(\bullet, l))$
2. $\delta(q_1, a, (l, -, -), \bullet) = (q_1, \mathbf{push}(\bullet, l))$

When the first b appears in the input, M starts a series of cycles. In the first cycle, the leaf of the degenerated tree gets a right descendant x which acts as a marker (Transition 3). Subsequently, the tree pointer is moved up to the marked node (Transition 4) and then to its ancestor (Transition 5).

3. $\delta(q_1, b, (l, t_2, -), \bullet) = (p_1, \mathbf{push}(x, r))$
4. $\delta(p_1, \lambda, (r, -, -), x) = (p_2, u)$
5. $\delta(p_2, \lambda, (l, t_2, +), \bullet) = (p_3, u)$

Next, the tree pointer is moved up towards the root whereby it reads two b's from the input for each level climbed up (Transitions 6 and 7).

6. $\delta(p_3, b, (l, t_2, -), \bullet) = (p_4, s)$
7. $\delta(p_4, b, (l, t_2, -), \bullet) = (p_3, u)$

When the tree pointer has reached the root, M has consumed $2n - 1$ symbols b. Once at the root, the tree pointer is moved down again with λ-steps until the first node marked with a right descendant appears (Transitions 8 and 9). From this node the tree pointer is moved up and M is sent into state q_1 again (Transition 10).

8. $\delta(p_3, \lambda, (-, +, -), \perp) = (s_1, d_l)$
9. $\delta(s_1, \lambda, (l, +, -), \bullet) = (s_1, d_l)$
10. $\delta(s_1, \lambda, (l, t_2, +), \bullet) = (q_1, u)$

If the current node is not the root now, a new cycle starts. That is, the current node gets a right descendant x as marker and so on. In the second cycle, M consumes $2n - 3$ symbols b. So, altogether M will read

$$2n - 1 + 2n - 3 + 2n - 5 + \cdots + 1 = n^2$$

many symbols b in these cycles. When M has passed through the last cycle, it will reach the root in state q_1 and accepts the input $a^n b^{n^2}$ if the next input symbol is the endmarker (Transition 11).

11. $\delta(q_1, \lhd, (-, t_2, -), \perp) = (q_+, s)$

If the next input symbol is a further b then M moves its tree pointer downwards with λ-steps whereby all right descendants (the markers) are removed Transitions 12 to 15). Once this process has reached the leaf of the tree-walking-storage the process of reading n^2 symbols b is repeated with new states, the first of these is $\bar{p}_1$ and the others are from the set Q_1 (Transition 16). At the end of this second process, M accepts if the last input symbol is c. We do not present the repetition of these transitions since they are straightforward.

12. $\delta(q_1, b, (-, +, -), \perp) = (r_1, d_l)$
13. $\delta(r_1, \lambda, (l, t_2, +), \bullet) = (r_2, d_r)$
14. $\delta(r_2, \lambda, (r, -, -), x) = (r_1, \texttt{pop})$
15. $\delta(r_1, \lambda, (l, +, -), \bullet) = (r_1, d_l)$
16. $\delta(r_1, \lambda, (l, -, -), \bullet) = (\bar{p}_1, \texttt{push}(x, r))$

We conclude that M accepts if and only if the input is $a^n b^{n^2}$ or $a^n b^{2n^2} c$, for some $n \geq 1$. That is, M accepts L.

For the time complexity of M, we consider the number of λ-steps. After having processed the a's from the input prefix without λ-step, M starts to run through cycles. In each cycle, it performs two λ-steps after pushing the marker and another at most $O(n)$ λ-steps to move the tree pointer downwards to the starting point of the next cycle. Since M runs through no more than $O(n)$ cycles in this phase, the number of λ-steps is bounded from above by $O(n^2)$. Preparing for and performing the second process of reading another n^2 symbols b takes another at most $O(n^2)$ λ-steps. We conclude that M takes a number of λ-steps that is in the order of the input length. So, it works in linear time. ■

3 Computational Capacity of General Deterministic Tree-Walking-Storage Automata

In this section we turn to consider the computational capacity of general deterministic tree-walking-storage automata, where *general* means that we do not impose a time limit on the machines a priori. We start at the lower end of the hierarchy and consider the weakest type of tree-walking-storage automata, that is, the checking variant. It turns out that twsDCA are no more powerful than classical checking stack automata.

Theorem 2. *A language is accepted by a twsDCA if and only if it is accepted by a DCSA.*

Proof. A DCSA is just a special case of a twsDCA. So, any language accepted by a DCSA is accepted by a twsDCA.

Conversely, since a twsDCA works in checking mode and thus cannot push any further nodes once the tree pointer has moved into the tree, it is evident that it cannot utilize the tree structure of its storage. In fact, it can never push a left *and* right descendant of some node. To this end, the twsDA had to come back to the node after pushing one of the descendants. But coming back means to move the tree pointer into the tree.

So, the storage of any twsDCA is a degenerated tree, that is, it can be seen as a linear structure as for classical DCSA.

With this observation it is straightforward to construct a DCSA that simulates a given twsDCA. □

A witness language separating the computational capacities of deterministic one-way checking stack automata and deterministic one-way non-erasing stack automata is $\{ a^{n^2} \mid n \geq 1 \}$ [9]. So, it is known that the language family $\mathscr{L}(\mathsf{DCSA}) = \mathscr{L}(\mathsf{twsDCA})$ is properly included in the family $\mathscr{L}(\mathsf{DNESA})$.

The next proposition yields a deterministic context-free language not accepted by any twsDCA. Since, for example, the non-context-free language $\{ a^n b^n a^n \mid n \geq 1 \}$ belongs to the family $\mathscr{L}(\mathsf{twsDCA})$, the family is incomparable with the family of deterministic context-free languages.

We define L_1 to be the deterministic context-free language

$$\{ a^m\#a^n\#v\#v^R\#a^{m-n} \mid m \geq n \geq 0, v \in \{a, b\}^* \}.$$

Proposition 3. *The language L_1 is not accepted by any twsDCA.*

We continue with the result that says that in the case of unlimited time deterministic tree-walking-storage automata are as powerful as Turing machines. This is in strong contrast to the classic models since it is known that all languages accepted by one-way stack automata are context sensitive [14].

Theorem 4. *Any recursively enumerable language is accepted by some twsDA.*

Theorem 4 suggests that we consider time constraints for deterministic tree-walking-storage automata. It is known that any classical DSA can effectively be transformed into an equivalent DSA that accepts in quadratic time (see [6,14] and [24] for a direct proof). Therefore, it is natural to consider deterministic tree-walking-storage automata that accept in polynomial time as well. Clearly, these devices are no longer computationally universal.

The family of languages accepted by a device of type X in polynomial time is denoted by $\mathscr{L}_{\text{poly}}(\mathsf{X})$.

4 Non-Erasing Tree-Walking-Storage Versus Non-Erasing Stack

As an intermediate goal, here we turn to separate the computational capacities of polynomial-time non-erasing tree-walking-storage automata and non-erasing stack automata. To this end, we provide a witness language L_2. In order to show that L_2 is not accepted by any DNESA we first show that it is not accepted by any twsDCA. In a second step, we reduce the problem in question to this result.

We define

$$L_2 = \{\, a^{n_1}\#a^{n_2}\#\cdots\#a^{n_k} \mid k \geq 3, n_i \geq 0, 1 \leq i \leq k, \text{ and } n_k = 2n_{k-2} \,\}.$$

Proposition 5. *The language L_2 is not accepted by any twsDCA.*

Proof. As in the proof of Proposition 3, we note that by Theorem 2, it is sufficient to show that L_2 is not accepted by any DCSA and assume in contrast to the assertion that L_2 is accepted by the DCSA $M = \langle Q, \Sigma, \Gamma, \delta, q_0, \lhd, \bot, F\rangle$.

Assume that there are some $\ell \geq 1$ and $n_1, n_2, \ldots, n_\ell \geq 0$ such that M moves the stack pointer into the stack at some step while processing the input $a^{n_1}\#a^{n_2}\#\cdots\#a^{n_\ell}$. Then, one can construct a DFA M' from M that accepts the language $\{\, a^n\#a^m\#a^{2n} \mid m, n \geq 0 \,\}$. To this end, the fixed stack storage of M after processing $a^{n_1}\#a^{n_2}\#\cdots\#a^{n_\ell}\#$ as well as the stack pointer position is maintained as part of the states of M'. Additionally, M' checks the input structure $a^*\#a^*\#a^*$. If M' processes the input $a^x\#a^y\#a^z$, it simulates M directly beginning with this stack storage. The DFA accepts if and only if the input structure is correct and the simulation ends accepting. That is, if and only if M accepts $a^{n_1}\#a^{n_2}\#\cdots\#a^{n_\ell}\#a^x\#a^y\#a^z$.

Since this input belongs to L_2 if and only if $z = 2x$, we derive that M' accepts in fact the language $\{\, a^n\#a^m\#a^{2n} \mid m, n \geq 0 \,\}$. But this language is not regular, which is a contradiction to the assumption that there are some $\ell \geq 1$ and $n_1, n_2, \ldots, n_\ell \geq 0$ such that M moves the stack pointer into the stack at some step while processing the input $a^{n_1}\#a^{n_2}\#\cdots\#a^{n_\ell}$.

If, however, M never moves the stack pointer into the stack on inputs of this form, it is essentially a DFA. Since L_2 is not regular, we obtain a contradiction also in this case. Together this implies a contradiction to the assumption that L_2 is accepted by some DCSA. □

By utilizing Proposition 5 we can show the next result.

Proposition 6. *The language L_2 is not accepted by any DNESA.*

Proof. In contrast to the assertion assume that language L_2 is accepted by the DNESA $M = \langle Q, \Sigma, \Gamma, \delta, q_0, \lhd, \bot, F\rangle$.

The first step of the proof is to show that M does not read more than $|Q|$ many consecutive symbols a without pushing. To this end, assume contrarily that, for some $\ell \geq 1$ and some $n_1, n_2, \ldots, n_\ell \geq 0$, on input $a^{n_1}\#a^{n_2}\#\cdots\#a^{n_\ell}$ automaton M reads $c > |Q|$ consecutive symbols a from the input factor a^{n_ℓ} without pushing.

The computation on the prefix has the following form, where the configuration with state q_1 is reached by a push operation and then the next c symbols a are read without push operations:

$$(q_0, a^{n_1}\#a^{n_2}\#\cdots\#a^{n_{\ell-1}}\#a^j, T_0, \lambda) \vdash^* (q_1, \lambda, T_1, P_1),$$

for some prefix a^j from a^{n_ℓ} and T_1 being a tree degenerated to a linked list, the stack, and P_1 being the leaf. Extending the input by a^i, for some $1 \leq i \leq c$, yields $(q_1, a^i, T_1, P_1) \vdash^* (q_i, \lambda, T_1, P_i)$.

Now, we consider an integer $h = |Q| + |T| + 1$, where $|T|$ denotes the number of nodes of the current stack T (the stack height). If M does not perform a push operation when processing at most h further input symbols a, then it runs into a loop and cannot accept L_2.

So, we extend the input by $\#a^h\#$ and consider the configuration that is reached just before the next push operation necessarily takes place:

$$(q_i, \#a^h\#, T_1, P_i) \vdash^* (r_i, u_i, T_1, P_1)$$

where the next step is a push operation, u_i is a suffix of $\#a^h\#$, and P_1 is the leaf of T_1. Since $1 \leq i \leq c$ and $c > |Q|$, there are at least two different values for i, say i_1 and i_2, such that $r_{i_1} = r_{i_2}$. So, we have the computations

$$(q_0, a^{n_1}\#a^{n_2}\#\cdots\#a^{n_{\ell-1}}\#a^{j+i_1}\#a^h\#, T_0, \lambda) \vdash^* (q_1, a^{i_1}\#a^h\#, T_1, P_1) \vdash^* (q_{i_1}, \#a^h\#, T_1, P_{i_1}) \vdash^* (r_{i_1}, u_{i_1}, T_1, P_1)$$

and

$$(q_0, a^{n_1}\#a^{n_2}\#\cdots\#a^{n_{\ell-1}}\#a^{j+i_2}\#a^h\#, T_0, \lambda) \vdash^* (q_1, a^{i_2}\#a^h\#, T_1, P_1) \vdash^* (q_{i_2}, \#a^h\#, T_1, P_{i_2}) \vdash^* (r_{i_2}, u_{i_2}, T_1, P_1).$$

If $u_{i_1} = u_{i_2}$, a contradiction follows since, for example, the acceptance of

$$a^{n_1}\#a^{n_2}\#\cdots\#a^{n_{\ell-1}}\#a^{j+i_1}\#a^h\#a^{2(j+i_1)} \in L_2$$

would imply the acceptance of

$$a^{n_1}\#a^{n_2}\#\cdots\#a^{n_{\ell-1}}\#a^{j+i_1}\#a^h\#a^{2(j+i_2)} \notin L_2.$$

Now, let $u_{i_1} \neq u_{i_2}$. Since u_{i_1} and u_{i_2} are suffixes of $\#a^h\#$, one of them must be the longer one, say it is u_{i_1}. So, we have $u_{i_1} = u''u_{i_2}$, for some non-empty u'', and $\#a^h\# = u'u''u_{i_2}$, for some u'.

Continuing from configuration $(r_{i_1}, u''u_{i_2}a^{2(j+i_1)}, T_1, P_1)$, automaton M will accept

$$a^{n_1}\#a^{n_2}\#\cdots\#a^{n_{\ell-1}}\#a^{j+i_1}\#a^h\#a^{2(j+i_1)} \in L_2$$

and continuing from configuration $(r_{i_2}, u_{i_2}a^{2(j+i_2)}, T_1, P_1)$, it will accept

$$a^{n_1}\#a^{n_2}\#\cdots\#a^{n_{m-1}}\#a^{j+i_2}\#a^h\#a^{2(j+i_2)} \in L_2.$$

Since $(r_{i_2}, u_{i_2}, T_1, P_1) = (r_{i_1}, u_{i_2}, T_1, P_1)$, removing the factor u'' from the input implies

$$(q_0, a^{n_1}\#a^{n_2}\#\cdots\#a^{n_{\ell-1}}\#a^{j+i_1}u'u_{i_2}a^{2(j+i_2)}, T_0, \lambda) \vdash^*$$
$$(q_1, a^{i_1}u'u_{i_2}a^{2(j+i_2)}, T_1, P_1) \vdash^* (q_{i_1}, u'u_{i_2}a^{2(j+i_2)}, T_1, P_{i_1}) \vdash^*$$
$$(r_{i_1}, u_{i_2}a^{2(j+i_2)}, T_1, P_1) = (r_{i_2}, u_{i_2}a^{2(j+i_2)}, T_1, P_1),$$

and, thus, $a^{n_1}\#a^{n_2}\#\cdots\#a^{n_{\ell-1}}\#a^{j+i_1}u'u_{i_2}a^{2(j+i_2)} \notin L_2$ is accepted, a contradiction. So, we have a contradiction to the assumption that M reads $c > |Q|$ consecutive input symbols a without pushing. In general, it may happen that M reads some # without pushing. But it cannot read more than c symbols before and more than c symbols after the #s. Otherwise the case from above applies. In total, there is a constant c_0 such that M reads at most c_0 symbols successively without pushing, in particular with the stack pointer inside the stack.

Next, M is simulated by a *checking* stack automaton M'. During its computation, M may move its stack pointer into the stack and back. During such excursions of the stack pointer, the stack storage cannot be changed, but M may read up to c_0 input symbols. The behavior of M in these excursions can entirely be described by a table that lists for every state in which M moves the stack pointer into the stack what happens. This can either be halting or moving the state pointer back to the leaf in some state. So, for every state, it is listed in which states the stack pointer may return to the leaf and which factors of the input have to be read to this end during the excursion. If M pushes when its stack pointer is at the leaf, the new table can be computed dependent on the old one and the symbol pushed. Altogether there are only finitely many such tables. At the end of the input, M' can simulate M directly by moving the stack pointer into the stack if necessary. Note, that at the end of the input M can only push a constant number of symbols without running into an infinite loop.

Finally, Proposition 5 gives us the contradiction to the assumption that L_2 is accepted by the DNESA M. □

By Proposition 5 and Proposition 6 we can now show the separation. Since, as mentioned before, it is known that any classical DSA can effectively be transformed into an equivalent DSA that accepts in quadratic time (see [6,14] and [24] for a direct proof), we conclude

$$\mathscr{L}(\mathsf{DNESA}) = \mathscr{L}_{\text{poly}}(\mathsf{DNESA}) \subseteq \mathscr{L}_{\text{poly}}(\mathsf{twsDNEA})$$

for structural reasons. So, it remains to be shown that the inclusion is proper. To this end, it is sufficient to prove that language L_2 is accepted by some twsDNEA in polynomial time.

Theorem 7. *The family of languages $\mathscr{L}$(DNESA) is properly included in the family of languages $\mathscr{L}_{\text{poly}}$(twsDNEA).*

5 A Non-Erasing Polynomial-Time Tree is Better Than Any Stack

By means of closure properties, in particular substitutions, it has been shown that the family of languages accepted by *nondeterministic* one-way non-erasing stack automata is properly included in the family of languages accepted by *nondeterministic one-way stack automata*. Moreover, the former family is closed under homomorphisms [9]. From these results, the proper inclusion

$$\mathscr{L}(\mathsf{DNESA}) \subset \mathscr{L}(\mathsf{DSA})$$

can be derived.

The situation for tree-walking-storage automata is quite different.

Theorem 8. *The families $\mathscr{L}(\mathsf{twsDNEA})$ and $\mathscr{L}(\mathsf{twsDA})$ coincide.*

The previous theorem reveals that, in fact, the computational power gained by allowing to pop from an ordinary stack is compensated by providing a tree-walking-storage.

Moreover, the simulation of a twsDA by a twsDNEA can be done in quadratic time. In particular, we have the identity

$$\mathscr{L}_{\text{poly}}(\mathsf{twsDNEA}) = \mathscr{L}_{\text{poly}}(\mathsf{twsDA}).$$

How about the relationship between $\mathscr{L}(\mathsf{DSA})$ and $\mathscr{L}_{\text{poly}}(\mathsf{twsDA})$? Since any classical DSA can effectively be transformed into an equivalent DSA that accepts in quadratic time (see [6,14] and [24] for a direct proof), we have the inclusion of $\mathscr{L}(\mathsf{DSA})$ in $\mathscr{L}_{\text{poly}}(\mathsf{twsDA})$. In fact, the inclusion turns out to be proper.

Theorem 9. *The family of languages $\mathscr{L}(\mathsf{DSA})$ is properly included in the family of languages $\mathscr{L}_{\text{poly}}(\mathsf{twsDNEA}) = \mathscr{L}_{\text{poly}}(\mathsf{twsDA})$.*

Proof. It remains to be shown that the inclusion is proper. A witness language is $L = \{\, a^n b^{n^2} \mid n \geq 1 \,\} \cup \{\, a^n b^{2n^2} c \mid n \geq 1 \,\}$ which does not belong to $\mathscr{L}(\mathsf{DSA})$ [22].

But by Example 1, language L is accepted by some twsDA in linear time. □

Finally, it is a trivial observation that polynomially time restricted stack and tree-walking-storage automata cannot be universal. That is,

$$\mathscr{L}_{\text{poly}}(\mathsf{twsDA}) \subset \mathscr{L}(\mathsf{twsDNEA}) = \mathsf{RE}.$$

References

1. Aho, A.V., Ullman, J.D.: Translations on a context-free grammar. Inform. Control **19**, 439–475 (1971)
2. Bensch, S., Björklund, J., Kutrib, M.: Deterministic stack transducers. Int. J. Found. Comput. Sci. **28**, 583–601 (2017)

3. Bojańczyk, M., Colcombet, T.: Tree-walking automata cannot be determinized. Theor. Comput. Sci. **350**, 164–173 (2006)
4. Bojańczyk, M., Colcombet, T.: Tree-walking automata do not recognize all regular languages. SIAM J. Comput. **38**, 658–701 (2008)
5. Denkinger, T.: An automata characterisation for multiple context-free languages. In: Brlek, S., Reutenauer, C. (eds.) DLT 2016. LNCS, vol. 9840, pp. 138–150. Springer, Heidelberg (2016). https://doi.org/10.1007/978-3-662-53132-7_12
6. Ginsburg, S., Greibach, S.A., Harrison, M.A.: Stack automata and compiling. J. ACM **14**, 172–201 (1967)
7. Ginsburg, S., Greibach, S.A., Harrison, M.A.: One-way stack automata. J. ACM **14**, 389–418 (1967)
8. Golubski, W., Lippe, W.: Tree-stack automata. Math. Syst. Theory **29**, 227–244 (1996)
9. Greibach, S.A.: Checking automata and one-way stack languages. J. Comput. Syst. Sci. **3**, 196–217 (1969)
10. Guessarian, I.: Pushdown tree automata. Math. Syst. Theory **16**, 237–263 (1983)
11. Gurari, E.M., Ibarra, O.H.: (Semi)alternating stack automata. Math. Syst. Theory **15**, 211–224 (1982)
12. Hopcroft, J.E., Ullman, J.D.: Nonerasing stack automata. J. Comput. Syst. Sci. **1**, 166–186 (1967)
13. Hopcroft, J.E., Ullman, J.D.: Deterministic stack automata and the quotient operator. J. Comput. Syst. Sci. **2**, 1–12 (1968)
14. Hopcroft, J.E., Ullman, J.D.: Sets accepted by one-way stack automata are context sensitive. Inform. Control **13**, 114–133 (1968)
15. Ibarra, O.H.: Characterizations of some tape and time complexity classes of Turing machines in terms of multihead and auxiliary stack automata. J. Comput. Syst. Sci. **5**, 88–117 (1971)
16. Ibarra, O.H., Jirásek, J., McQuillan, I., Prigioniero, L.: Space complexity of stack automata models. Int. J. Found. Comput. Sci. **32**, 801–823 (2021)
17. Ibarra, O.H., McQuillan, I.: Variations of checking stack automata: obtaining unexpected decidability properties. Theor. Comput. Sci. **738**, 1–12 (2018). https://doi.org/10.1016/j.tcs.2018.04.024
18. Ibarra, O.H., McQuillan, I.: Generalizations of checking stack automata: characterizations and hierarchies. Int. J. Found. Comput. Sci. **32**, 481–508 (2021)
19. Kosaraju, S.R.: 1-way stack automaton with jumps. J. Comput. Syst. Sci. **9**, 164–176 (1974)
20. Kutrib, M., Malcher, A., Wendlandt, M.: Tinput-driven pushdown, counter, and stack automata. Fund. Inform. **155**, 59–88 (2017)
21. Lange, K.: A note on the P-completeness of deterministic one-way stack language. J. UCS **16**, 795–799 (2010)
22. Ogden, W.F.: Intercalation theorems for stack languages. In: Proceedings of the First Annual ACM Symposium on Theory of Computing (STOC 1969), pp. 31–42. ACM Press, New York (1969)
23. Shamir, E., Beeri, C.: Checking stacks and context-free programmed grammars accept p-complete languages. In: Loeckx, J. (ed.) ICALP 1974. LNCS, vol. 14, pp. 27–33. Springer, Heidelberg (1974). https://doi.org/10.1007/978-3-662-21545-6_3
24. Wagner, K., Wechsung, G.: Computational Complexity. Reidel, Dordrecht (1986)

Rewriting Rules for Arithmetics in Alternate Base Systems

Zuzana Masáková[(✉)], Edita Pelantová, and Katarína Studeničová

FNSPE, Czech Technical University in Prague, Prague, Czechia
{zuzana.masakova,edita.pelantova,studekat}@fjfi.cvut.cz

Abstract. For alternate Cantor real base numeration systems we generalize the result of Frougny and Solomyak on arithmetics on the set of numbers with finite expansion. We provide a class of alternate bases which satisfy the so-called finiteness property. The proof uses rewriting rules on the language of expansions in the corresponding numeration system. The proof is constructive and provides a method for performing addition of expansions in Cantor real bases.

Keywords: Rewriting rule · Cantor base · Parry condition · Finiteness property

1 Introduction

Cantor real numeration systems represent a generalization of the – nowadays well-known – Rényi numeration systems [8]. They have been introduced independently by two groups of authors, Caalim and Demeglio [2], and Charlier and Cisternino [3]. In [3] a Cantor real numeration system is given by an infinite sequence $\boldsymbol{\mathcal{B}} = (\beta_n)_{n\geq1}$, $\beta_n > 1$, where the $\boldsymbol{\mathcal{B}}$-expansions of numbers in the interval $[0, 1)$ are found by the greedy algorithm. The authors then provide a description of the language of the $\boldsymbol{\mathcal{B}}$-expansions using a set of lexicographic inequalities. They focus on the case of purely periodic Cantor real bases $(\beta_n)_{n\geq1}$, which they call alternate bases. In analogy with the result of Bertrand-Mathis [1] for Rényi numeration systems, which are alternate bases with period 1, they define a symbolic dynamical system called $\boldsymbol{\mathcal{B}}$-shift, and show under what conditions the $\boldsymbol{\mathcal{B}}$-shift is sofic. In [4] it is proven that this happens precisely for the so-called Parry alternate bases. It is also shown under what algebraic assumptions on the base $\boldsymbol{\mathcal{B}}$ a normalization function is computable by a finite Büchi automaton.

In this paper, we focus on the arithmetical properties of Cantor numeration systems. We are in particular interested in addition of numbers whose $\boldsymbol{\mathcal{B}}$-expansions are sequences with finite support. We define the finiteness and positive finiteness property and provide some necessary and sufficient conditions on an alternate base $\boldsymbol{\mathcal{B}}$ for these properties. By that, we generalize the results

The work was supported by projects CZ.02.1.01/0.0/0.0/16_019/0000778 and SGS23/187/OHK4/3T/14.

F. Drewes and M. Volkov (Eds.): DLT 2023, LNCS 13911, pp. 195–207, 2023.
https://doi.org/10.1007/978-3-031-33264-7_16

for Rényi numeration systems obtained by Frougny and Solomyak [6]. Among other, we prove that the finiteness property of an alternate base $\boldsymbol{\mathcal{B}} = (\beta_n)_{n\geq 1}$ with period p implies that $\delta = \prod_{i=1}^{p} \beta_i$ is a Pisot or a Salem number and $\beta_i \in \mathbb{Q}(\delta)$ for $i \in \{1, 2, \ldots, p\}$, see Theorem 2. On the other hand, we provide a sufficient condition for the finiteness and positive finiteness property (Theorems 4 and 5). The latter result is proven using a system of rewriting rules on non-admissible strings. We then give a family of alternate bases with finiteness property (Theorem 3).

The rewriting system provides a method for performing addition on $\boldsymbol{\mathcal{B}}$-expansions. It is conceivable that for some classes of bases this tool will allow us to design an addition automaton, as it was described by Frougny [5] for the Rényi numeration systems with confluent Parry bases $\beta > 1$. An analogue of confluent bases in the frame of Cantor real bases is yet to be explored.

In this contribution, we announce the results, but include only certain proofs. Full version with all the demonstrations can be found in the forthcoming paper [7].

2 Preliminaries

2.1 Finiteness in Cantor Real Bases

A *Cantor real base* is a sequence $\boldsymbol{\mathcal{B}} = (\beta_n)_{n\geq 1}$ of real numbers $\beta_n > 1$ such that $\prod_{n\geq 1} \beta_n = +\infty$. A *$\boldsymbol{\mathcal{B}}$-representation* of a number $x \in [0, 1]$ is a sequence $\mathbf{x} = x_1x_2x_3\ldots$ of non-negative integers such that

$$x = \sum_{n\geq 1} \frac{x_n}{\beta_1 \cdots \beta_n} =: \operatorname{val} \mathbf{x}.$$

Note that $\operatorname{val} \mathbf{x}$ can be defined for any (not only non-negative) sequence $\mathbf{x}$ of integers. We say that the representation is *finite*, if its support $\operatorname{supp} \mathbf{x} = \{n \in \mathbb{N} : a_n \neq 0\}$ is finite. Denote

$$\mathcal{F} = \{\mathbf{x} = (x_n)_{n\geq 1} : \operatorname{supp} \mathbf{x} \text{ is finite and } x_n \in \mathbb{N} \text{ for each } n\}$$

the set of sequences of non-negative integers with finite support.

A particular $\boldsymbol{\mathcal{B}}$-representation can be obtained by the *greedy algorithm*: For a given $x \in [0, 1)$, set $x_1 = \lfloor \beta_1 x \rfloor$, $r_1 = \beta_1 x - x_1$, and for $n \geq 2$ set $x_n = \lfloor \beta_n r_{n-1} \rfloor$, $r_n = \beta_n r_{n-1} - x_n$. The obtained $\boldsymbol{\mathcal{B}}$-representation is denoted by $d_{\boldsymbol{\mathcal{B}}}(x) = x_1x_2x_3\cdots$ and called the *$\boldsymbol{\mathcal{B}}$-expansion of x*. It follows from the greedy algorithm that the digits in the $\boldsymbol{\mathcal{B}}$-expansion satisfy $0 \leq x_i < \beta_i$ and that $d_{\boldsymbol{\mathcal{B}}}(x)$ is lexicographically greatest among all $\boldsymbol{\mathcal{B}}$-representations of x. Note that the algorithm works also for $x = 1$; the sequence $d_{\boldsymbol{\mathcal{B}}}(1)$ is called the *$\boldsymbol{\mathcal{B}}$-expansion of 1* and it is lexicographically greatest among all $\boldsymbol{\mathcal{B}}$-representations of 1.

If a $\boldsymbol{\mathcal{B}}$-expansion is finite, i.e. it is of the form $w_1w_2\cdots w_k0^\omega$, where 0^ω stands for infinite repetition of zeros, we sometimes omit the suffix 0^ω. We define

$$\operatorname{Fin}(\boldsymbol{\mathcal{B}}) = \{x \in [0, 1) : d_{\boldsymbol{\mathcal{B}}}(x) \in \mathcal{F}\}. \tag{1}$$

Definition 1. *We say that a Cantor real base* $\boldsymbol{\mathcal{B}}$ *satisfies Property (PF) (the positive finiteness property), if for every* $x, y \in \mathrm{Fin}(\boldsymbol{\mathcal{B}})$ *we have*

$$x + y \in [0,1) \implies x + y \in \mathrm{Fin}(\boldsymbol{\mathcal{B}}). \tag{2}$$

We say that a Cantor real base $\boldsymbol{\mathcal{B}}$ *satisfies Property (F) (the finiteness property), if for every* $x, y \in \mathrm{Fin}(\boldsymbol{\mathcal{B}})$ *we have*

$$x+y \in [0,1) \implies x+y \in \mathrm{Fin}(\boldsymbol{\mathcal{B}}) \quad \textit{and} \quad x-y \in [0,1) \implies x-y \in \mathrm{Fin}(\boldsymbol{\mathcal{B}}). \tag{3}$$

Property (F) expresses the fact that the set of finite expansions is closed under addition and subtraction of its elements, provided the sum belong to the interval $[0,1)$. Property (PF) focuses only on addition.

For a given sequence of real numbers $\beta_n > 1$, $n \geq 1$ denote by $\boldsymbol{\mathcal{B}}^{(k)}$ the Cantor real base $\boldsymbol{\mathcal{B}}^{(k)} = (\beta_n)_{n \geq k}$. Realize the following relation between these bases.

Lemma 1. *Let* $\boldsymbol{\mathcal{B}} = \boldsymbol{\mathcal{B}}^{(1)} = (\beta_n)_{n\geq 1}$ *be a Cantor real base. If* $\boldsymbol{\mathcal{B}}^{(1)}$ *satisfies Property (PF) or Property (F), then, for every* $k \geq 1$, *the base* $\boldsymbol{\mathcal{B}}^{(k)}$ *satisfies Property (PF) or Property (F), respectively.*

Proof. Take $z \in [0,1)$. From the greedy algorithm, it is not difficult to show that

$$d_{\boldsymbol{\mathcal{B}}^{(2)}}(z) = z_2 z_3 z_4 \cdots \iff d_{\boldsymbol{\mathcal{B}}^{(1)}}(z/\beta_1) = 0 z_2 z_3 z_4 \cdots . \tag{4}$$

Consequently,

$$\frac{1}{\beta_1} \mathrm{Fin}\left(\boldsymbol{\mathcal{B}}^{(2)}\right) \subset \mathrm{Fin}\left(\boldsymbol{\mathcal{B}}^{(1)}\right).$$

Take $x, y \in \mathrm{Fin}\left(\boldsymbol{\mathcal{B}}^{(2)}\right)$ such that $x + y < 1$. Then $\frac{x}{\beta_1}, \frac{y}{\beta_1} \in \mathrm{Fin}\left(\boldsymbol{\mathcal{B}}^{(1)}\right)$, and, moreover, $z = \frac{x}{\beta_1} + \frac{y}{\beta_1} < \frac{1}{\beta_1} < 1$. By Property (PF) of the base $\boldsymbol{\mathcal{B}}^{(1)}$, necessarily $z \in \mathrm{Fin}\left(\boldsymbol{\mathcal{B}}^{(1)}\right)$. Moreover, since $z < \frac{1}{\beta_1}$, the greedy algorithm implies that the $\boldsymbol{\mathcal{B}}^{(1)}$-expansion of z/β_1 is of the form $d_{\boldsymbol{\mathcal{B}}^{(1)}}(z/\beta_1) = 0 z_2 z_3 z_4 \cdots \in \mathcal{F}$. Therefore $d_{\boldsymbol{\mathcal{B}}^{(2)}}(\beta_1 z) = z_2 z_3 z_4 \cdots \in \mathcal{F}$, which means that $x + y \in \mathrm{Fin}\left(\boldsymbol{\mathcal{B}}^{(2)}\right)$. This proves that $\boldsymbol{\mathcal{B}}^{(2)}$ satisfies Property (PF). The proof for subtraction is analogous. For $\boldsymbol{\mathcal{B}}^{(k)}$, $k \geq 3$, we proceed by induction. □

2.2 Admissibility

Let $\boldsymbol{\mathcal{B}} = (\beta_n)_{n\geq 1}$ be a Cantor real base. We say that a sequence $\mathbf{x} = (x_n)_{n\geq 1}$ of integers is *admissible* in base $\boldsymbol{\mathcal{B}}$, if there exists a real number $x \in [0,1)$ such that $\mathbf{x} = d_{\boldsymbol{\mathcal{B}}}(x)$. The admissible sequences have been characterized in [3] in terms of the quasi-greedy expansions of 1. The *quasi-greedy expansion of 1* in base $\boldsymbol{\mathcal{B}}$ is defined as

$$d^*_{\boldsymbol{\mathcal{B}}}(1) = \lim_{x \to 1-} d_{\boldsymbol{\mathcal{B}}}(x),$$

where the limit is taken over the product topology. Note that the quasi-greedy expansion of 1 is lexicographically greatest among all $\boldsymbol{\mathcal{B}}$-representations of 1 with infinitely many non-zero digits. We obviously have

$$d^*_{\boldsymbol{\mathcal{B}}}(1) \preceq_{lex} d_{\boldsymbol{\mathcal{B}}}(1),$$

with equality precisely if $d_{\boldsymbol{\mathcal{B}}}(1) \notin \mathcal{F}$.

Theorem 1. *[3, Theorem 26] Let $\boldsymbol{\mathcal{B}} = (\beta_n)_{n\geq 1}$ be a Cantor real base. A sequence of non-negative integers $z_1z_2z_3\cdots$ is admissible in base $\boldsymbol{\mathcal{B}}$ if and only if for each $n \geq 1$*

$$z_n z_{n+1} z_{n+2} \cdots \prec_{lex} d^*_{\boldsymbol{\mathcal{B}}^{(n)}}(1).$$

For recognizing admissibility of finite digit strings, we have a condition formulated using the $\boldsymbol{\mathcal{B}}$-expansions of 1 instead of the quasi-greedy expansions. It is a direct consequence of the previous theorem.

Proposition 1. *Let $\boldsymbol{\mathcal{B}} = (\beta_n)_{n\geq 1}$ be a Cantor real base. A sequence of non-negative integers $z_1z_2z_3\cdots \in \mathcal{F}$ is admissible in base $\boldsymbol{\mathcal{B}}$ if and only if for each $n \geq 1$*

$$z_n z_{n+1} z_{n+2} \cdots \prec_{lex} d_{\boldsymbol{\mathcal{B}}^{(n)}}(1).$$

Proof. The implication $\Rightarrow$ follows from Theorem 1, since

$$z_n z_{n+1} z_{n+2} \cdots \prec_{lex} d_{\boldsymbol{\mathcal{B}}^*(n)}(1) \preceq_{lex} d_{\boldsymbol{\mathcal{B}}(n)}(1).$$

For the opposite implication $\Leftarrow$, we will prove an equivalent statement:

Let $\mathbf{z} = z_1z_2z_3\cdots$ belong to $\mathcal{F}$. If $\mathbf{z}$ *is not admissible in base $\boldsymbol{\mathcal{B}}$, then there exists index $\ell \in \mathbb{N}, \ell \geq 1$, such that $d_{\boldsymbol{\mathcal{B}}^{(\ell)}}(1) \preceq z_\ell z_{\ell+1} z_{\ell+2}\cdots$.*

By Theorem 1, if $\mathbf{z} = z_1z_2z_3\cdots$ is not admissible in $\boldsymbol{\mathcal{B}}$, then there exists an index $i \geq 1$ such that $d^*_{\boldsymbol{\mathcal{B}}^{(i)}}(1) \preceq_{lex} z_i z_{i+1} z_{i+2}\cdots$. Obviously, $i \leq \max \operatorname{supp} \mathbf{z}$. Since $\mathbf{z} \in \mathcal{F}$ and $d^*_{\boldsymbol{\mathcal{B}}^{(i)}}(1) \notin \mathcal{F}$, we have a strict inequality

$$d^*_{\boldsymbol{\mathcal{B}}^{(i)}}(1) \prec_{lex} z_i z_{i+1} z_{i+2}\cdots. \tag{5}$$

If $d_{\boldsymbol{\mathcal{B}}^{(i)}}(1) \notin \mathcal{F}$, then $d^*_{\boldsymbol{\mathcal{B}}^{(i)}}(1) = d_{\boldsymbol{\mathcal{B}}^{(i)}}(1)$ and the proof is finished. Consider the other case, namely that $d_{\boldsymbol{\mathcal{B}}^{(i)}}(1) = w_1w_2\cdots w_{n-1}w_n0^\omega \in \mathcal{F}$, where $w_n \geq 1$. It has been shown in [3, Equation (4.1)] that $d^*_{\boldsymbol{\mathcal{B}}^{(i)}}(1) = w_1\cdots w_{n-1}(w_n-1)d^*_{\boldsymbol{\mathcal{B}}^{(i+n)}}(1)$. Inequality (5) implies

1. either $w_1\cdots w_{n-1}(w_n-1) \prec_{lex} z_i z_{i+1}\cdots z_{i+n-1}$,
2. or $d^*_{\boldsymbol{\mathcal{B}}^{(i+n)}}(1) \prec_{lex} z_{i+n} z_{i+n+1}\cdots$.

Case 1. The lexicographically smallest word of length n, strictly greater than $w_1\cdots w_{n-1}(w_n-1)$ is $w_1\cdots w_{n-1}w_n$. Hence we have $w_1\cdots w_{n-1}w_n \preceq_{lex} z_i z_{i+1}\cdots z_{i+n-1}$. Therefore,

$$d_{\boldsymbol{\mathcal{B}}^{(i)}}(1) \preceq_{lex} z_i z_{i+1}\cdots z_{i+n-1}0^\omega \preceq_{lex} z_i z_{i+1} z_{i+2}\cdots,$$

as we wanted to show.

Case 2. We have found $i_{new} = i + n > i$ so that (5) is satisfied when substituting i_{new} in place of i. The same idea can be repeated. As $\operatorname{supp} \mathbf{z}$ is bounded, finitely many steps lead to Case 1. □

2.3 Alternate Bases

In case that the Cantor real base $\boldsymbol{\mathcal{B}} = (\beta_n)_{n\geq 1}$ is a purely periodic sequence with period p, we call it an *alternate base* and denote

$$\boldsymbol{\mathcal{B}} = (\beta_1, \beta_2, \ldots, \beta_p), \qquad \delta = \prod_{i=1}^{p} \beta_i.$$

For the Cantor real bases $\boldsymbol{\mathcal{B}}^{(n)} = (\beta_n, \beta_{n+1}, \ldots, \beta_{n+p-1})$, we have $\boldsymbol{\mathcal{B}}^{(n)} = \boldsymbol{\mathcal{B}}^{(n+p)}$ for every n. The indices will therefore be considered mod p. Let us stress that throughout this paper the representatives of the congruence classes will be taken in the set $\mathbb{Z}_p = \{1, 2, \ldots, p\}$.
For the Cantor real base $\boldsymbol{\mathcal{B}}^{(n)}$, $n \geq 1$, denote the $\boldsymbol{\mathcal{B}}^{(n)}$-expansions of 1 by

$$\mathbf{t}^{(n)} = t_1^{(n)} t_2^{(n)} t_3^{(n)} \cdots = d_{\boldsymbol{\mathcal{B}}^{(n)}}(1). \tag{6}$$

Note that $\mathbf{t}^{(n)} = \mathbf{t}^{(n+p)}$ for each $n \in \mathbb{N}$. In fact, we only work with $\mathbf{t}^{(\ell)}$, $\ell \in \mathbb{Z}_p$.

The specific case when $p = 1$ yields the well-known numeration systems in one real base $\beta > 1$ as defined by Rényi [8]. These systems define a sofic β-shift precisely when the base β is a Parry number. In analogy to this case we say that an alternate base $\boldsymbol{\mathcal{B}}$ with period $p \geq 1$ is a *Parry alternate base*, if all the sequences $\mathbf{t}^{(\ell)}$, $\ell \in \mathbb{Z}_p$, are eventually periodic. If, moreover, $\mathbf{t}^{(\ell)} \in \mathcal{F}$ for all $\ell \in \mathbb{Z}_p$, we speak about *simple Parry alternate base*.

Note that (4) for an alternate base $\boldsymbol{\mathcal{B}}$ implies

$$d_{\boldsymbol{\mathcal{B}}}(z) = z_1 z_2 z_3 \cdots \iff d_{\boldsymbol{\mathcal{B}}}(z/\delta) = 0^p z_1 z_2 z_2 \cdots \tag{7}$$

and hence $\frac{1}{\delta}\,\mathrm{Fin}(\boldsymbol{\mathcal{B}}) \subset \mathrm{Fin}(\boldsymbol{\mathcal{B}})$.

3 Main Results

We focus on the finiteness property of alternate bases. We first provide necessary conditions on an alternate base in order to satisfy Property (PF) and Property (F). For that, we need to recall some number theoretical notions. An algebraic integer $\delta > 1$ is a *Pisot number*, if all its conjugates other than δ belong to the interior of the unit disc; an algebraic integer $\delta > 1$ is a *Salem number*, if all its conjugates other than δ belong to the unit disc and at least one lies on its boundary. If δ is an algebraic number of degree d, then the minimal subfield of $\mathbb{C}$ containing δ is of the form

$$\mathbb{Q}(\delta) = \left\{c_0 + c_1\delta + \cdots + c_{d-1}\delta^{d-1} \,:\, c_i \in \mathbb{Q}\right\}.$$

The number field $\mathbb{Q}(\delta)$ has precisely d embeddings into $\mathbb{C}$, namely the field isomorphisms $\psi : \mathbb{Q}(\delta) \to \mathbb{Q}(\gamma)$ induced by $\psi(\delta) = \gamma$, where γ is a conjugate of δ.

The necessary condition is formulated as Theorem 2. Its proof follows the same ideas as are given in [6] for the case of Rényi numeration systems, although it needs some rather technical considerations. It can be found in the full version of this paper [7].

Theorem 2. *Let* $\boldsymbol{\mathcal{B}} = (\beta_1, \beta_2, \ldots, \beta_p)$ *be an alternate base.*

1. *If* $\boldsymbol{\mathcal{B}}$ *satisfies Property (PF), then* $\delta = \prod_{i=1}^{p} \beta_i$ *is a Pisot or a Salem number and* $\beta_i \in \mathbb{Q}(\delta)$ *for every* $i \in \mathbb{Z}_p$.
2. *If, moreover,* $\boldsymbol{\mathcal{B}}$ *satisfies Property (F), then* $\boldsymbol{\mathcal{B}}$ *is a simple Parry alternate base, and for any non-identical embedding* ψ *of* $\mathbb{Q}(\delta)$ *into* $\mathbb{C}$ *the vector* $(\psi(\beta_1), \ldots, \psi(\beta_p))$ *is not positive.*

As the main result of this paper we provide a class of alternate Parry bases satisfying Property (F).

Theorem 3. *Let* $\boldsymbol{\mathcal{B}}$ *be an alternate base with period p such that the corresponding expansions* $\mathbf{t}^{(\ell)}$ *of 1 satisfy*

$$t_1^{(\ell)} \geq t_2^{(\ell-1)} \geq t_3^{(\ell-2)} \geq \cdots \quad \textit{for every } \ell \in \mathbb{Z}_p. \tag{8}$$

Then $\boldsymbol{\mathcal{B}}$ *is a Parry alternate base and has Property (PF).*

Moreover, if $\boldsymbol{\mathcal{B}}$ *is a simple Parry alternate base, then it has Property (F).*

Note that in inequalities (8) we have upper indices in $\mathbb{Z}$, but they are always counted mod p. Condition (8) is a generalization of the class given for $p = 1$ in [6], where Frougny and Solomyak demonstrate finiteness property for the bases $\beta > 1$, which are Pisot roots of the polynomials $x^d - t_1 x^{d-1} - \cdots - t_{d-1}x - t_d$ satisfying the following condition

$$t_1 \geq t_2 \geq \cdots \geq t_{d-1} \geq t_d, \quad t_i \in \mathbb{Z}.$$

Theorem 3 is a consequence of a sufficient condition for Properties (PF) and (F) which we present and prove in Sect. 4. The sufficient condition is formulated in terms of rewriting rules of non-admissible sequences (Theorem 4). At the end of this contribution we explain the application of the sufficient condition for the demonstration of Theorem 3. The full proof is non-trivial; we present it in [7].

Let us illustrate all the above results.

Example 1. Consider the favourite example of [3], namely the alternate base $\boldsymbol{\mathcal{B}} = (\beta_1, \beta_2) = \left(\frac{1+\sqrt{13}}{2}, \frac{5+\sqrt{13}}{6}\right)$. Expansions of 1 are of the form

$$d_{\boldsymbol{\mathcal{B}}^{(1)}}(1) = t_1^{(1)} t_2^{(1)} t_3^{(1)} 0^\omega = 2010^\omega \quad \text{and} \quad d_{\boldsymbol{\mathcal{B}}^{(2)}}(1) = t_1^{(2)} t_2^{(2)} 0^\omega = 110^\omega.$$

Consequently, $\boldsymbol{\mathcal{B}}$ is a simple Parry alternate base. The inequalities (8) for $p = 2$ are of the form

$$t_1^{(1)} \geq t_2^{(2)} \geq t_3^{(1)} \geq \cdots, \qquad \text{in our case} \qquad 2 \geq 1 \geq 1 \geq 0 \geq \cdots,$$
$$t_1^{(2)} \geq t_2^{(1)} \geq t_3^{(2)} \geq \cdots, \qquad \text{in our case} \qquad 1 \geq 0 \geq 0 \geq 0 \geq \cdots.$$

Therefore, according to Theorem 3, the base $\boldsymbol{\mathcal{B}}$ has Property (F).

Note that the necessary conditions of Property (F) as presented in Theorem 2 are satisfied. Indeed, we have $\delta = \frac{1+\sqrt{13}}{2} \cdot \frac{5+\sqrt{13}}{6} = \frac{3+\sqrt{13}}{2}$, which is the positive root of $x^2 - 3x - 1$. The other root of this polynomial is $\gamma = \frac{3-\sqrt{13}}{2} \approx -0.303$, thus δ is a Pisot number.

Obviously, $\beta_1, \beta_2 \in \mathbb{Q}(\delta) = \mathbb{Q}\left(\sqrt{13}\right)$. The only non-identical embedding ψ of the field $\mathbb{Q}\left(\sqrt{13}\right)$ into $\mathbb{C}$ is the Galois automorphism induced by $\sqrt{13} \mapsto -\sqrt{13}$. Thus we have

$$\psi(\beta_1) = \frac{1-\sqrt{13}}{2} \approx -1.303, \qquad \psi(\beta_2) = \frac{5-\sqrt{13}}{6} \approx 0.232,$$

therefore indeed $(\psi(\beta_1), \psi(\beta_2))$ is not a positive vector.

4 Sufficient Condition

In this section we state and prove a sufficient condition for an alternate base to satisfy Property (PF). It can be reformulated as stated in the following lemma.

Lemma 2. *A Cantor real base $\boldsymbol{\mathcal{B}}$ satisfies Property (PF) if and only if for any string $\mathbf{z} \in \mathcal{F}$ such that $\operatorname{val}(\mathbf{z}) < 1$ we have $\operatorname{val}(\mathbf{z}) \in \operatorname{Fin}(\boldsymbol{\mathcal{B}})$, i.e. $d_{\boldsymbol{\mathcal{B}}}(\operatorname{val}\mathbf{z}) \in \mathcal{F}$.*

In other words, for each string $\mathbf{z} \in \mathcal{F}$ with $\operatorname{val}(\mathbf{z}) < 1$, we will find a string $\mathbf{x} \in \mathcal{F}$ admissible in base $\boldsymbol{\mathcal{B}}$ such that $\operatorname{val}\mathbf{x} = \operatorname{val}\mathbf{z}$. We will use the fact that admissible strings are lexicographically the greatest among all strings representing the same value.

Definition 2. *Let $\boldsymbol{\mathcal{B}} = (\beta_1, \beta_2, \ldots, \beta_p)$ be an alternate base and denote the corresponding $\boldsymbol{\mathcal{B}}^{(\ell)}$-expansions of 1 by $\mathbf{t}^{(\ell)}$, as in (6). Denote the following set of strings with finite support*

$$\mathcal{S} = \left\{0^{p+\ell-1} t_1^{(\ell)} t_2^{(\ell)} \cdots t_{k-1}^{(\ell)} \left(t_k^{(\ell)}+1\right) 0^\omega : \ell \in \mathbb{Z}_p, k \geq 1\right\} \cup \\ \cup \left\{0^{p+\ell-1}\mathbf{t}^{(\ell)} \in \mathcal{F} : \ell \in \mathbb{Z}_p\right\}.$$

We say that the alternate base $\boldsymbol{\mathcal{B}}$ has the rewriting property, if for any string $\mathbf{a} \in \mathcal{S}$ there exists a string $T(\mathbf{a})$, such that

$$T(\mathbf{a}) \in \mathcal{F}, \quad \operatorname{val} T(\mathbf{a}) = \operatorname{val} \mathbf{a}, \quad \textit{and} \quad T(\mathbf{a}) \succ_{lex} \mathbf{a}. \tag{9}$$

The following lemma shows that given an alternate base $\boldsymbol{\mathcal{B}}$ with rewriting property, a non-admissible $\mathbf{z} \in \mathcal{F}$ can always be replaced by a lexicographically greater string in $\mathcal{F}$ of the same value. For that, let us introduce the notation for digit-wise addition of strings. Let $\mathbf{x} = x_1x_2x_3\cdots$ and $\mathbf{y} = y_1y_2y_3\cdots$ be two sequences with $x_n, y_n \in \mathbb{Z}$ for every $n \in \mathbb{N}, n \geq 1$. Then $\mathbf{x} \oplus \mathbf{y}$ stands for the sequence $z_1z_2z_3\cdots$, where $z_n = x_n + y_n$ for every $n \in \mathbb{N}, n \geq 1$. Obviously, in this notation, we have

- $\operatorname{val}(\mathbf{x} \oplus \mathbf{y}) = \operatorname{val}\mathbf{x} + \operatorname{val}\mathbf{y}$;
- if $\mathbf{x} \prec_{lex} \tilde{\mathbf{x}}$, then for every $\mathbf{y}$ the inequality $\mathbf{x} \oplus \mathbf{y} \prec_{lex} \tilde{\mathbf{x}} \oplus \mathbf{y}$ holds true.

Lemma 3. *Let $\boldsymbol{\mathcal{B}}$ be an alternate base with rewriting property. Then for every $\mathbf{z} \in \mathcal{F}$ non-admissible in $\boldsymbol{\mathcal{B}}$ with value $\operatorname{val}\mathbf{z} < \frac{1}{\delta}$ there exist $\mathbf{x}, \mathbf{y} \in \mathcal{F}$ and $j \in \mathbb{N}$ such that $\mathbf{z} = (0^{pj}\mathbf{x}) \oplus \mathbf{y}$ and $\mathbf{x} \in \mathcal{S}$. Consequently, $(0^{pj}T(\mathbf{x})) \oplus \mathbf{y} \in \mathcal{F}$ is a $\boldsymbol{\mathcal{B}}$-representation of $\operatorname{val}\mathbf{z}$ lexicographically strictly greater than $\mathbf{z}$.*

Proof. Consider $\mathbf{z} = z_1 z_2 \cdots \in \mathcal{F}$ non-admissible in base $\boldsymbol{\mathcal{B}}$. Since $\operatorname{val}\mathbf{z} < \frac{1}{\delta}$, necessarily $z_1 = z_2 = \cdots = z_p = 0$. By Proposition 1, there exists $i \in \mathbb{N}$, $i \geq p+1$, such that $z_i z_{i+1} \cdots \succeq_{lex} d_{\boldsymbol{\mathcal{B}}^{(i)}}(1)$. We distinguish two cases how to define the string $\mathbf{x}$:

a) Suppose that $z_i z_{i+1} \cdots = d_{\boldsymbol{\mathcal{B}}^{(i)}}(1)$.
As $\mathbf{z} \in \mathcal{F}$, we have that $d_{\boldsymbol{\mathcal{B}}^{(i)}}(1) \in \mathcal{F}$. We can take $\mathbf{x} \in \mathcal{S}$ and $j \in \mathbb{N}$ such that $0^{pj}\mathbf{x} = 0^{i-1} d_{\boldsymbol{\mathcal{B}}^{(i)}}(1)$.
b) Suppose that $z_i z_{i+1} \cdots \succ_{lex} d_{\boldsymbol{\mathcal{B}}^{(i)}}(1)$.
Choose $n \in \mathbb{N}$ minimal such that $z_i z_{i+1} \cdots z_{i+n} \succ_{lex} d_{\boldsymbol{\mathcal{B}}^{(i)}}(1)$. Then necessarily $z_i z_{i+1} \cdots z_{i+n-1} = t_1^{(i)} t_2^{(i)} \cdots t_n^{(i)}$ and $z_{i+n} > t_{n+1}^{(i)}$. In this case we can define $\mathbf{x} \in \mathcal{S}$ and $j \in \mathbb{N}$ so that

$$0^{pj}\mathbf{x} = 0^{i-1} t_1^{(i)} t_2^{(i)} \cdots t_n^{(i)} \left(t_{n+1}^{(i)} + 1\right) 0^{\omega}.$$

It is obvious that the choice of $\mathbf{x}$ allows us to find the string $\mathbf{y}$ with non-negative digits such that $(0^{pj}\mathbf{x}) \oplus \mathbf{y} = \mathbf{z}$. Now replacing $0^{pj}\mathbf{x}$ by $0^{pj}T(\mathbf{x})$ yields $(0^{pj}T(\mathbf{x})) \oplus \mathbf{y} \succ_{lex} (0^{pj}\mathbf{x}) \oplus \mathbf{y} = \mathbf{z}$. □

In rewriting the non-admissible string $\mathbf{z} \in \mathcal{F}$, we need to ensure that after finitely many steps, the procedure yields the lexicographically maximal string, i.e. the $\boldsymbol{\mathcal{B}}$-expansion of $\operatorname{val}\mathbf{z}$.

Definition 3. *We say that the base $\boldsymbol{\mathcal{B}}$ has the weight property, if it has the rewriting property and there exist positive integers w_n, $n \geq 1$, such that the weight function $g : \mathcal{F} \to \mathbb{N}$, $g(\mathbf{a}) = \sum_{n \geq 1} w_n a_n$, satisfies*

1. *$g(0^p\mathbf{a}) = g(\mathbf{a})$ for any $\mathbf{a} \in \mathcal{F}$;*
2. *$g(\mathbf{a}) \geq g(T(\mathbf{a}))$ for any $\mathbf{a} \in \mathcal{S}$.*

The weight property is sufficient to guarantee the positive finiteness property. We will use the obvious fact that the weight function satisfies $g(\mathbf{x} \oplus \mathbf{y}) = g(\mathbf{x}) + g(\mathbf{y})$ for any digit strings $\mathbf{x}, \mathbf{y} \in \mathcal{F}$.

Theorem 4. *Let $\boldsymbol{\mathcal{B}} = (\beta_1, \beta_2, \ldots, \beta_p)$ be an alternate base satisfying the weight property. Then $\boldsymbol{\mathcal{B}}$ has Property (PF).*

Lemma 4. *Let $g : \mathcal{F} \mapsto \mathbb{N}$ be a weight function and let $\left(\mathbf{a}^{(k)}\right)_{k \geq 1}$ be a sequence of infinite strings satisfying for every $k \geq 1$*

- *$\mathbf{a}^{(k)} \in \mathcal{F}$;*

- $\mathbf{a}^{(k)} \prec_{lex} \mathbf{a}^{(k+1)}$.

Then the integer sequence $\big(g(\mathbf{a}^{(k)})\big)_{k\geq 1}$ *is not bounded.*

Proof. Let $w_1, w_2, \ldots$ be positive integers and $g(\mathbf{a}) = \sum_{n\geq 1} w_n a_n$ for every sequence $\mathbf{a} = (a_n)_{n\geq 1} \in \mathcal{F}$.

We proceed by contradiction. Assume that there exists $H \in \mathbb{N}$ such that $g(\mathbf{a}^{(k)}) \leq H$ for every $k \geq 1$. Since all coefficients w_n are positive integers, we derive

$$a_n^{(k)} \in \{0, 1, \ldots, H\} \text{ and } \#\operatorname{supp}\mathbf{a}^{(k)} \leq H \quad \text{for all } n, k \geq 1. \tag{10}$$

The set $\{0, 1, \ldots, H\}^{\mathbb{N}\setminus\{0\}}$ equipped with the product topology is a compact space and thus the increasing sequence $\big(\mathbf{a}^{(k)}\big)_{k\geq 1}$ has a limit, say $\mathbf{b} = \lim_{k\to+\infty} \mathbf{a}^{(k)}$. Obviously, $\mathbf{a}^{(k)} \preceq_{lex} \mathbf{b}$ for each $k \geq 1$.

Let us at first show that $\operatorname{supp}\mathbf{b}$ is infinite. Suppose the contrary, i.e. that $\mathbf{b} = b_1 b_2 \cdots b_N 0^\omega$ for some $N \geq 1$. Since $\mathbf{b} = \lim_{k\to+\infty} \mathbf{a}^{(k)}$, there exists $k_0 \geq 1$ such that $\mathbf{a}^{(k)}$ has a prefix $b_1 b_2 \cdots b_N$ for each $k > k_0$. The inequality $\mathbf{a}^{(k)} \preceq_{lex} \mathbf{b} = b_1 b_2 \cdots b_N 0^\omega$ implies $\mathbf{a}^{(k)} = \mathbf{b}$ for all $k > k_0$, and that is a contradiction with the fact that $\big(\mathbf{a}^{(k)}\big)_{k\geq 1}$ is strictly increasing.

Since $\operatorname{supp}\mathbf{b}$ is infinite, we can choose $M \geq 1$ such that the set

$$S_M := \{n \geq 1 : b_n \neq 0 \text{ and } n \leq M\} \subset \operatorname{supp}\mathbf{b}$$

has cardinality $\#S_M > H$. Since for all sufficiently large k the string $\mathbf{a}^{(k)}$ has prefix $b_1 b_2 \cdots b_M$, it has to be $S_M \subset \operatorname{supp}\mathbf{a}^{(k)}$, and that is a contradiction with (10). □

Proof (Proof of Theorem 4). By Lemma 2, it is sufficient to show that for $\mathbf{z} \in \mathcal{F}$ such that $z = \operatorname{val}\mathbf{z} < 1$, the $\boldsymbol{\mathcal{B}}$-expansion of $z = \operatorname{val}\mathbf{z}$ belongs to $\mathcal{F}$ as well. We distinguish two cases:

a) Suppose that $\operatorname{val}\mathbf{z} < \frac{1}{\delta}$.
If $\mathbf{z}$ is admissible in base $\boldsymbol{\mathcal{B}}$, there is nothing left to discuss. On the other hand, if $\mathbf{z}$ is not admissible in base $\boldsymbol{\mathcal{B}}$, then Lemma 3 allows us to find a lexicographically greater $\boldsymbol{\mathcal{B}}$-representation of z which also has a finite support. Let us show that after finitely many applications of Lemma 3 we get the $\boldsymbol{\mathcal{B}}$-expansion of z.
Let us proceed by contradiction. Assume that applying Lemma 3 repeatedly yields an infinite sequence of strings $\mathbf{z} = \mathbf{z}^{(0)}, \mathbf{z}^{(1)}, \mathbf{z}^{(2)}, \ldots$ such that for all $n \in \mathbb{N}$

$$\operatorname{val}\mathbf{z}^{(n)} = z, \quad \mathbf{z}^{(n)} \text{ is not admissible in} \boldsymbol{\mathcal{B}}, \quad \text{and} \quad \mathbf{z}^{(n)} \prec_{lex} \mathbf{z}^{(n+1)}.$$

Let g be a weight function satisfying properties 1 and 2 from Definition 3. To obtain $\mathbf{z}^{(n+1)}$ we use Lemma 3, i.e., if $\mathbf{z}^{(n)} = \big(0^{pj}\mathbf{x}\big) \oplus \mathbf{y}$, then $\mathbf{z}^{(n+1)} = \big(0^{pj}T(\mathbf{x})\big) \oplus \mathbf{y}$.
Properties of g guarantee that $g\big(\mathbf{z}^{(n+1)}\big) \leq g\big(\mathbf{z}^{(n)}\big)$ for all $n \in \mathbb{N}$. In particular, the values $g\big(\mathbf{z}^{(n)}\big) \in \mathbb{N}$ are bounded by $g\big(\mathbf{z}^{(0)}\big) = g(\mathbf{z})$, and that is a contradiction with Lemma 4.

b) Suppose that $\frac{1}{\delta} \le \operatorname{val} \mathbf{z} < 1$.
Obviously, the string $\mathbf{z}' := 0^p\mathbf{z}$ belongs to $\mathcal{F}$ and $z' = \operatorname{val}\mathbf{z}' = \frac{1}{\delta} z < \frac{1}{\delta}$. By a), the $\boldsymbol{\mathcal{B}}$-expansion of z' has finite support, and must be of the form $0^p z_{p+1}z_{p+2}\cdots \in \mathcal{F}$. Relation (7) yields $d_{\boldsymbol{\mathcal{B}}}(z) = z_{p+1}z_{p+2}\cdots \in \mathcal{F}$. □

The above theorem gives us a sufficient condition for Property (PF). As follows from Item 2 of Theorem 2, when searching for bases with Property (F), it suffices to limit our considerations to simple Parry alternate bases. The following proposition shows that for such bases Properties (F) and (PF) are equivalent.

Proposition 2. *Let $\boldsymbol{\mathcal{B}} = (\beta_1, \beta_2, \ldots, \beta_p)$ be a simple Parry alternate base. Then $\boldsymbol{\mathcal{B}}$ has (F) if and only if $\boldsymbol{\mathcal{B}}$ has (PF).*

We first prove an auxiliary statement.

Lemma 5. *Let $\boldsymbol{\mathcal{B}}$ be a simple Parry alternate base. For all $j, k \in \mathbb{N}$, $j \ge 1$, there exists a $\boldsymbol{\mathcal{B}}$-representation of the number $\operatorname{val}\left(0^{j-1}10^\omega\right)$ of the form*

$$0^{j-1}u_ju_{j+1}u_{j+2}\cdots \in \mathcal{F}$$

such that $u_{j+k} \ge 1$.

Proof. We proceed by induction on k. For $k = 0$, the statement is obvious. Let $k \ge 0$. By induction hypothesis, there exists a string $\mathbf{u} = 0^{j-1}u_ju_{j+1}u_{j+2}\cdots \in \mathcal{F}$, such that $\operatorname{val}(0^{j-1}10^\omega) = \operatorname{val}\mathbf{u}$ and $u_{j+k} \ge 1$. From the definition of $\mathbf{t}^{(j+k+1)}$, we derive that the string $0^{j+k-1}(-1)\mathbf{t}^{(j+k+1)}$ of integer digits has value 0, i.e. $\operatorname{val}\left(0^{j+k-1}(-1)\mathbf{t}^{(j+k+1)}\right) = 0$. The string $\mathbf{u} \oplus 0^{j+k-1}(-1)\mathbf{t}^{(j+k+1)}$ has all digits non-negative and has finite support. Its digit at position $j + k + 1$ is equal to $u_{j+k+1} + t_1^{(j+k+1)} \ge 1$. □

Proof (Proof of Proposition 2). Clearly, if $\boldsymbol{\mathcal{B}}$ satisfies (F), then it satisfies (PF). For the opposite implication, assume that a simple Parry alternate base $\boldsymbol{\mathcal{B}}$ has Property (PF). Therefore, according to Lemma 2, if a number $z \in [0, 1)$ has a $\boldsymbol{\mathcal{B}}$-representation in $\mathcal{F}$, then the $\boldsymbol{\mathcal{B}}$-expansion of z belongs to $\mathcal{F}$ as well.

In order to show Property (F), it is thus sufficient to verify that for any $x, y \in \operatorname{Fin}(\boldsymbol{\mathcal{B}}) \cap [0, 1)$ such that $z := x - y \in [0, 1)$ we can find a $\boldsymbol{\mathcal{B}}$-representation of z with finite support. Let $\mathbf{x}, \mathbf{y} \in \mathcal{F}$ be the $\boldsymbol{\mathcal{B}}$-expansions of numbers $x, y \in [0, 1)$ such that $x = \operatorname{val}\mathbf{x} > y = \operatorname{val}\mathbf{y} > 0$.

We proceed by induction on the sum of digits in $\mathbf{y}$. Since $\operatorname{val}\mathbf{x} > \operatorname{val}\mathbf{y}$, there exist $j, k \in \mathbb{N}$, $j \ge 1$, such that $\mathbf{x} = \mathbf{x}' \oplus \mathbf{x}''$ and $\mathbf{y} = \mathbf{y}' \oplus \mathbf{y}''$, where $\mathbf{x}'' = 0^{j-1}10^\omega$ and $\mathbf{y}'' = 0^{j+k-1}10^\omega$. It follows directly from Lemma 5 that $\operatorname{val}\mathbf{x}'' - \operatorname{val}\mathbf{y}''$ has a finite $\boldsymbol{\mathcal{B}}$-representation, say $\mathbf{z}'' \in \mathcal{F}$. Hence

$$z = \operatorname{val}\mathbf{x} - \operatorname{val}\mathbf{y} = \operatorname{val}\mathbf{x}' + \operatorname{val}\mathbf{z}'' - \operatorname{val}\mathbf{y}'.$$

Property (PF) guarantees that $\operatorname{val}\mathbf{x}' + \operatorname{val}\mathbf{z}''$ has a finite $\boldsymbol{\mathcal{B}}$-expansion, say $\mathbf{x}_{new}$. Thus

$$z = \operatorname{val}\mathbf{x} - \operatorname{val}\mathbf{y} = \operatorname{val}\mathbf{x}_{new} - \operatorname{val}\mathbf{y}'.$$

The sum of digits in $\mathbf{y}'$ is smaller by 1 than the sum of digits in $\mathbf{y}$. Induction hypothesis implies that z has a finite $\boldsymbol{\mathcal{B}}$-representation, and, by Property (PF), also a finite $\boldsymbol{\mathcal{B}}$-expansion. □

As a consequence of Proposition 2, we formulate a sufficient condition for Property (F).

Theorem 5. *Let $\boldsymbol{\mathcal{B}} = (\beta_1, \beta_2, \ldots, \beta_p)$ be a simple Parry alternate base satisfying the weight property. Then $\boldsymbol{\mathcal{B}}$ has Property (F).*

Theorems 4 and 5 provide a sufficient condition for positive finiteness and finiteness property of alternate bases. In order to prove Theorem 3, it is sufficient to demonstrate that alternate bases satisfying (8) verify the assumption for the sufficient condition, namely the weight property. For that we need to find rewriting rules for each string in the set $\mathcal{S}$ from Definition 2. The rules are of two types:

Type 1: For $\ell, i \in \mathbb{Z}_p$ and $k \in \mathbb{N}$ set

$$\mathbf{x}_{\ell,i,k} = 0^{p+\ell-1} t_1^{(\ell)} t_2^{(\ell)} \cdots t_{pk+i-1}^{(\ell)} \left(t_{pk+i}^{(\ell)} + 1\right) 0^\omega,$$

$$T(\mathbf{x}_{\ell,i,k}) = 0^{p+\ell-2} 1 0^{pk+i} \left(t_1^{(\ell+i)} - t_{pk+i+1}^{(\ell)}\right) \left(t_2^{(\ell+i)} - t_{pk+i+2}^{(\ell)}\right) \cdots .$$

Type 2: For $\ell \in \mathbb{Z}_p$ set

$$\mathbf{x}_\ell = 0^{p+\ell-1} \mathbf{t}^{(\ell)}, \qquad T(\mathbf{x}_\ell) = 0^{p+\ell-2} 1 0^\omega.$$

The full proof that these rewriting rules are suitable for the rewriting property of the base $\boldsymbol{\mathcal{B}}$, and that they are suitable for defining a weight function with desired features can be found in [7].

Example 2. Let us present rewriting rules for the base as in Example 1. First realize that for an arbitrary alternate base we have the following implication: If $d_{\boldsymbol{\mathcal{B}}^{(\ell)}}(1) \in \mathcal{F}$, say $d_{\boldsymbol{\mathcal{B}}^{(\ell)}}(1) = t_1^{(\ell)} \cdots t_n^{(\ell)} 0^\omega$, then $\mathbf{x}_{\ell,i,k} = \mathbf{x}_\ell \oplus \left(0^{p(k+1)+\ell+i-2} 1 0^\omega\right)$ for ℓ, i, k such that $pk + i \geq n$. In other words, for the string $\mathbf{x}$ in Lemma 3 we can take $\mathbf{x}_\ell$ instead of $\mathbf{x}_{\ell,i,k}$. As a consequence, in such case we do not need to use the rewriting rule for the string $\mathbf{x}_{\ell,i,k}$.

Since for the base as in Example 1 the expansions of 1 are $d_{\boldsymbol{\mathcal{B}}}(1) = 2010^\omega$ and $d_{\boldsymbol{\mathcal{B}}^{(2)}}(1) = 110^\omega$, the necessary rewriting rules are the following.

$\mathbf{x}_{1,1,0}$	$\mathbf{x}_{1,2,0}$	$\mathbf{x}_1$	$\mathbf{x}_{2,1,0}$	$\mathbf{x}_2$
0030^ω	00210^ω	002010^ω	00020^ω	000110^ω
$T(\mathbf{x}_{1,1,0})$	$T(\mathbf{x}_{1,2,0})$	$T(\mathbf{x}_1)$	$T(\mathbf{x}_{2,1,0})$	$T(\mathbf{x}_2)$
01010^ω	01001010^ω	010^ω	00101010^ω	0010^ω

Let us also demonstrate an algorithm for addition using rewriting rules. Consider for example numbers $x, y \in [0, 1)$ with $d_{\boldsymbol{\mathcal{B}}}(x) = 000020010^\omega$ and

$d_{\boldsymbol{\mathcal{B}}}(y) = 000020^\omega$. Denote $z = x+y$. The digit-wise sum of the two $\boldsymbol{\mathcal{B}}$-expansions gives a non-admissible representation of z, namely $d_{\boldsymbol{\mathcal{B}}}(x) \oplus d_{\boldsymbol{\mathcal{B}}}(y) = 000040010^\omega$. Using rewriting rules for strings in the set $\mathcal{S}$, as shown in the above table, we can write

$$\underbrace{000040010^\omega}_{non-admissible} = (000010010^\omega) \oplus (\underbrace{000030^\omega}_{00\mathbf{x}_{1,1,0}})$$
$$\downarrow$$
$$(000010010^\omega) \oplus (\underbrace{0001010^\omega}_{00T(\mathbf{x}_{1,1,0})}) = \underbrace{000111010^\omega}_{non-admissible}.$$

In this way we have obtained another representation of the same number z. Since the latter is still non-admissible, we continue by rewriting

$$\underbrace{000111010^\omega}_{non-admissible} = (000001010^\omega) \oplus (\underbrace{000110^\omega}_{\mathbf{x}_2})$$
$$\downarrow$$
$$(000001010^\omega) \oplus (\underbrace{0010^\omega}_{T(\mathbf{x}_2)}) = \underbrace{001001010^\omega}_{admissible}.$$

The resulting digit string is admissible, and therefore it is the $\boldsymbol{\mathcal{B}}$-expansion of z, i.e. $d_{\boldsymbol{\mathcal{B}}}(z) = d_{\boldsymbol{\mathcal{B}}}(x+y) = 001001010^\omega$.

In the above example we have illustrated the algorithm for addition which can be used for any alternate base $\boldsymbol{\mathcal{B}}$ satisfying the weight property: If a representation of a number z is not admissible, by Lemma 3 one can use a rewriting rule to find a lexicographically greater representation of z. Note that the choice of a rewriting rule may not be unique. The proof of Theorem 4 ensures that after finitely many steps, we obtain the $\boldsymbol{\mathcal{B}}$-expansion of z. The choice of rewriting rules applied in the individual steps may influence the number of needed rewrites.

References

1. Bertrand, A.: Développements en base de Pisot et répartition modulo 1. Comptes Rendus Hebdomadaires des Séances de l'Académie des Sciences. Séries A et B **285**(6), A419–A421 (1977)
2. Caalim, J., Demegillo, S.: Beta Cantor series expansion and admissible sequences. Acta Polytechnica **60**(3), 214–224 (2020). https://doi.org/10.14311/ap.2020.60.0214
3. Charlier, É., Cisternino, C.: Expansions in Cantor real bases. Monatshefte für Mathematik **195**(4), 585–610 (2021). https://doi.org/10.1007/s00605-021-01598-6
4. Charlier, É., Cisternino, C., Masáková, Z., Pelantová, E.: Spectrum, algebraicity and normalization in alternate bases. J. Number Theory (2023, to appear). https://doi.org/10.48550/ARXIV.2202.03718
5. Frougny, C.: Confluent linear numeration systems. Theoret. Comput. Sci. **106**(2), 183–219 (1992). https://doi.org/10.1016/0304-3975(92)90249-F
6. Frougny, C., Solomyak, B.: Finite β-expansions. Ergodic Theory Dynam. Syst. **12**(4), 713–723 (1992). https://doi.org/10.1017/s0143385700007057

7. Masáková, Z., Pelantová, E., Studeničová, K.: Rewriting rules for arithmetics in alternate base systems (2023). https://doi.org/10.48550/arXiv.2302.10708
8. Rényi, A.: Representations for real numbers and their ergodic properties. Acta Math. Acad. Scientiarum Hungaricae **8**, 477–493 (1957). https://doi.org/10.1007/BF02020331

Synchronizing Automata with Coinciding Cycles

Jakub Ruszil(✉)

Doctoral School of Exact and Natural Sciences, Faculty of Mathematics and Computer Science, Jagiellonian University, Cracow, Poland
ruszil@ii.uj.edu.pl

Abstract. We present a new class of automata for which we prove the bound of $\Theta(n^2)$ on the length of the shortest synchronizing words, where n is the number of states. Furthermore we show that a subset of letters of an alphabet of those automata has synchronization property as defined by Arnold and Steinberg in [1].

Keywords: Synchronization · Černý conjecture · Reset word

1 Introduction

The concept of synchronization of finite automata is essential in various areas of computer science. It consists in regaining control over a system by applying a specific set of input instructions. These instructions lead the system to a fixed state no matter which state it was in at the beginning. The idea of synchronization has been studied for many classes of complete deterministic finite automata (DFA) [2,3,8,14,15,17,18,21–25] and non-deterministic finite automata [11,12]. One of the most famous longstanding open problems in automata theory, known as Černý Conjecture, states that for a given synchronizing DFA with n states one can always find a synchronizing word of length at most $(n-1)^2$. This conjecture was proven for numerous classes of automata, but the problem is still not solved in general case. The concept of synchronization has been also considered in coding theory [4,13], parts orienting in manufacturing [8,16], testing of reactive systems [19] and Markov Decision Processes [5,6].

Example 1 depicts the Černý automaton with four states. The shortest synchronizing word for this automaton is *baaabaaab*.

Amongst many classes of automata investigated by researchers, there is a number of them defined utilizing the properties of cycles induced by the letters - see for example [7,10,20]. Continuing in that spirit we define the class of automata with a certain conditions placed on the intersections of the cycles and then investigate the synchronization properties of such automata.

F. Drewes and M. Volkov (Eds.): DLT 2023, LNCS 13911, pp. 208–218, 2023.
https://doi.org/10.1007/978-3-031-33264-7_17

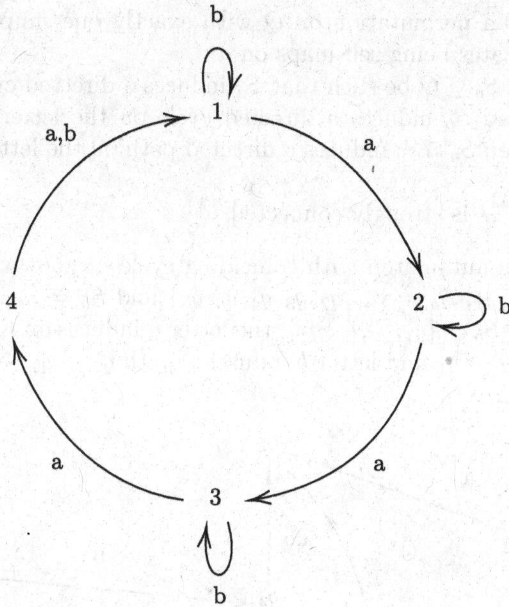

Fig. 1. Černý automaton with four states

2 Preliminaries

Deterministic finite automaton (DFA) is an ordered tuple $\mathcal{A} = (Q, \Sigma, \delta)$ where Σ is a finite set of letters, Q is a finite set of states and $\delta : Q \times \Sigma \to Q$ is a transition function. For $w \in \Sigma^*$ and $q \in Q$ we define $\delta(q, w)$ inductively: $\delta(q, \epsilon) = q$ and $\delta(q, aw) = \delta(\delta(q, a), w)$ for $a \in \Sigma$, where ϵ is the empty word and $\delta(q, a)$ is defined. We write $q.w$ instead of $\delta(q, w)$ wherever it does not cause ambiguity. We also define $q.w^{-1} = \{p \in Q : p.w = q\}$. If $w \in \Sigma^*$ then denote $|w|$ as a length of a word $|w|$. A word $w \in \Sigma^*$ is called *synchronizing* if there exists $\overline{q} \in Q$ such that for every $q \in Q$, $\delta(q, w) = \overline{q}$. A DFA is called *synchronizing* if it admits any synchronizing word. We state here useful characterization of synchronizing automata obtained by Černý in [23] (Theorem 2).

Theorem 1. *A DFA $\mathcal{A}$ is synchronizing if and only if for any $p, q \in Q$ there exists a word $w \in \Sigma^*$ such that $p.w = q.w$.*

Before moving further we provide a couple of definitions. Let $a \in \Sigma$, $S \subset Q$ and $G = (V, E)$ be a digraph. We say that S *induces* G *on the letter* a *in* $\mathcal{A}$, if a digraph with a vertex set S and edge set created from corresponding edges labelled with a is isomorphic to G. For the sake of simplicity we consider an isolated vertex as a directed path. Let $\Pi \subset \Sigma$. We define an automaton $\mathcal{A}$ *restricted to* Π as a DFA $\mathcal{B} = (Q, \Pi, \gamma)$ where γ is defined as δ but only on letters from Π. We say that a DFA is with *coincinding cycles* if there is a set of letters $\Pi \subseteq \Sigma$ and $|\Pi| > 1$ such that:

- every $a \in \Pi$ is a permutation on Q with exactly one simple directed cycle and all other states being self-maps on a
- let $S_a \subseteq Q$ and $S_b \subseteq Q$ be such that S_a induces a directed cycle on the letter $a \in \Pi$ in $\mathcal{A}$ and S_b induces a directed cycle on the letter $b \in \Pi$ in $\mathcal{A}$. If $S_a \cap S_b \neq \varnothing$ then $S_a \cap S_b$ induces a directed path on the letter a in $\mathcal{A}$ and on a letter b in $\mathcal{A}$
- $\mathcal{A}$ restricted to Π is strongly connected

The example of an automaton with coinciding cycles is shown on the Fig. 2. In that case $\Pi = \{a, b\}$, $S_a = \{q_1, q_2, q_3, q_4, q_5, q_6\}$ and $S_b = \{q_1, q_2, q_3, q_4, q_7, q_8\}$. Observe, that $S_a \cap S_b = \{q_1, q_2, q_3, q_4\}$, the letter a induces on that set a directed path $q_4 \to q_3 \to q_2 \to q_1$ and letter b induces a path $q_2 \to q_1 \to q_4 \to q_3$.

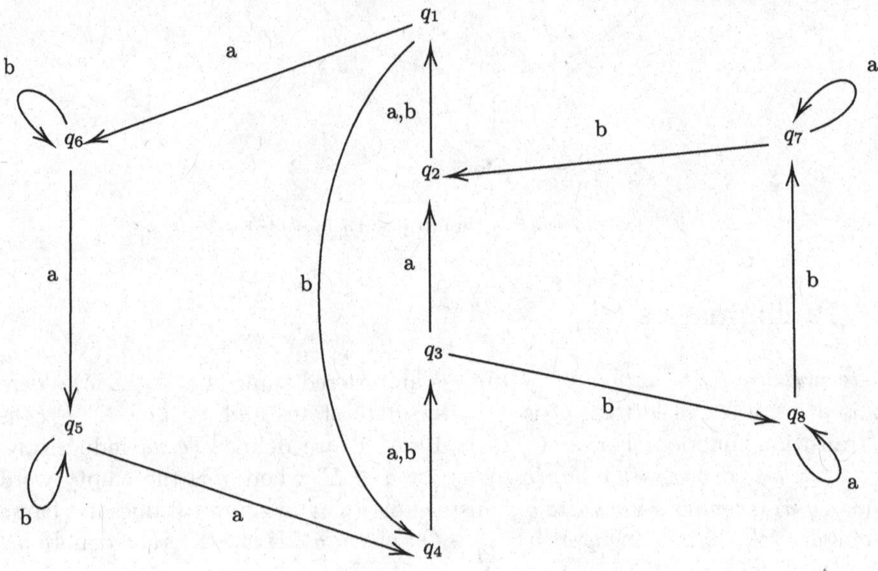

Fig. 2. Example of automaton with coinciding cycles

Let $a, b \in \Pi, p, q \in S_a$. Define $\text{dist}(p, q, a) = \min\{l \in \mathbb{N} : p.a^l = q\}$. Observe that since S_a induces a cycle then for any $k \in \mathbb{N}$ it holds $\text{dist}(p, q, a) = \text{dist}(p.a^k, q.a^k, a)$. Define also $\text{lastdist}_{a,b}(p)$ as the length of the path from p to the last state in the path induced by a in $S_a \cap S_b$. If $p \notin S_a \cap S_b$ then we set $\text{lastdist}_{a,b}(p) = \infty$. For example, if we take the automaton from the Fig. 2, then $\text{lastdist}_{a,b}(q_2) = 1$ (the last element on the path induced by a is q_1) and $\text{lastdist}_{b,a}(q_2) = 3$ (the last element on the path induced by b is q_3). Finally for $a, b \in \Pi$ define functions $\text{mindist}_{\text{a,b}} : Q \times Q \to Q$ and $\text{maxdist}_{\text{a,b}} : Q \times Q \to Q$ as

$$\text{mindist}_{\text{a,b}}(p, q) = \begin{cases} p \text{ if } \text{lastdist}_{a,b}(p) < \text{lastdist}_{a,b}(q), \\ q \text{ otherwise} \end{cases}$$

and

$$\mathrm{maxdist_{a,b}}(p,q) = \begin{cases} p \text{ if } \mathrm{lastdist}_{a,b}(p) > \mathrm{lastdist}_{a,b}(q), \\ q \text{ otherwise.} \end{cases}$$

First we make useful observations.

Observation 1. *For every $a \in \Pi$, there exists $b \in \Pi$ such that $S_a \cap S_b \neq \varnothing$.*

Proof. It is obvious, since $\mathcal{A}$ restricted to Π is strongly connected and $|\Pi| > 1$.

Observation 2. *For every $a \in \Pi$, we have $|S_a| < |Q|$.*

Proof. Suppose for the sake of contradiction that this is not the case. So there exists a letter a that is a directed cycle on Q. From Observation 1 we know that there exists $b \in \Pi$ such that $S_a \cap S_b \neq \varnothing$. But then we have $Q \cap S_b = S_a \cap S_b = S_b$ so it is a contradiction with the second condition of definition of automata with coinciding cycles since $S_a \cap S_b$ should induce a path on S_b.

Observe that we can identify a letter a of an automaton as a function $a : Q \to Q$. Let Π be a set of permutation letters. As defined in [1], we say that Π has a *synchronization property* if for any non-permutation letter x, automaton induced by letters $\Pi \cup \{x\}$ has a synchronizing word. Now we are ready to move forward to Sect. 3.

3 Upper Bound

In this section we provide quadratic upper bound on the length of the shortest synchronizing word in automata with coinciding cycles as well as necessary and sufficient condition for such automata to be synchronizing. From that point we assume that $|Q| = n$, where $n \in \mathbb{N}$ and if $a \in \Pi \subset \Sigma$ then $S_a \subset Q$ is as in the second condition of automata with coinciding cycles. First we prove a couple of lemmas.

Lemma 1. *Let X be a finite set, and $A_1, \ldots, A_k \subset X$ be a sequence of sets such that $A_i \cap A_{i+1} \neq \varnothing$. Then there exist a subsequence of this sequence $B_1, \ldots, B_l$ such that $B_1 = A_1, B_l = A_k$, $B_i \cap B_{i+1} \neq \varnothing$ and for $i, j \in \{1, \ldots, l\}$ such that $j \notin \{i-1, i, i+1\}$ we have that $B_i \cap B_j = \varnothing$.*

Proof. We construct desired subsequence inductively. Let $B_1 = A_1$. Having B_i we choose B_{i+1} to be A_j such that $A_j \cap B_i \neq \varnothing$, $j > i$ and for any $j' > j$ we have that $A_{j'} \cap B_i = \varnothing$. Since for any i holds $A_i \cap A_{i+1} \neq \varnothing$ then the construction ends with B_l equal to A_k and, since in the i-th step we have chosen maximal j for which $B_i \cap A_j \neq \varnothing$, we know that for $i, j \in \{1, ..., l\}$ such that $j \notin \{i-1, i, i+1\}$ holds $B_i \cap B_j = \varnothing$ and that concludes the proof.

Lemma 2. *Let $p, q \in Q$, $p \neq q$ and $a, b, c \in \Pi$ such that $S_b \cap S_c \neq \varnothing$ and $S_a \cap S_b = \varnothing$. If $p \in S_a$ and $q \in S_b$, then there exists a word $w \in \Pi^*$ such that $p.w \in S_a$, $q.w \in S_c$ and $|w| \leq |S_b| - |S_b \cap S_c|$.*

Proof. Since $S_a \cap S_b = \varnothing$, we have $p \notin S_b$. Notice, that $p.b = p$ and since $S_b \cap S_c \neq \varnothing$, there exists a word b^l such that $q.b^l \in S_b \cap S_c$ and $l \leq |S_b| - |S_b \cap S_c|$. We set $w = b^l$ and that concludes the proof.

Lemma 3. *Let $a, b, c \in \Pi$, $p, q \in Q$ and $p \neq q$. If $p \in S_a$, $q \in S_b$, $S_a \cap S_b \neq \varnothing$ and $S_a \cap S_c \neq \varnothing$ then there exists a word $w \in \Pi^*$ such that either $p.w \in S_a \setminus S_c$ and $q.w \in S_c$ or $q.w \in S_a \setminus S_c$ and $p.w \in S_c$ and $|w| \leq |S_b| + |S_a|$.*

Proof. If $p \notin S_a \cap S_b$ then there exists a word b^l such that $p.b^l = p$, $q.b^l \in S_a \cap S_b$ and $l \leq |S_b| - |S_a \cap S_b|$. We set $u = b^l$. Otherwise set

$$r_1 = \mathrm{mindist}_{\mathrm{a,b}}(p, q)$$

and

$$r_2 = \mathrm{maxdist}_{\mathrm{a,b}}(p, q).$$

Observe that then there exists a^k such that $r_1.a^k \in S_a \setminus S_b$, $r_1.a^{k-1} \in S_a \cap S_b$, $k \leq |S_a \cap S_b|$ and by definition of r_2 it follows that $r_2.a^k \in S_b$. So there exists b^l such that $r_1.a^k b^l = r_1.a^k$ and $r_2.a^k b^l \in S_a \cap S_b$ and $l \leq |S_b| - |S_a \cap S_b|$. We set $u = a^k b^l$. By this time we have $p.u \in S_a \setminus S_b$ and $q.u \in S_a \cap S_b$ or vice versa and $|u| \leq |S_b|$.

If $r_1.u \notin S_c$ and $r_2.u \notin S_c$ then we know that, since $r_2.u \neq r_1.u$ and S_a is a cycle, there exists a word a^k such that $r_1.ua^k \in S_a \setminus S_c$, $r_2.ua^k \in S_a \cap S_c$ or vice versa and $k < |S_a| - |S_a \cap S_c|$. Then we set $w = ua^k$ and we have $|w| \leq |S_b| + |S_a| - |S_a \cap S_c|$. Otherwise $r_1.u \in S_a \cap S_c$ or $r_2.u \in S_a \cap S_c$. If also $r_2.u \in S_a \setminus S_c$ or $r_1.u \in S_a \setminus S_c$ then we set $w = u$ and we know that $|w| \leq |S_b|$. If $r_1.u \in S_a \cap S_c$ and $r_2.u \in S_a \cap S_c$ then define

$$s_1 = \mathrm{mindist}_{a,c}(r_1.u, r_2.u)$$

and

$$s_2 = \mathrm{maxdist}_{a,c}(r_1.u, r_2.u).$$

Using a similar argument as above we obtain that there exists a word a^k such that $s_1.a^k \in S_a \setminus S_c$, $s_1.a^{k-1} \in S_a \cap S_c$, $s_2.a^k \in S_a \cap S_c$ and $k \leq |S_a \cap S_c|$. We set $w = ua^k$ with $|w| \leq |S_b| + |S_a \cap S_c|$.

Observe that the word w satisfies $p.w \in S_a \setminus S_c$ and $q.w \in S_c$ or vice versa and $|w| \leq \max\{|S_b|, |S_b| + |S_a| - |S_a \cap S_c|, |S_b| + |S_a \cap S_c|\}$ and from that inequality we deduce $|w| \leq |S_a| + |S_b|$ and the result holds.

Using the three former lemmas we are ready to prove the main proof of that section. First we prove lemma stating, that choosing two states and two cycles, we can always find a linear length word which sends the chosen states to the states belonging to the chosen cycles.

Lemma 4. *Let $a, b \in \Pi$, and $S_a \subset Q$ induces a directed cycle in $\mathcal{A}$ on a and $S_b \subset Q$ induces a directed cycle in $\mathcal{A}$ on b. Let also $p, q \in Q$. Then there exists a word $w \in \Pi^*$ such that either $p.w \in S_a$ and $q.w \in S_b$ or $q.w \in S_a$ and $p.w \in S_b$ and $|w| \leq 3n - |S_b|$.*

Proof. Let $\mathcal{B}$ be an automaton $\mathcal{A}$ restricted to Π. We are going to construct the desired word. Let $w = \epsilon$. If $p \in S_a$ and $q \in S_b$ then we are done. Assume without loss of generality that $q \notin S_b$. If also $p \notin S_a$, then, since $\mathcal{B}$ is strongly connected, we know that there exists a word $v \in \Sigma^*$ such that $p.v \in S_a$ and $|v| \leq n - |S_a|$. If also $q.v \in S_b$ then we set $w = v$ and we are done. Assume that $q.v \notin S_b$. Then, since $\mathcal{B}$ is strongly connected, there exists $c_1 \in \Pi$ such that $q.v \in S_{c_1}$. Since $\mathcal{B}$ is strongly connected, there exists a sequence of pairwise different letters $(c_2, \ldots, c_k)$, such that $S_{c_i} \cap S_{c_{i+1}} \neq \varnothing$ for $1 \leq i \leq k-1$ and $S_{c_k} \cap S_b \neq \varnothing$. Using Lemma 1 we choose the subsequence of sequence $(c_2, \ldots, c_k, b)$, say $(c'_1, \ldots, c'_l)$, such that $c'_1 = c_2$ and $c'_l = b$. Investigate three cases.

Case 1: $S_a \cap S_{c'_i} \neq \varnothing$ and $S_a \cap S_{c'_j} \neq \varnothing$ for $0 < i \neq j < l+1$

Find the first and the last index of sequence $(c'_1, \ldots, c'_l)$, respectively i and j, such that $S_{c'_i} \cap S_a \neq \varnothing$ and $S_{c'_j} \cap S_a \neq \varnothing$ Note $w_1 = \epsilon$. We apply Lemma 2 $i-1$ times inductively with, states $p.vw_1 \ldots w_{m-1}$ and $q.vw_1 \ldots w_{m-1}$, and letters a, c'_{m-1}, c'_m for $m \in \{2, .., i\}$ to find a word w_m such that $p.vw_1 \ldots w_{m-1}w_m \in S_a$, $q.vw_1 \ldots w_{m-1}w_m \in S_{c'_m}$ and $|w_m| \leq |S_{c'_{m-1}}| - |S_{c'_{m-1}} \cap S_{c'_m}|$. We obtain word $u = w_1 \ldots w_i$, such that $p.vu \in S_a$ and $q.vu \in S_{c'_i}$. Now we use Lemma 3 to obtain a word u' such that either $p.vuu' \in S_a \setminus S_{c'_j}$ and $q.vuu' \in S_{c'_j}$ or vice versa and $|u'| \leq |S_a| + |S_{c'_i}|$. Without loss of generality assume that first option holds. Notice that there exists a word c'^k_j such that $p.vuu'c'^k_j \in S_a$, $q.vuu'c'^k_j \in S_{c'_j} \cap S_{c'_{j+1}}$ and $k \leq |S_{c'_j}| - |S_{c'_j} \cap S_{c'_{j+1}}|$. Now we can again apply Lemma 2 in the same manner to obtain a word $u'' = w_{j+2}..w_l$ such that $p.vuu'c'^k_j u'' \in S_a$ and $q.vuu'c'^k_j u'' \in S_b$. We set $w = vuu'c'^k_j u''$ and we only need to bound its length from above. We know that

$$|w| = |v| + |u| + |u'| + |c'^k_j| + |u''| = n - |S_a| + |w_1 w_2 \ldots w_i| + |u'| + k + |w_{j+2}..w_l| \leq$$

$$n - |S_a| + \sum_{m=1}^{i-1} (|S_{c'_m}| - |S_{c'_m} \cap S_{c'_{m+1}}|) + |S_{c'_i}| + |S_a| +$$

$$|S_{c'_j}| - |S_{c'_j} \cap S_{c'_{j+1}}| + \sum_{m=j+1}^{l-1} (|S_{c'_m}| - |S_{c'_m} \cap S_{c'_{m+1}}|) =$$

$$n + \sum_{m=1}^{i-1} (|S_{c'_m}| - |S_{c'_m} \cap S_{c'_{m+1}}|) + |S_{c_i}| + \sum_{m=j}^{l-1} (|S_{c'_m}| - |S_{c'_m} \cap S_{c'_{m+1}}|)$$

On the other hand we know that

$$|S_{c'_1} \cup \ldots \cup S_{c'_i}| = \bigcup_{\varnothing \neq K \subseteq \{1, \ldots, i\}} (-1)^{|K|+1} |\bigcap_{k \in K} S_{c'_k}|$$

and

$$|S_{c'_j} \cup \ldots \cup S_{c'_l}| = \bigcup_{\varnothing \neq K \subseteq \{j, \ldots, l\}} (-1)^{|K|+1} |\bigcap_{k \in K} S_{c'_k}|$$

and also

$$|S_{c'_1} \cup \ldots \cup S_{c'_i}| + |S_{c'_j} \cup \ldots \cup S_{c'_l}| \leq 2n.$$

But by Lemma 1 it follows, that for any $K \subseteq \{1, \ldots, l\}$ different than $\{i, i+1\}$ for $i \in \{1, \ldots, l-1\}$ we have $|\bigcap_{k \in K} S_{c'_k}| = \varnothing$. So we know that $\sum_{m=1}^{i-1}(|S_{c'_m}| - |S_{c'_m} \cap S_{c'_{m+1}}|) + |S_{c'_i}| = |S_{c'_1} \cup \ldots \cup S_{c'_i}|$ and $\sum_{m=j}^{l-1}(|S_{c'_m}| - |S_{c'_m} \cap S_{c'_{m+1}}|) = |S_{c'_j} \cup \ldots \cup S_{c'_l}| - |S_{c'_l}|$, and since $c'_l = b$, the lemma holds for that case.

Case 2: $S_a \cap S_{c'_i} = \varnothing$ for every $0 < i < l+1$

Since for every $0 < i < l+1$ we have $S_a \cap S_{c'_i} = \varnothing$ then we can apply Lemma 2 $l-1$ times to obtain a word $u = w_1 w_2 \ldots w_{l-1}$ such that $p.vu \in S_a$, $q.vu \in S_{c'_l} = S_b$ and each $|w_i| \leq |S_{c'_i}| - |S_{c'_i} \cap S_{c'_{i+1}}|$ what implies $|vu| = n - |S_a| + \sum_{i=1}^{l-1}(|S_{c'_i}| - |S_{c'_i} \cap S_{c'_{i+1}}|)$ and using Lemma 1 one can immediately deduce that lemma holds for that case in the similar manner as in the former case.

Case 3: $S_a \cap S_{c'_i} \neq \varnothing$ for exactly one $0 < i < l+1$

As in the Case 1 we can construct the word u such that $p.vu \in S_a$, $q.vu \in S_{c'_i}$ and $|u| \leq \sum_{m=1}^{i-1}(|S_{c'_m}| - |S_{c'_m} \cap S_{c'_{m+1}}|)$. We are now looking for the word u' such that either $p.vuu' \in S_a \setminus S_{c'_i}$ and $q.vuu' \in S_{c'_i}$ or $q.vuu' \in S_a \setminus S_{c'_i}$ and $p.vuu' \in S_{c'_i}$. Using similar argument as in the proofs of the former lemmas we can find such a word with $|u'| \leq |S_a \cap S_{c'_i}|$. As in the previous cases, using Lemma 2 we now construct a word u'' such that either $p.vuu'u'' \in S_a$ and $q.vuu'u'' \in S_{c'_l}$ or $q.vuu'u'' \in S_a$ and $p.vuu'u'' \in S_{c'_l}$ with $|u''| \leq \sum_{m=i}^{l-1}(|S_{c'_m}| - |S_{c'_m} \cap S_{c'_{m+1}}|)$. The rest of the argument is analogous to the ones in the former cases.

The next lemma states, that for any two pairs of states we can find a word that sends the first pair to the second one.

Lemma 5. *Let $p, q \in Q$. Then for any $r, s \in Q$ there exists a word $w \in \Pi^*$ such that either $r.w = p$ and $s.w = q$ or $s.w = p$ and $r.w = q$ and $|w| \leq 6n$.*

Proof. We investigate two cases.

Case 1: There exists $a \in \Pi$ such that $p, q \in S_a$.

The idea of the proof in this case is to "push" one of the states r, s "outside" the cycle induced by a to the state right behind the "entrance" to this cycle, then to set the other state in a right distance after this entrance, push the first state back to the cycle of a and then rotate those two states to obtain p and q.

Using Observation 1 we can find $b \in \Pi$ such that $S_a \cap S_b \neq \varnothing$. Now we use Lemma 4 to obtain a word $u \in \Pi^*$ of length at most $3n$ such that $r.u \in S_a$ and $s.u \in S_b$ or $s.u \in S_a$ and $r.u \in S_b$. Without lose of generality assume that the first option holds. Now we are looking for a word $v \in \Pi^*$ such that $r.uv \in S_a$, $s.uv \in S_b \setminus S_a$ and $s.uvb \in S_a$. If $r.u \notin S_a \cap S_b$, then it suffices to take $v = b^l$, where $l < |S_b|$. If $r.u \in S_a \cap S_b$ then define

$$t_1 = \text{mindist}_{a,b}(r.u, s.u)$$

and

$$t_2 = \text{maxdist}_{a,b}(r.u, s.u).$$

Observe that there exists a word a^k where $k \leq |S_a \cap S_b|$ such that $t_1.a^k \in S_a \setminus S_b$ and, by definition of t_2 it follows that $t_2.a^k \in S_b$. Now we can find a word b^l in the same manner as before and obtain that $v = a^k b^l \leq |S_b| + |S_a \cap S_b|$.

Now we are looking for a word $z \in \Pi^*$ such that $\text{dist}(r.uvzb, s.uvzb, a) = \text{dist}(p, q, a)$. It is obvious that, since $|S_a \cap S_b|$ induces a path on letter b and for $t \in S_a \setminus S_b$ we have that $t.b = t$, there exists $t' \in S_a$ such that $\text{dist}(t'.b, s.uvb, a) = \text{dist}(p, q, a)$. Moreover, there exists a word a^m where $m < |S_a|$ such that $r.uvza^k = t'$, and since $s.uv \in S_b \setminus S_a$ then $s.uva^k = s.uv$. We set $u = a^m$. Since $\text{dist}(r.uvzb, s.uvzb, a) = \text{dist}(p, q, a)$ we know that there exists a word a^k where $k < |S_a|$ such that $r.uvzba^k = p$ and $s.uvzba^k = q$. Set $w = wvuba^k$ and notice that $|w| = |uvzba^k| \leq 3n - |S_b| + |S_b| + |S_a \cap S_b| + 2|S_a| \leq 6n$ and that concludes this case.

Case 2: There exist $a, b \in \Pi$ such that $p \in S_a$ and $q \in S_b$

If also $p \in S_a \cap S_b$ or $q \in S_a \cap S_b$ then we can apply Case 1 of this proof to obtain the required word. Assume that $p \in S_a \setminus S_b$ and $q \in S_b \setminus S_a$. Using Lemma 4 we obtain a word $u \in \Pi^*$ of length at most $3n$ such that $r.u \in S_a$ and $s.u \in S_b$. If $S_a \cap S_b = \varnothing$ then there exists a word $a^k b^l$ such that $k < |S_a|$ and $l < |S_b|$, and $r.ua^k b^l = p$ and $s.ua^k b^l = q$. Set $w = ua^k b^l$ and observe that $|w| < 3n$. So let us assume $S_a \cap S_b \neq \varnothing$. If also $r.u \in S_a \setminus S_b$ and $s.u \in S_b \setminus S_a$ then $w = ua^k b^l$ for some $k < |S_a|$ and $l < |S_b|$ and we are done. Otherwise define

$$t_1 = \text{mindist}_{a,b}(r.u, s.u)$$

and

$$t_2 = \text{maxdist}_{a,b}(r.u, s.u)$$

and we can find word a^k where $k \leq |S_a \cap S_b|$ such that $t_1.a^k \in S_a \setminus S_b$ and $t_2.a^k \in S_b$. Then we apply word a^l, where $l < |S_a|$ to obtain $t_1.a^l = p$, and, since $p \in S_a \setminus S_b$, then $p.b = p$. Now we apply word b^m, where $m < |S_b|$ to obtain $t_2.a^{k+l}b^m = q$. Set $w = ua^{k+l}b^m$ and observe that $k+l+m \leq |S_a|+|S_b|+|S_a \cap S_b|$ what concludes the proof.

Theorem 2. *Let $\mathcal{A}$ be an automaton with coinciding cycles. Then $\mathcal{A}$ is synchronizing if and only if there exist $x \in \Sigma$ and $p, q \in Q$ such that $p.x = q.x$. The shortest synchronizing word for such automaton is of length at most $1 + (6n + 1)(n - 2)$.*

Proof. First we prove "$\Rightarrow$" implication. For the sake of contradiction suppose, that $\mathcal{A}$ is synchronizing, and there is no such letter. But that would mean that every letter in Σ is a bijection, so for any $a \in \Sigma$ we have that $Q.a = Q$ what is a contradiction.

To prove "$\Leftarrow$" implication assume that there exist $x \in \Sigma$ and $p, q \in Q$ such that $p.x = q.x$. Notice that then $|Q.x| < n$. Then we can act as in proof of Theorem 1 by using Lemma 5 to obtain words $w_i x$ such that $|Q.w_1 x \ldots w_{i-1} x w_i x| < |Q.w_1 x \ldots w_{i-1} x|$. Using the bound from Lemma 5 concludes the proof.

Now we can state a straightforward conclusion from a previous theorem.

Corollary 1. *Any automaton with coinciding cycles has synchronization property.*

Proof. Immediate from Theorem 2, since any non-permutation letter x has $p, q \in Q$ such that $p.x = q.x$.

4 Worst-Case Lower Bound

We use the result from [9] to show that we cannot significantly improve the result from the previous section. First we recall the construction and a theorem showing the lower bound for the shortest synchronizing word of a family of automata arisen from that construction and then we shall prove that the family fulfills the conditions of automata with coninciding cycles. We start with the construction.

Let $Q_n = \{q_0, .., q_{n-1}\}$ and $\Sigma = \{a_1, ..., a_n\}$ for $n \in \mathbb{N}$. Define an automaton $\mathscr{V}_n = (Q_n, \Sigma_n, \delta_n)$ with transition function δ_n given as follows:

- $\delta_n(q_i, a_j) = q_i$ for $0 \leq i < n$, $1 \leq j < n$,$i \neq j$, $j \neq i+1$, $i \neq n$
- $\delta_n(q_i, a_i) = q_{i-1}$ for $0 < i \leq n-1$
- $\delta_n(q_i, a_{i+1}) = q_{i+1}$ for $0 \leq i < n-1$
- $\delta_n(q_0, a_n) = \delta_n(q_1, a_n) = q_0$
- $\delta_n(q_i, a_n) = q_i$ for $2 \leq i \leq n-1$

We recall the result from [9] establishing the length of the shortest synchronizing word for the automaton $\mathscr{V}_n$. The $\mathscr{V}_5$ automaton is depicted on the Fig. 3 (self-maps $a_1, \ldots, a_4$ are omited).

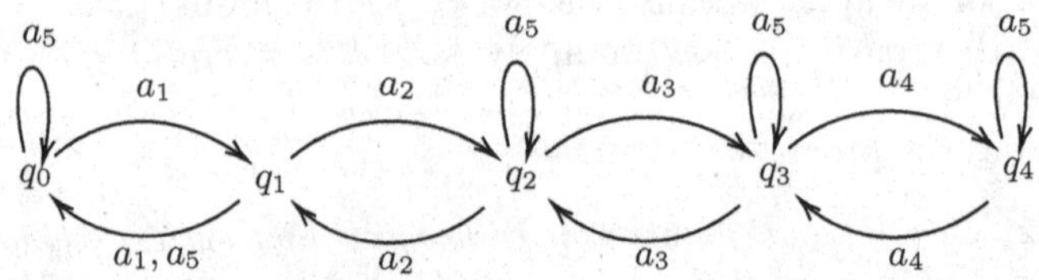

Fig. 3. The $\mathscr{V}_5$ automaton

Theorem 3 (Theorem 4 in [9]). *The length of the shortest synchronizing word for the automaton $\mathscr{V}_n$ is $\frac{(n-1)n}{2}$.*

Having that we are ready to prove the following lemma:

Lemma 6. *For $n > 2$ automaton $\mathscr{V}_n$ is the automaton with coinciding cycles.*

Proof. First notice that letter a_n is a non-permutation letter. Define $\Pi_n = \Sigma_n \setminus \{a_n\}$. Since $n > 2$ then, by definition of $\mathscr{V}_n$, we have that $|\Pi_n| > 1$. Consider letter $a_i \in \Pi_n$. By definition we obtain that for any $j \notin \{i-1, i\}$ letter a_i acts as a self-map on q_j and acts as a cyclic permutation on a set $\{q_{i-1}, q_i\}$ and that proves the first condition. Define $S_i = \{q_{i-1}, q_i\}$ for $1 \leq i \leq n$. We have already proven, that a_i induces a directed cycle on S_i. It is obvious that for any $i \neq j$ we have that $S_i \cap S_j \neq \varnothing$ if, and only if $j = i-1$ or $j = i+1$. In any of those cases $|S_i \cap S_j| = 1$ so the intersection of those sets must be an isolated vertex and that proves the second condition. For the proof of the third condition it suffices to notice, that for any $i \neq j$ we have that $q_i.a_{i+1}a_{i+2}..a_j = q_j$ and $q_j.a_j a_{j-1}..a_{i+1} = q_i$ what concludes the proof.

Since we have shown the existence of a family of automata in the class of automata with coinciding cycles for which the shortest synchronizing word are of length $\Theta(n^2)$, we are certain that we cannot improve the bound from the previous section significantly i.e. the bound proven in the Sect. 3 is asymptotically strict.

5 Conclusions and Further Work

We have presented a class of automata with coinciding cycles and we have proven the quadratic upper bound for the length of the shortest synchronizing word for them. What is more we have shown that every automaton from this class has synchronization property and found the evidence that the upper bound for synchronizing word that we have established is asymptotically strict.

In the further work we would like to relax the second condition of the automata with coinciding cycles in a way to abandon the necessity of inducing the directed paths by intersections of cycles. One can also try to improve the upper bound established in this paper by a linear factor. Utilizing avoiding words as for example in [21] could lead to the major improvements. Another interesting question is if one can achieve $\Theta(n^2)$ length of the shortest synchronizing word using constant number of letters in the alphabet.

References

1. Arnold, F., Steinberg, B.: Synchronizing groups and automata. Theor. Comput. Sci. **359**(1), 101–110 (2006)
2. Berlinkov, M.V.: On two algorithmic problems about synchronizing automata. In: Shur, A.M., Volkov, M.V. (eds.) DLT 2014. LNCS, vol. 8633, pp. 61–67. Springer, Cham (2014). https://doi.org/10.1007/978-3-319-09698-8_6
3. Berlinkov, M., Szykuła, M.: Algebraic synchronization criterion and computing reset words. Inf. Sci. **369**, 718–730 (2016)
4. Biskup, M.T., Plandowski, W.: Shortest synchronizing strings for Huffman codes. Theor. Comput. Sci. **410**, 3925–3941 (2009)

5. Doyen, L., Massart, T., Shirmohammadi, M.: Robust synchronization in Markov decision processes. In: Baldan, P., Gorla, D. (eds.) CONCUR 2014. LNCS, vol. 8704, pp. 234–248. Springer, Heidelberg (2014). https://doi.org/10.1007/978-3-662-44584-6_17
6. Doyen, L., Massart, T., Shirmohammadi, M.: The complexity of synchronizing Markov decision processes. J. Comput. Syst. Sci. **100**, 96–129 (2019)
7. Dubuc, L.: Sur les automates circulaires et la conjecture de Černý. RAIRO Theor. Inf. Appl. **32**(1–3), 21–34 (1998)
8. Eppstein, D.: Reset sequences for monotonic automata. SIAM J. of Comput. **19**, 500–510 (1990)
9. Gonze, F., Gusev, V.V., Gerencsér, B., Jungers, R.M., Volkov, M.V.: On the interplay between Babai and Černý's conjectures. In: Charlier, É., Leroy, J., Rigo, M. (eds.) Developments in Language Theory, pp. 185–197. Springer International Publishing, Cham (2017). https://doi.org/10.1007/978-3-319-62809-7_13
10. Gusev, V.V., Pribavkina, E.V.: Reset thresholds of automata with two cycle lengths. In: Holzer, M., Kutrib, M. (eds.) Implementation and Application of Automata, pp. 200–210. Springer International Publishing, Cham (2014). https://doi.org/10.1142/S0129054115400080
11. Imreh, B., Steinby, M.: Directable nondeterministic automata. Acta Cybern. **14**, 105–115 (1999)
12. Ito, M., Shikishima-Tsuji, K.: Some results on directable automata. In: Karhumäki, J., Maurer, H., Păun, G., Rozenberg, G. (eds.) Theory Is Forever. LNCS, vol. 3113, pp. 125–133. Springer, Heidelberg (2004). https://doi.org/10.1007/978-3-540-27812-2_12
13. Jürgensen, H.: Synchronization. Inf. Comput. **206**, 1033–1044 (2008)
14. Kari, J.: A counter example to a conjecture concerning synchronizing word in finite automata. EATCS Bull. **73**, 146–147 (2001)
15. Kari, J.: Synchronizing finite automata on Eulerian digraphs. Theor. Comput. Sci. **295**, 223–232 (2003)
16. Natarajan, B.K.: An algorithmic approach to the automated design of parts orienters. In: 27th Annual Symposium on Foundations of Computer Science, pp. 132–142 (1986)
17. Pin, J.E.: On two combinatorial problems arising from automata theory. Proc. Int. Colloquium Graph Theory Comb. **75**, 535–548 (1983)
18. Rystsov, I.K.: Reset words for commutative and solvable automata. Theor. Comput. Sci. **172**, 273–279 (1997)
19. Sandberg, S.: Homing and synchronizing sequences. Model-Based Test. React. Syst. **3472**, 5–33 (2005)
20. Steinberg, B.: The Černý conjecture for one-cluster automata with prime length cycle. Theor. Comput. Sci. **412**(39), 5487–5491 (2011)
21. Szykuła, M.: Improving the Upper Bound on the Length of the Shortest Reset Word. In: STACS 2018, pp. 56:1–56:13 (2018)
22. Trahtman, A.: The Černý conjecture for aperiodic automata. Discrete Math. Theor. Comput. Sci. **9**, 3–10 (2007)
23. Černý, J.: Poznámka k homogénnym eksperimentom s konečnými automatami. Mat.-Fyz. Cas. Slovens. Akad. Vied. **14**, 208–216 (1964)
24. Volkov, M.: Synchronizing automata and the Černý conjecture. Lang. Automata Theory Appl. **5196**, 11–27 (2008)
25. Volkov, M.: Slowly synchronizing automata with idempotent letters of low rank. J. Automata, Lang. Comb. **24**, 375–386 (2019)

Approaching Repetition Thresholds via Local Resampling and Entropy Compression

Arseny M. Shur(✉)

Bar Ilan University, Ramat Gan, Israel
shur@datalab.cs.biu.ac.il

Abstract. We analyze a simple algorithm, transforming an input word into a word avoiding certain repetitions such as fractional powers and undirected powers. This transformation can be made reversible by adding the log of the run of the algorithm to the output. We introduce a compression scheme for the logs; its analysis proves that $(1+\frac{1}{d})$-powers and undirected $(1+\frac{1}{d})$-powers can be avoided over $d+O(1)$ letters. These results are closer to the optimum than it is usually expected from a purely information-theoretic considerations. In the second part, we present experimental results obtained by the mentioned algorithm in the extreme case of $(d+1)$-ary words avoiding $(1+\frac{1}{d})^{+}$-powers.

Keywords: Power-free word · Repetition threshold · Entropy compression · Local resampling

1 Introduction

Many combinatorial problems on words (strings) fit into a simple frame: given a type of repetition, find the minimum alphabet allowing to construct an arbitrary long (or infinite) word containing no repetitions of this type. This type of problems dates back to Thue [18], who proved that squares and cubes can be avoided over the ternary and binary alphabets, respectively. Since then, all *optimal* results for such problems were proved by explicitly constructing words without repetitions of the required sort. While some constructions of such words are simple and mathematically beautiful, like the Thue–Morse word [19], some others have complicated and not intuitively clear structure, and often require the aid of computer to be found. Two examples of such complicated constructions are Keränen's quaternary abelian-square-free word [9] and the morphic constructions found by Currie and Rampersad [2] and Rao [14] to prove the remaining cases of Dejean's conjecture [5]. Up to now, in both cases no "simpler" explicit constructions were proposed, and it may well happen that no such

Supported by the grant MPM no. ERC 683064 under the EU's Horizon 2020 Research and Innovation Programme and by the State of Israel through the Center for Absorption in Science of the Ministry of Aliyah and Immigration.

F. Drewes and M. Volkov (Eds.): DLT 2023, LNCS 13911, pp. 219–232, 2023.
https://doi.org/10.1007/978-3-031-33264-7_18

constructions exist. In such a situation, it is natural to study the performance of non-constructive methods on these problems, and this is what we try to do.

The constructive proof of the Lovász Local Lemma by Moser and Tardos [10] boosted the popularity of non-constructive proofs based on local resampling algorithms and entropy compression. The first result about words was proved by such an argument in [7]: given an arbitrary infinite sequence of four-letter alphabets $\{A_n\}_1^\infty$, it is possible to construct a square-free word $w = w[1..\infty]$ such that $w[n] \in A_n$ for all n. It is conjectured that the same result can be obtained with the alphabets of size 3, but up to now this was proved only in a special case [15]. Further results obtained by this technique concern the avoidance of long patterns [12] and shuffle squares [8]. In this paper we apply this technique to *fractional powers*.

An $(\frac{m}{n})$-*power* of a length-n word u is the length-m prefix v of the infinite word $uuu\cdots$ ($m > n$ is assumed). A word is said to be α-*free* if it contains no $(\frac{m}{n})$-powers with $\frac{m}{n} \geq \alpha$. By $F(\alpha)$ we denote the minimum alphabet of an infinite α-free word, also called the *avoidability index* (of α). Thus, the mentioned results by Thue can be written as $F(2) = 3$ and $F(3) = 2$; moreover, in [19] he proved that $F(\alpha) = 2$ for all $\alpha > 2$. An "inverse" of F, called *repetition threshold*, was defined by Dejean [5] as $\mathsf{RT}(k) = \inf\{\alpha : F(\alpha) = k\}$. Dejean showed that $\mathsf{RT}(3) = 7/4$ and conjectured the remaining values $\mathsf{RT}(4) = 7/5$ (proved by Pansiot [13]) and $\mathsf{RT}(k) = \frac{k}{k-1}$ for $k \geq 5$ (proved by efforts of many authors [1,2,11,14]).

If $\alpha < 2$, then α-powers are words of the form vxv, where α is the ratio between the lengths of the words vxv and vx. Currie and Mol [3] defined *undirected α-powers* as the words of the form vxv', where v' is either v or the reversal of v; the meaning of α remains the same. They defined $F_u(\alpha)$ and $\mathsf{RT}_u(k)$ as the "undirected" analogs of $F(\alpha)$ and $\mathsf{RT}(k)$, conjectured that $\mathsf{RT}_u(k) = \frac{k-1}{k-2}$ for all $k \geq 4$, and confirmed this conjecture for $k \leq 21$.

In this paper we apply entropy compression to obtain simple upper bounds for $F(\alpha)$ and $F_u(\alpha)$. It is rather surprising that the obtained bounds are only a small constant away from the actual values of these functions. Our main bounds are as follows:

$$\text{if } 1 + \frac{1}{d+1} < \alpha \leq 1 + \frac{1}{d}, \text{ then } F(\alpha) \leq d+6 \text{ and } F_u(\alpha) \leq d+10,$$

with better (but still suboptimal) bounds for small values of d. In addition, we show that the local resampling algorithm we used shows very good practical results in constructing words with "threshold" restrictions from Dejean's conjecture. So the algorithm can possibly be used for an alternative "computer-free" proof of Dejean's conjecture.

The paper is organized as follows. Section 2 contains necessary preliminaries, the local resampling algorithm and its basic properties. In Sect. 3 we define compression schemes associated with the algorithm and prove the bounds on the avoidability index for α-powers and undirected α-powers. In Sect. 4 we describe the results of experiments with the algorithm and propose a finite-state Markov model to explain the results.

2 Definitions and Notation

We study words over finite alphabets. For a length-n word u we write $u = u[1..n]$; the elements of the range $[1..n]$ are *positions* in u. The length of u is denoted by $|u|$; the empty word λ has length 0. The *reversal* of u is the word $u^R = u[n] \cdots u[1]$. A word w is a *factor* of u if $u = vwz$ for some (possibly empty) words v and z; the condition $v = \lambda$ (resp., $z = \lambda$) means that w is a *prefix* (resp., *suffix*) of u. If v is a suffix of u, then uv^{-1} denotes the word obtained from u by deleting this suffix. Any factor w of u can be represented as $w = u[i..j]$ for some i and j; a representation of this form specifies the *occurrence of w at position i*.

For any α, $1 < \alpha \leq 2$, an *α-power* is a word of the form vxv such that $\frac{|vxv|}{|vx|} = \alpha$, and an *undirected α-power* is either an α-power or a word of the form vxv^R satisfying the same length restrictions. We also say that vxv is an *ℓ-repetition of period p and exponent α*, where $\ell = |v|$, $p = |vx|$. A word is *α-free* if no one of its factors is a β-power with $\beta \geq \alpha$. It is convenient to extend the set of reals, defining, for each α, the "number" α^+ by postulating that $\alpha^+ \leq \beta$ iff $\alpha < \beta$. Then α^+-free words are defined in the same way as α-free words. We say that an extended real α is *k-avoidable* if there exists an infinite k-ary α-free word. *Undirected α-free* words and *undirected k-avoidable* powers are defined in the same way. Throughout the paper, log() denotes the binary logarithm.

Consider the following simple algorithm, transforming a k-ary word w to a k-ary α-free word $u = \mathsf{RF}(w, \alpha)$, where $\alpha \leq 2$ is an extended real.

Algorithm 1. Constructing an α-free word $\mathsf{RF}(w, \alpha)$ from a k-ary word w

1: $u \leftarrow \lambda$
2: **for** $t = 1$ to $|w|$ **do**
3: $\quad u \leftarrow u{\cdot}w[t]$
4: $\quad$ **if** u has a suffix with exponent at least α **then**
5: $\quad\quad$ take the shortest v such that u has a suffix vxv with $\frac{|vxv|}{|vx|} \geq \alpha$
6: $\quad\quad u \leftarrow uv^{-1}$
7: $\mathsf{RF}(w, \alpha) = u$

Remark 1. The deletion of v in line 6 cannot delete all occurrences of some letter because uv^{-1} contains v as a factor. In particular, the word $\mathsf{RF}(w, \alpha)$ contains all letters occurring in w.

Remark 2. If w is α-free, then $\mathsf{RT}(w, \alpha) = w$. Hence any k-ary α-free word is a valid output of Algorithm 1.

For each run of Algorithm 1 we associate its *log* $\mathsf{Log}(w, \alpha)$. The log is the set of all triples (t, ℓ, p) such that at the t'th iteration of the main cycle the suffix v of an ℓ-repetition vxv was deleted; p is the period of this repetition. The log allows one to reconstruct the input word w, as the following lemma shows.

Lemma 1. *An arbitrary word w can be uniquely reconstructed from the pair* $(\mathsf{RF}(w, \alpha), \mathsf{Log}(w, \alpha))$.

Proof. Let u_t be the current α-free word after the t'th iteration of Algorithm 1. Given u_t and $\mathsf{Log}(w,\alpha)$, we can find $w[t]$ and u_{t-1}. Indeed, if the log contains no triple (t,ℓ,p), then u_t was obtained from u_{t-1} by appending $w[t]$, so $w[t] = u_t[|u|]$, and $u_{t-1} = u_t{\cdot}u_t[|u|]^{-1}$. If a triple (t,ℓ,p) belongs to the log, then after appending $w[t]$ to u_{t-1}, some length-ℓ suffix v of the obtained word was deleted. According to the choice of v in Algorithm 1, u_t has the suffix vx of length p. Hence we can recover v, $w[t] = v[|v|]$, and $u_{t-1} = u_t v{\cdot}v[|v|]^{-1}$.

As $\mathsf{RF}(w,\alpha)$ equals the word $u_{|w|}$, the whole word w can be reconstructed by applying the above procedure for all t's in reversed order. □

3 Logs Compression and Repetition Thresholds

The use of Algorithm 1 (and its variations for different types of repetitions) for proving a given repetition k-avoidable is based on the following lemma.

Lemma 2. *Suppose that* $\mathsf{Enc}()$ *is an injective function which maps each set* $S = \mathsf{Log}(w,\alpha)$ *to a bit sequence* $\mathsf{Enc}(S)$*. Let* $E(n,k,\alpha)$ *be the expected length of* $\mathsf{Enc}(S)$ *if* w *is chosen uniformly at random among all* k*-ary words of length* n*.*

1) If $E(n,k,\alpha) \le n\log k - \delta(n)$*, for some infinitely growing function* $\delta(n)$*, then the exponent* α *is* k*-avoidable; moreover,*
2) If $\lim_{n\to\infty} \frac{E(n,k,\alpha)}{n} \le \log k - \varepsilon$ *for some constant* $\varepsilon > 0$*, then the number of* k*-ary* α*-free words grows exponentially with length.*

Proof. By Lemma 1, Algorithm 1 builds an injection $w \mapsto (\mathsf{RF}(w,\alpha), \mathsf{Log}(w,\alpha))$. As w is chosen from a distribution with the entropy $n\log k$, Shannon's theorem guarantees that under any coding scheme the expected length of the code for the pair $(\mathsf{RF}(w,\alpha), \mathsf{Log}(w,\alpha))$ is at least $n\log k$. So in case 1) we need, on expectation, at least $\delta(n)$ bits to encode $\mathsf{RF}(w,\alpha)$. Hence the set of possible output words of Algorithm 1 (which is precisely the set of all k-ary α-free words; see Remark 2) infinitely grows with the growth of n. Similarly, in case 2) we need at least $\varepsilon - o(1)$ bits per symbol of w to encode $\mathsf{RF}(w,\alpha)$. Hence the number of k-ary α-free words of length at most n is not smaller than $2^{(\varepsilon-o(1))n}$. □

We define three encoding schemes for logs. Fix $\alpha = 1 + \frac{1}{d}$, where d is an arbitrary positive integer; later we adapt the schemes to the exponents $(1+\frac{1}{d})^+$. In the descriptions of all schemes we assume d, k, and $n = |w|$ to be fixed. As we are going to apply Lemma 2(2), we do not care about $O(\log n)$ additional memory costs. In particular, we omit rounding to simplify calculations.

First encoding scheme. We denote $m = |\mathsf{Log}(w,\alpha)|$, $\gamma = \frac{m}{n}$. The bit sequence $\mathsf{Enc}_1(w,\alpha)$, encoding $\mathsf{Log}(w,\alpha)$, consists of four blocks:

- first block of $\log n$ bits stores m in binary;
- second block of $\log\binom{n}{m}$ bits encodes all times t;
- third block of $\log\binom{n}{m}$ bits encodes all lengths ℓ;
- fourth block of $m\log d$ bits encodes all periods p.

Since the deletions were made at m iterations, we can represent the set of these iterations as a "characteristic" binary sequence of length n with exactly m 1's; the number of such sequences is $\binom{n}{m}$. We enumerate them, say, in lexicographic order and store the number of the required sequence in binary, using $\log\binom{n}{m}$ bits. Note that we know n in advance and m from the first block; hence we know where the second block ends and can decode the number into a characteristic sequence. Next, let $\ell_1, \ldots, \ell_m$ be the lengths of the words deleted at the iterations $t_1 < \cdots < t_m$. We pack these lengths into another binary sequence of length n with exactly m 1's such that the i'th (from the left) 1 in this sequence is at the position $\ell_1 + \cdots + \ell_i$. This sequence is encoded and decoded in the same way as the second block.

Finally, to encode periods we observe that

($\star$) each triple (t, ℓ, p) satisfies $d(\ell - 1) + 1 \le p \le d\ell$.

To see this, consider the deletion by Algorithm 1 of the suffix v of an ℓ-repetition vxv. The period p of vxv is at most $d\ell$, because $\alpha = \frac{|vxv|}{|vx|} = \frac{\ell+p}{p} \ge 1 + \frac{1}{d}$. On the other hand, all proper prefixes of the current word u are α-free. Hence so is the word $vxv{\cdot}v[\ell]^{-1}$, which is an $(\ell-1)$-repetition of period p. Therefore, $\frac{\ell-1+p}{p} < 1 + \frac{1}{d}$, implying $p > (\ell - 1)d$. Using ($\star$), we store the number $p' = p - d(\ell-1) - 1 \in \{0, \ldots, d-1\}$ instead of p. We view the sequence of all numbers p' as a d-ary number with m digits. Converting this number into binary, we obtain the fourth block of $m \log d$ bits.

Lemma 3. *Let $E(n, k, \alpha)$ be the expectation of $|\mathsf{Enc}_1(w, \alpha)|$ over all k-ary words of length n. Then $\lim_{n\to\infty} \frac{E(n,k,\alpha)}{n} \le 2\log(\sqrt{d} + 1)$.*

Proof. We recall that $m = \gamma n$ and apply Stirling's formula to get

$$\log\binom{n}{m} = \log \frac{(1 + o(1)) \cdot \sqrt{2\pi n}\left(\frac{n}{e}\right)^n}{\sqrt{2\pi\gamma n} \cdot \left(\frac{\gamma n}{e}\right)^{\gamma n} \cdot \sqrt{2\pi(1-\gamma)n} \cdot \left(\frac{(1-\gamma)n}{e}\right)^{(1-\gamma)n}} = n \cdot \left(\gamma \log \frac{1}{\gamma} + (1-\gamma) \log \frac{1}{1-\gamma}\right) + O(\log n) \quad (1)$$

Thus for each log $\mathsf{Log}(w, \alpha)$ of a fixed cardinality γn we have

$$|\mathsf{Enc}_1(w, \alpha)| = n \cdot \left(\gamma \log \frac{1}{\gamma} + (1-\gamma) \log \frac{1}{1-\gamma} + \gamma \log d\right) + O(\log n) \quad (2)$$

The function $f(\gamma) = \gamma \log \frac{1}{\gamma} + (1-\gamma)\log\frac{1}{1-\gamma} + \gamma \log d$ has the derivative $f'(\gamma) = 2\log\frac{1-\gamma}{\gamma} + \log d$ which decreases inside the interval $(0, 1)$ and has the only zero at $\gamma_0 = \frac{\sqrt{d}}{\sqrt{d}+1}$. Hence $f(\gamma)$ reaches the maximum at γ_0. Since $f(\gamma_0) = 2\log(\sqrt{d}+1)$, the lemma follows. □

Corollary 1. *The exponent $1 + \frac{1}{d}$ is k-avoidable if $k > d + 2\sqrt{d} + 1$.*

An alternative encoding of periods, based on the following lemma, is more economical in the case $k < 2d$.

Lemma 4. *Suppose that Algorithm 1 runs for a pair* (w, α), *where* w *is* k*-ary and* $\alpha = 1 + \frac{1}{d}$. *Let* $(t, \ell, p) \in \mathsf{Log}(w, \alpha)$, $\ell \geq 2$. *Given* t, ℓ, *the word* u *obtained at the* t*'th iteration, and at most* $\log(k-d)$ *bits of additional information, one can restore* p.

Proof. Note that

(∗) every $d+1$ subsequent letters of an α-free word are distinct.

At t'th iteration, Algorithm 1 detected some ℓ-repetition vxv and deleted its suffix v, getting $u = u'vx$ for some word u'. Since $p = |vx|$, p is determined by the position of the suffix vx in u. By (⋆), there are d subsequent candidate positions; they contain different letters by (∗). Hence we can encode the first letter of v to identify the position. Again by (∗), the first letter of v is distinct from the last d letters of u. Thus, there are $k - d$ candidates for this first letter (and not necessarily all of them occur at candidate positions). Given u, one can determine the set of candidate letters. Additional $\log(k-d)$ bits allow one to identify the correct candidate. □

Second Encoding Scheme. As in the first scheme, $m = |\mathsf{Log}(w, \alpha)|$, $\gamma = \frac{m}{n}$. Let m_1 be the number of triples in $\mathsf{Log}(w, \alpha)$ with $\ell = 1$ and denote $\beta = \frac{d}{k}$, $c = k - d$. The bit sequence $\mathsf{Enc}_2(w, \alpha)$, encoding $\mathsf{Log}(w, \alpha)$, consists of six blocks:

- first block of $2 \log n$ bits stores m and m_1 in binary;
- second block of $\log \binom{n}{m}$ bits encodes all times t;
- third block of $\log \binom{m}{m_1}$ bits encodes all lengths $\ell = 1$;
- fourth block of $\log \binom{n-m}{m-m_1}$ bits encodes all lengths $\ell \geq 2$;
- fifth block of $m_1 \log d$ bits encodes all periods p for $\ell = 1$;
- sixth block of $(m - m_1) \log c$ bits encodes all periods p for $\ell \geq 2$.

The first block is trivial, the second block is the same as in $\mathsf{Enc}_1()$. Next, we have m triples in $\mathsf{Log}(w, \alpha)$, naturally ordered by t's. A binary sequence of length m with m_1 1's indicates which triples have $\ell = 1$. The third block stores a binary number of $\log \binom{m}{m_1}$ bits, encoding the choice among these sequences. Let $(t_1, \ell_1, p_1), \ldots, (t_{m-m_1}, \ell_{m-m_1}, p_{m-m_1})$ be the remaining triples, ordered by t's, in $\mathsf{Log}(w, \alpha)$. We describe all lengths ℓ_i with a bit sequence containing $m - m_1$ 1's such that the i'th from the left 1 is at the position $\ell_1 + \cdots + \ell_i - i$. Since the total length of all deleted suffixes is less than n, and m_1 of these suffixes have length 1, we obtain $\sum_{i=1}^{m-m_1} \ell_i < n - m_1$. Hence the position of the rightmost 1 in the analyzed sequence is less than $(n - m_1) - (m - m_1) = n - m$, so we can take the sequence of length $n - m$. The choice among these sequences can be encoded in $\log \binom{n-m}{m-m_1}$ bits, constituting the fourth block of $\mathsf{Enc}_2(w, \alpha)$.

In the last two blocks we encode, similar to the fourth block of $\mathsf{Enc}_1()$, the periods of all processed 1-repetitions and the periods of all processed ℓ-repetitions with $\ell \geq 2$, respectively. Since the length of the first block is known in advance, the lengths of remaining five blocks can be computed from n, m, and m_1.

Lemma 5. *Let $E(n,k,\alpha)$ be the expectation of $|\mathsf{Enc}_2(w,\alpha)|$ over all k-ary words of length n. Then $\lim_{n\to\infty}\frac{E(n,k,\alpha)}{n} < \log k$ whenever $k \geq d+6$.*

Proof. First we observe that the deletion of a single symbol ends an iteration of Algorithm 1 with the probability β. Then m_1 is the result of n Bernoulli trials with the success probability β. Hence m_1 is close to βn; namely, w.h.p. $m_1 = (\beta + O(n^{-\varepsilon}))n$ for any $\varepsilon \in (0, \frac{1}{2})$. Using Stirling's formula and elementary O-transformations we get the following analog of (1):

$$\begin{aligned}
\log\binom{m}{m_1} &= m\cdot\Big(\frac{\beta+O(n^{-\varepsilon})}{\gamma}\log\frac{\gamma}{\beta+O(n^{-\varepsilon})}\\
&\quad+\frac{\gamma-\beta+O(n^{-\varepsilon})}{\gamma}\log\frac{\gamma}{\gamma-\beta+O(n^{-\varepsilon})}\Big)+O(\log n)\\
&= m\cdot\Big(\frac{\beta}{\gamma}\log\frac{\gamma}{\beta}+\frac{\gamma-\beta}{\gamma}\log\frac{\gamma}{\gamma-\beta}+O(n^{-\varepsilon})\Big)+O(\log n)\\
&= n\cdot\Big(\beta\log\frac{\gamma}{\beta}+(\gamma-\beta)\log\frac{\gamma}{\gamma-\beta}\Big)+O(n^{1-\varepsilon})\\
&= n\cdot\Big(\gamma\log\gamma+\beta\log\frac{1}{\beta}+(\gamma-\beta)\log\frac{1}{\gamma-\beta}\Big)+O(n^{1-\varepsilon}) \qquad (3)
\end{aligned}$$

In the same way, we get

$$\begin{aligned}
\log\binom{n-m}{m-m_1} &= n\cdot\Big((\gamma-\beta)\log\frac{1-\gamma}{\gamma-\beta}+(1-2\gamma+\beta)\log\frac{1-\gamma}{1-2\gamma+\beta}\Big)+O(n^{1-\varepsilon})\\
&= n\cdot\Big((1-\gamma)\log(1-\gamma)+(\gamma-\beta)\log\frac{1}{\gamma-\beta}+(1-2\gamma+\beta)\log\frac{1}{1-2\gamma+\beta}\Big)+O(n^{1-\varepsilon}) \qquad (4)
\end{aligned}$$

Summing up (1), (3), and (4) and observing that (1) is annihilated by the first terms in the parentheses of (3) and (4), we get the expression for the total length of the second, third, and fourth blocks of $\mathsf{Enc}_2(w,\alpha)$:

$$\begin{aligned}
&\log\binom{n}{m}+\log\binom{m}{m_1}+\log\binom{n-m}{m-m_1}\\
&= n\cdot\Big(\beta\log\frac{1}{\beta}+2(\gamma-\beta)\log\frac{1}{\gamma-\beta}+(1-2\gamma+\beta)\log\frac{1}{1-2\gamma+\beta}\Big)+O(n^{1-\varepsilon}) \qquad (5)
\end{aligned}$$

The fifth and sixth blocks are, respectively, $m_1\log d$ and $(m-m_1)\log c$ bits long. W.h.p., this gives $n\cdot(\beta\log d+(\gamma-\beta)\log c)+O(n^{1-\varepsilon})$ bits in total. Thus, the total code length of a log $\mathsf{Log}(w,\alpha)$ of a fixed cardinality γn is, w.h.p.,

$$\begin{aligned}
|\mathsf{Enc}_2(w,\alpha)| &= n\cdot\Big(\beta\log k+2(\gamma-\beta)\log\frac{1}{\gamma-\beta}\\
&\quad+(1-2\gamma+\beta)\log\frac{1}{1-2\gamma+\beta}+(\gamma-\beta)\log c\Big)+O(n^{1-\varepsilon}), \qquad (6)
\end{aligned}$$

where the first term in parentheses is the sum of $\beta \log d$ and $\beta \log \frac{1}{\beta}$. As in the proof of Lemma 3, we denote the expression in parentheses in (6) by $f(\gamma)$ and find its maximum; note that $\beta \le \gamma \le \frac{1+\beta}{2}$. We get the derivative $f'(\gamma) = 2\log\frac{1-2\gamma+\beta}{\gamma-\beta} + \log c$ which decreases inside the interval $(\beta, \frac{1+\beta}{2})$ and has the only zero at $\gamma_0 = \frac{(\beta+1)\sqrt{c}+\beta}{2\sqrt{c}+1}$. Hence $f(\gamma)$ reaches the maximum at γ_0. One has $\gamma_0 - \beta = \frac{(1-\beta)\sqrt{c}}{2\sqrt{c}+1}$, $1 - 2\gamma_0 + \beta = \frac{1-\beta}{2\sqrt{c}+1}$, and then

$$\begin{aligned} f(\gamma_0) = \beta \log k + 2\frac{(1-\beta)\sqrt{c}}{2\sqrt{c}+1}\log\frac{2\sqrt{c}+1}{(1-\beta)\sqrt{c}} \\ + \frac{1-\beta}{2\sqrt{c}+1}\log\frac{2\sqrt{c}+1}{1-\beta} + \frac{(1-\beta)\sqrt{c}}{2\sqrt{c}+1}\log c \\ = \beta\log k + (1-\beta)\log\frac{2\sqrt{c}+1}{1-\beta} \end{aligned} \quad (7)$$

(the terms $2\frac{(1-\beta)\sqrt{c}}{2\sqrt{c}+1}\log\frac{1}{\sqrt{c}}$ and $\frac{(1-\beta)\sqrt{c}}{2\sqrt{c}+1}\log c$ annihilate). From (7) we see that $f(\gamma_0) < \log k$ iff $\frac{2\sqrt{c}+1}{1-\beta} < k$. As $1-\beta = \frac{c}{k}$, this condition simplifies to

$$2\sqrt{c} + 1 < c, \quad (8)$$

which is true for all $c \ge 6$. Thus, for $k \ge d+6$ the expression in parentheses in (6) is less then $\log k$. Then w.h.p. $|\mathsf{Enc}_2(w,\alpha)| < n(\log k - \varepsilon)$ for some $\varepsilon > 0$. The reference to Lemma 2 finishes the proof. □

Corollary 2. *The exponent* $1 + \frac{1}{d}$ *is* k*-avoidable if* $k \ge d + 6$.

The third encoding is of very different nature; it generalizes the idea from [7] and allows us to replace the strict inequality in Corollary 1 with a non-strict one, improving the bound for $d = 1$ and $d = 4$.

Third Encoding Scheme. For $\alpha = 1 + \frac{1}{d}$, we represent $\mathsf{Log}(w,\alpha)$ by a $(d+1)$-ary word $y = \mathsf{Enc}_3(w,\alpha)$ as follows. We start with the empty word and process all iterations in order. At t'th iteration, if $\mathsf{Log}(w,\alpha)$ contains a triple (t,ℓ,p), then we add to y the word $0p'1^{\ell-1}$, where $p' = p - d(\ell-1) \in \{1,\ldots,d\}$; if no such triple exists, we add 0.

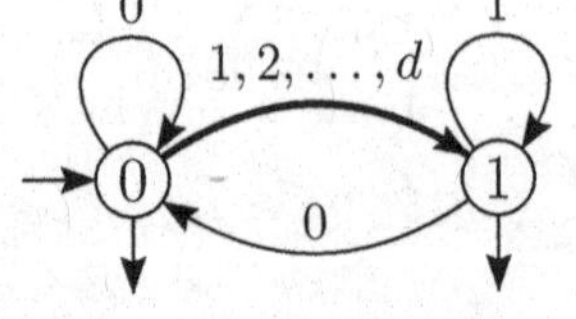

Fig. 1. A two-state DFA accepting all encoded logs (third scheme). The bold arrow depicts a multiple edge.

The resulting word y has length less than $2n$, contains exactly n zeroes, and is accepted by a two-state partial DFA $\mathcal{A}_d$ depicted in Fig. 1. Thus, the number of words of length $< 2n$ accepted by $\mathcal{A}_d$ is the upper bound for the number of logs of length-n words. Each accepted word corresponds to a unique walk in $\mathcal{A}_d$. A basic statement on counting walks in digraphs (see, e.g., [4]) says that a strongly connected digraph contains $\Theta(\mu^n)$ distinct walks of length n, where μ is the maximum positive root of the characteristic polynomial of its adjacency matrix. For $\mathcal{A}_d$, the matrix is $\left(\begin{smallmatrix}1 & d\\ 1 & 1\end{smallmatrix}\right)$, and its characteristic polynomial $(\mu-1)^2-d$ has the maximum root $\mu=\sqrt{d}+1$. As the number of walks of length $<2n$ is Θ-equivalent to the number of walks of length exactly $2n$, we see that the number of logs for length-n words is $O((d+2\sqrt{d}+1)^n)$. This estimate gives us exactly the result of Corollary 1. To slightly improve it, we use another basic fact from [4]: if W is a finite walk in a strongly connected digraph G containing $\Theta(\mu^n)$ distinct walks of length n, then the number of length-n walks in G, *avoiding the walk* W, is $\Theta((\mu-\varepsilon)^n)$ for some $\varepsilon>0$. Therefore, the lemma below proves that the number of logs is $O((d+2\sqrt{d}+1-\varepsilon)^n)$ for some $\varepsilon>0$.

Lemma 6. *For any w, the word* $\mathsf{Enc}_3(w,\alpha)$ *has no factor* 0110110.

Proof. The factor 0110110 in $\mathsf{Enc}_3(w,\alpha)$ means that $\mathsf{Log}(w,\alpha)$ contains, for some t, the triples $(t,d+1,2)$ and $(t+1,d+1,2)$. Then at t'th iteration Algorithm 1 added some letter b to the current α-free word $u_{t-1}=u'abxa$ (here a is a letter, x is a factor of length $d-1$) and deleted the suffix ab to get $u_t=u'abx$. As u_{t-1} is α-free, the last letters of u' and x are distinct. But then it is impossible to get a 2-repetition with period $d{+}1$ by adding a letter to u_t. Hence $\mathsf{Log}(w,\alpha)$ cannot contain the triples $(t,d+1,2)$ and $(t+1,d+1,2)$ simultaneously. The lemma now follows. □

Thus we improved the result of Corollary 1 to

Corollary 3. *The exponent $1+\frac{1}{d}$ is k-avoidable if $k\geq d+2\sqrt{d}+1$.*

Summarizing Corollaries 2 and 3, we obtain, *by purely information-theoretic argument*, the following result.

Theorem 1. *The exponent $1+\frac{1}{d}$ is k-avoidable if $k\geq d+\min\{6,2\sqrt{d}+1\}$.*

Adapting the Second Encoding Scheme to Exponents with a Plus. Consider $\alpha=\left(1+\frac{1}{d+1}\right)^+$. For this α, there are d possible periods of 1-repetitions (the condition $(*)$ remains valid) and $d{+}1$ periods of ℓ-repetitions for each $\ell\geq 2$: the condition $(\star)$ is changed to $(\ell-1)(d+1)\leq p\leq \ell(d+1)-1$. Because of $(*)$, Lemma 4 holds. Then we can use the second scheme without adaptation: we need $\log d$ (resp., $\log c$) bits to encode the period of 1-repetition (resp., of an ℓ-repetition for $\ell\geq 2$). The proof of Lemma 5 also remains exactly the same, as the meaning of β has not changed (the only difference is that now α equals $\left(1+\frac{1}{d+1}\right)^+$, not $1+\frac{1}{d}$). Thus, the conclusion of Lemma 5 remains true, and the alphabet of $d{+}6$ letters suffices to avoid α-powers.

For small d, we can adapt the second scheme, encoding the periods of ℓ-repetitions with $\ell\geq 2$ by $\log(d{+}1)$ bits. Again, the whole analysis from Lemma 5 remains the same, with the replacement of the constant c by the constant $d+1$. The inequality (8) becomes $2\sqrt{d+1}+1<c$, which is true whenever $k>d+1+2\sqrt{d+1}$. Thus, the analog of Theorem 1 we can obtain by our methods for exponents with a plus is

Theorem 2. *The exponent* $\left(1+\frac{1}{d}\right)^{+}$ *is* k*-avoidable if* $k \geq d+\min\{5, 2\sqrt{d}+1\}$.

Avoiding Undirected Powers. Algorithm 1 can be adjusted in an obvious way to construct undirected α-free words. When logging a run of the modified algorithm, one has to add to each triple (t, ℓ, p) with $\ell \geq 2$ a 1-bit flag indicating whether the repetition found was of the form vxv or vxv^R. We apply the second encoding scheme to the obtained logs, with a single change: in the sixth block, instead of spending $\log c$ bits for the period of a triple, we spend $(\log c+1)$ bits to store both the period and the flag. As $\log c+1 = \log 2c$, and $(*)$ is valid for both undirected $(1+\frac{1}{d})$-powers and undirected $(1+\frac{1}{d+1})^{+}$-powers, we can follow the analysis of Lemma 5 to the final inequality (8), which now takes the form $2\sqrt{2c}+1 < c$; it holds for $c \geq 10$. This gives us

Theorem 3. *The undirected exponents* $\left(1+\frac{1}{d}\right)$ *and* $\left(1+\frac{1}{d+1}\right)^{+}$ *are* k*-avoidable if* $k \geq d+10$.

4 Constructing Threshold Words: Experimental Results

To understand whether Algorithm 1 can help to prove optimal results (in particular, to get an alternative proof of Dejean's conjecture) it is natural to run the algorithm for threshold exponents and answer some obvious questions:

- Whether the algorithm apparently builds arbitrary long α-free words?
- If yes, how does the ratio between the lengths of the input and output behave? Is it stable for the same alphabet? For different alphabets?
- What repetitions (with $\ell > 1$) the algorithm finds? Is there a model that explains the observed statistics of repetitions?

A similar experiment with square-free words over different alphabets is described in [17]. Let $\alpha = \left(1+\frac{1}{d+1}\right)^{+}$ and $k = d+2$, $k \geq 5$. We slightly modify Algorithm 1 using the following two observations.

1. Since the word u reaches the length d, the prefix $u[1..d]$, consisting of d distinct letters by $(*)$, is never changed (see Remark 1). The number of iterations needed to build this prefix depends on d only; it is easy to show that w.h.p. this number is $o(d^{1+\varepsilon})$ for any $\varepsilon > 0$.
   - We replace $u = \lambda$ by $u = a_1 \cdots a_d$ in line 1
2. We call an iteration of Algorithm 1 *void*, if it leaves u unchanged (i.e., the current letter $w[t]$ is first added to u and then deleted). As we have seen in the proof of Lemma 5, the number of void iterations is, w.h.p., $n \cdot (\frac{d}{k} + O(n^{-\varepsilon}))$. The output word will remain the same if we delete all letters, corresponding to void iterations, from w, leaving a word w' of length close to $\frac{2}{k}n$. In turn, each letter of w' can be encoded by one bit: given u, we know the only two letters which will produce a non-void iteration.

– Instead of w, the input is a binary word, interpreted as the *Pansiot encoding* [13] of the word w' described above: 0 and 1 encode the letters which are not among the last d letters of u (0 encodes the letter which had occurred in u more recently)

This modification of Algorithm 1 minimizes the dependence on d and allows us to compare results over different alphabets. The parameters of our experiment were as follows: for each $k = 5, \ldots, 30$ we ran the algorithm multiple times, constructing w letter by letter uniformly at random and stopping when the prescribed length of u was reached (500 "short" runs for $|u| = 20000$ plus 500 "long" runs for $|u| = 100000$). The statistics, gathered from a run, consists of the number $n = |w|$ and the number of deletions for each period. For each alphabet, the distributions of periods of deletions for short and long runs look statistically indistinguishable, so below we analyze the aggregate data.

1. As the alphabet increases, the average "conversion ratio" between the lengths of inputs and outputs shows fast convergence to the limit $C \approx 3.57$; the ratios for $k \geq 10$ are shown in Fig. 2. Thus, the conversion ratio for the original Algorithm 1 is close to $\frac{Ck}{2}$.

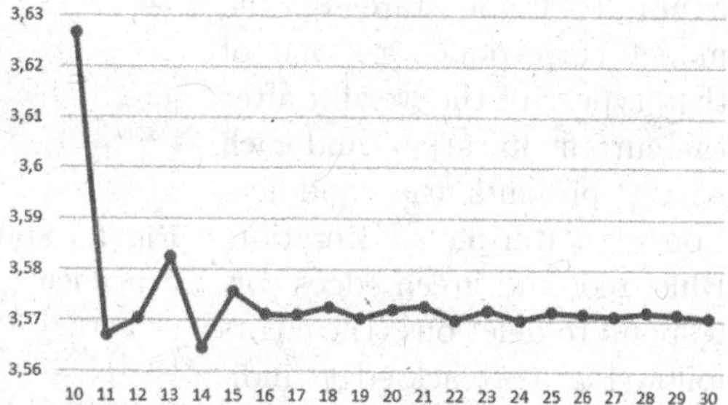

Fig. 2. Statistics of modified Algorithm 1: the dependence of average ratios between the lengths of inputs and outputs on the size of the alphabet.

2. If an ℓ-repetition vxv is found by our algorithm, then $\ell > 1$ as we excluded void iterations, all prefixes of vxv are α-free by construction, and all suffixes of vxv are α-free because of the choice in line 5. Hence vxv is a *minimal* ℓ-repetition. For $\alpha = \left(1+\frac{1}{d+1}\right)^+$ and $k = d+2$, such repetitions for small ℓ were described in [16]. In particular,

- if $\ell = 2$, the period is $k-1$ or k;
- if $\ell = 3$ or $6 \leq \ell \leq (k+2)/2$, the period is $(\ell-1)k$;
- no minimal 4- and 5-repetitions exist for $k \geq 8$.

By *frequency* of a repetition we mean the ratio $\frac{\text{number of detections}}{\text{total number of iterations}}$ computed over all experiments for fixed k. As k grows, the frequencies stabilize very quickly. Almost all detected repetitions are either 2-repetitions of period $k-1$, or 2-repetitions of period k, or 3-repetitions of period $2k$, with the frequencies ≈ 0.1578, ≈ 0.1385, and ≈ 0.0424, respectively (see Fig. 3 for $k \geq 10$). The frequency of 6-repetitions is about $5 \cdot 10^{-6}$, and, as k grows, the total frequency of all other repetitions goes below 10^{-6}.

The results of experiments suggest that the alternative proof of Dejean's conjecture is close. Indeed, there are "just" two things to do:

- develop a model for 2- and 3-repetitions to prove the observed frequencies;
- prove some (even loose) upper bound on the frequencies of longer repetitions.

While we have no idea how to prove the upper bound, we are sure that the required model can be found. As was proved in [6], the words over different alphabets, avoiding only 2- and 3-repetitions, have similar structure and can be described by a "limit" set of 2-dimensional words as $k \to \infty$.

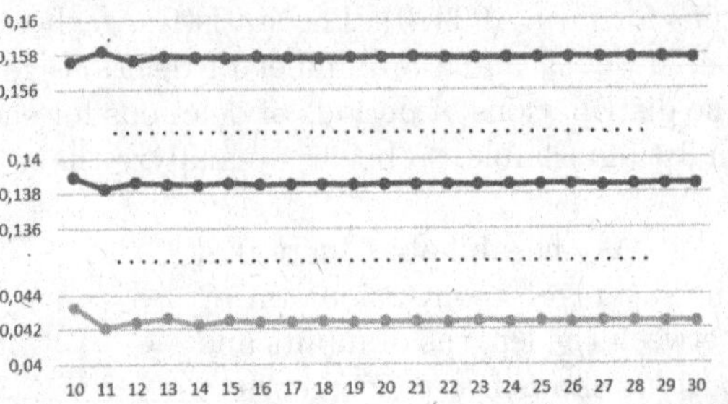

Fig. 3. Statistics of modified Algorithm 1: the dependence of average frequencies of 2-repetitions of period $k-1$ (blue), 2-repetitions of period k (red), and 3-repetitions of period $2k$ (green) on the size of the alphabet. (Color figure online)

Based on the structural results from [6], we built a 12-state Markov model (Fig. 4) which approximately represents the work of the modified Algorithm 1. Each state of the model corresponds to one of the "types" of the word u after the current iteration, and each edge represents the modification of u during an iteration. Blue, red, and green edges correspond to deletions (the colors follow Fig. 3); black edges indicate that no deletion was made. An edge with a single (resp., double) arrow has the probability 0.25 (resp., 0.5). The graph of the model is strongly connected. Hence the stationary distribution can be obtained by standard eigenvector computations for the matrix of the model. Given the stationary distribution, one immediately gets the probabilities of encountering each of three considered repetitions at the present iteration. These probabilities, rounded to four decimal digits, are 0.1600, 0.1360 and 0.0399 for 2-repetition of period $k-1$, 2-repetition of period k, and 3-repetition of period $2k$, respectively. This is close (but not equal!) to the frequencies observed in the experiments. It looks possible that some adjustment of edge probabilities will allow the model to predict precisely the observed values.

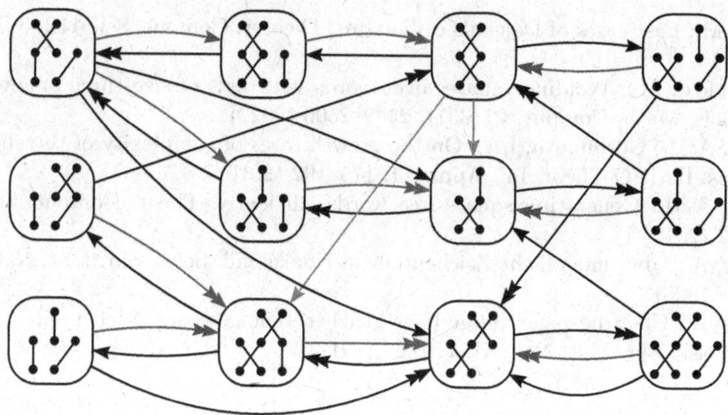

Fig. 4. A Markov model approximating the work of modified Algorithm 1. States correspond to 12 types of suffixes of the current α-free word u, edges represent the change of u at one iteration. Coloured edges correspond to repetitions (the colours follow Fig. 3). Each edge with a double arrow represents two parallel edges. The graph is 4-out-regular, so all (single) edges have the same associated probability 0.25.

References

1. Carpi, A.: On Dejean's conjecture over large alphabets. Theoret. Comput. Sci. **385**, 137–151 (1999)
2. Currie, J.D., Rampersad, N.: A proof of Dejean's conjecture. Math. Comp. **80**, 1063–1070 (2011)
3. Currie, J.D., Mol, L.: The undirected repetition threshold and undirected pattern avoidance. Theor. Comput. Sci. **866**, 56–69 (2021)
4. Cvetković, D.M., Doob, M., Sachs, H.: Spectra of graphs, 3rd edn. Theory and applications. Johann Ambrosius Barth, Heidelberg (1995)
5. Dejean, F.: Sur un théorème de Thue. J. Combin. Theory. Ser. A **13**, 90–99 (1972)
6. Gorbunova, I.A., Shur, A.M.: On Pansiot words avoiding 3-repetitions. Int. J. Found. Comput. Sci. **23**(8), 1583–1594 (2012)
7. Grytczuk, J., Kozik, J., Micek, P.: New approach to nonrepetitive sequences. Random Struct. Algorithms **42**(2), 214–225 (2013)
8. Guégan, G., Ochem, P.: A short proof that shuffle squares are 7-avoidable. RAIRO Theor. Inf. Appl. **50**(1), 101–103 (2016)
9. Keränen, V.: Abelian squares are avoidable on 4 letters. In: Kuich, W. (ed.) ICALP 1992. LNCS, vol. 623, pp. 41–52. Springer, Heidelberg (1992). https://doi.org/10.1007/3-540-55719-9_62
10. Moser, R.A., Tardos, G.: A constructive proof of the general Lovász local lemma. J. ACM 57(2), 11:1–11:15 (2010)
11. Moulin-Ollagnier, J.: Proof of Dejean's conjecture for alphabets with 5, 6, 7, 8, 9, 10 and 11 letters. Theoret. Comput. Sci. **95**, 187–205 (1992)
12. Ochem, P., Pinlou, A.: Application of entropy compression in pattern avoidance. Electron. J. Comb. **21**(2), 2 (2014)
13. Pansiot, J.J.: A propos d'une conjecture de F. Dejean sur les répétitions dans les mots. Discr. Appl. Math. 7, 297–311 (1984)

14. Rao, M.: Last cases of Dejean's conjecture. Theoret. Comput. Sci. **412**, 3010–3018 (2011)
15. Rosenfeld, M.: Avoiding squares over words with lists of size three amongst four symbols. Math. Comput. **91**(337), 2489–2500 (2022)
16. Shur, A.M., Gorbunova, I.A.: On the growth rates of complexity of threshold languages. RAIRO Theor. Inf. Appl. **44**, 175–192 (2010)
17. Shur, A.M.: Generating square-free words efficiently. Theor. Comput. Sci. **601**, 67–72 (2015)
18. Thue, A.: Über unendliche Zeichenreihen. Norske vid. Selsk. Skr. Mat. Nat. Kl. **7**, 1–22 (1906)
19. Thue, A.: Über die gegenseitige Lage gleicher Teile gewisser Zeichenreihen. Norske vid. Selsk. Skr. Mat. Nat. Kl. **1**, 1–67 (1912)

Languages Generated by Conjunctive Query Fragments of FC[REG]

Sam M. Thompson(✉) and Dominik D. Freydenberger

Loughborough University, Loughborough, UK
{s.thompson4,d.d.freydenberger}@lboro.ac.uk

Abstract. FC is a finite-model variant on the theory of concatenation, FC[REG] extends FC with regular constraints. This paper considers the languages generated by their conjunctive query fragments, FC-CQ and FC[REG]-CQ. We compare the expressive power of FC[REG]-CQ to that of various related language generators, such as regular expressions, patterns, and typed patterns. We then consider decision problems for FC-CQ and FC[REG]-CQ, and show that certain static analysis problems (such as equivalence and regularity) are undecidable. While this paper defines FC-CQ based on the logic FC, they can equally be understood as synchronized intersections of pattern languages, or as systems of restricted word equations.

Keywords: Word equations · Conjunctive queries · Expressive power · Decision problems · Descriptional complexity

1 Introduction

This paper studies the languages that are generated by fragments of FC, a variant of the theory of concatenation, that has applications in information extraction.

Word equations and the Theory of Concatenation. We begin with *word equations*, which have the form $\alpha_L \doteq \alpha_R$, where both α_L and α_R are words over an alphabet Σ of terminal symbols and an alphabet of variables. Each variable represents some word from Σ^*. These equations can be used to define languages or relations (see Karhumäki, Mignosi, and Plandowski [16]), by choosing one or more variables. For example, consider the equation $(xy \doteq \mathtt{ab})$, where x and y are variables. Then, the variable x defines the language $\{\varepsilon, \mathtt{a}, \mathtt{ab}\}$, and the pair (x, y) defines the relation $\{(\varepsilon, \mathtt{ab}), (\mathtt{a}, \mathtt{b}), (\mathtt{ab}, \varepsilon)\}$.

The *theory of concatenation* uses word equations as atomic formulas, and combines them with the usual connectives and first-order quantifiers that range over words in Σ^*. For example, the variable x in the formula $\neg\exists y\colon (x \doteq yy)$ defines the language of all words that are not a square.

Usually, only its existential-positive fragment is studied, as *satisfiability* is decidable (see e.g. [4]); and even venturing only a little beyond that quickly leads to undecidability (see e.g. Durnev [6]). Also relevant for the current paper is an extension that allows the use of *regular constraints*, that is, atoms of the form $(x \mathrel{\dot\in} \alpha)$ for any regular expression α, which express that $x \in \mathcal{L}(\alpha)$.

F. Drewes and M. Volkov (Eds.): DLT 2023, LNCS 13911, pp. 233–245, 2023.
https://doi.org/10.1007/978-3-031-33264-7_19

The Logic FC. Freydenberger and Peterfreund [11] introduced the logic FC as a finite-model version of the theory of concatenation. FC fixes one word w as the *input word* (the word for which membership in the language is tested, or the word from which information is extracted), and considers factors of w for the variables. This is in contrast to the "standard" theory of concatenation, where variables range over the infinite set Σ^*.

In FC formulas, the input word w is represented by w. For example, the FC-formula $\exists x\colon (\mathsf{w} \doteq xx)$ accepts those words w that are squares, and

$$\forall x\colon (\neg(x \doteq \varepsilon) \rightarrow \neg\exists y\colon (x \doteq yy))$$

accepts those words that do not contain a non-empty square.

The logic FC[REG] extends FC by allowing regular constraints. For example,

$$\exists x, y\colon ((\mathsf{w} \doteq xyxy) \wedge (x \mathrel{\dot\in} \mathtt{a}^+) \wedge (y \mathrel{\dot\in} \mathtt{b}^+))$$

defines the language $\{\mathtt{a}^m\mathtt{b}^n\mathtt{a}^m\mathtt{b}^n \mid m, n \geq 1\}$.

FC[REG] is motivated by information extraction. Fagin, Kimelfeld, Reiss, and Vansummeren [7] introduced *core spanners* to formalize the core of the query language AQL that is used in IBM's SystemT. Core spanners use relational algebra on top of regular expressions with variables; this allows one to query text like one queries a relational database. But for various reasons, restrictions from database theory and algorithmic finite-model theory that make relational queries tractable do not translate to spanners. In contrast to this, the canonical tractability restrictions from relational first-order logic can be adapted to FC[REG]. Moreover, its definition is much simpler than that of core spanners (see [11]), and existential-positive FC[REG] has the same expressive power as core spanners (with polynomial-time transformations in both directions).

Conjunctive Queries. One of the aforementioned tractability restrictions is based on *conjunctive queries (CQs)*, a model that has been studied extensively in database theory (see Abiteboul, Hull, and Vianu [1] as a starting point). From the logic point of view, a CQ is a series of existential quantifiers applied to a conjunction of atoms[1]. While evaluating these is still NP-hard in general, there is a tractable subclass, the so-called *acyclic* CQs. Moreover, determining whether a CQ is acyclic can be decided in polynomial time.

Freydenberger and Thompson [13] adapted this to FC and FC[REG] by introducing FC-CQs and FC[REG]-CQs. These are defined analogously to CQs, and they have the same desirable algorithmic properties. Furthermore, they allow for additional optimization, as certain queries can be split into acyclic FC-CQs. As a step towards answering the question what can (and cannot) be expressed with acyclic FC-CQs and FC[REG]-CQs, the present paper looks into the properties of FC-CQ and FC[REG]-CQ in general.

Two reasons led the authors to believe that approaching this based on languages is a natural choice. First, even though database theory usually sees

[1] In terms of SQL, it is `SELECT` applied to a `JOIN` of (potentially many) tables.

inexpressibility more as the question which relations cannot be expressed, all relation inexpressibility results for fragments of FC[REG] and equivalent models (like spanners) rely on language inexpressibility (see [7,9–11]). Second, as we shall see, FC-CQs (and FC[REG]-CQs) are conceptually very close to (typed) pattern languages and system of word equations (with regular constraints). Hence, they define rich and natural classes of languages, which makes them interesting beyond their background and application in logic.

Contributions. In Sect. 3, we first consider the closure properties of FC-CQ and FC[REG]-CQ, and show that FC-CQ and FC[REG]-CQ are both closed under intersection, but are not closed under complement. Regarding the closure under union, we show that FC-CQ is not closed under union, but whether FC[REG]-CQ is closed under union is left open. We then compare the expressive power of FC-CQ and FC[REG]-CQ to that of related language generators, such as patterns and regularly typed patterns. The results regarding expressive power are summarized in Fig. 1. To conclude this section, we show that a large and natural class of xregex can be represented by FC[REG]-CQ.

Section 4 considers the complexity bounds and decidability of various decision problems. The main result of this section is that the universality problem for FC[REG]-CQ is undecidable, and the regularity problem for FC-CQ is undecidable. In Sect. 5, we use Hartmanis' meta theorem to show non-recursive trade-offs from FC-CQs to regular expressions.

2 Preliminaries

For $n \geq 1$, let $[n] := \{1, 2, \dots, n\}$. Let $\emptyset$ denote the *empty set*. We use $|S|$ for the *cardinality* of S. If S is a subset of T then we write $S \subseteq T$ and if $S \neq T$ also holds, then $S \subset T$. The difference of two sets S and T is denoted as $S \setminus T$. We use $\mathbb{N}$ to denote the set $\{0, 1, \dots\}$ and $\mathbb{N}_+ := \mathbb{N} \setminus \{0\}$.

Let A be an alphabet. We use $|w|$ to denote the length of some word $w \in A^*$, and the word of length zero (the *empty word*) is denoted ε. The number of occurrences of some $\mathtt{a} \in A$ within w is $|w|_{\mathtt{a}}$. We write $u \cdot v$ or just uv for the concatenation of words $u, v \in A^*$. If we have words $w_1, w_2, \dots, w_n \in A^*$, then we use $\Pi_{i=1}^{n} w_i$ as shorthand for $w_1 \cdot w_2 \cdots w_n$. If $u = p \cdot v \cdot s$ for $p, s \in A^*$, then v is a *factor* of u, denoted $v \sqsubseteq u$. If $u \neq v$ also holds, then $v \sqsubset u$. Let Σ be a fixed and finite alphabet of *terminal symbols* and let Ξ be a countably infinite alphabet of *variables*. We assume that $\Sigma \cap \Xi = \emptyset$ and $|\Sigma| \geq 2$. A *language* $L \subseteq \Sigma^*$ is a set of words. If A and B are alphabets, then a function $\sigma \colon A^* \to B^*$ is a *morphism* if $\sigma(w \cdot u) = \sigma(w) \cdot \sigma(v)$ holds for all $w, u \in A^*$.

2.1 Patterns

A *pattern* is a word $\alpha \in (\Sigma \cup \Xi)^*$. A *pattern substitution* (or just *substitution*) is a partial morphism $\sigma \colon (\Sigma \cup \Xi)^* \to \Sigma^*$ where $\sigma(\mathtt{a}) = \mathtt{a}$ must hold for all $\mathtt{a} \in \Sigma$. Let $\mathsf{vars}(\alpha)$ be the set of variables in α. We always assume that the domain of σ is a superset of $\mathsf{vars}(\alpha)$. The language α generates is defined as:

$$\mathcal{L}(\alpha) := \{\sigma(\alpha) \mid \sigma \text{ is a substitution}\}.$$

Note that some literature, such as Angluin [2], defines the language of a pattern based on *non-erasing substitutions*, whereas we always assume variables can be mapped to the empty word.

Example 2.1. Let $\Sigma := \{\mathtt{a}, \mathtt{b}\}$. Consider the pattern $\alpha := \mathtt{ab}x\mathtt{ba}xyx$ and the pattern substitution $\sigma\colon (\Sigma \cup \Xi)^* \to \Sigma^*$, where $\sigma(x) = \mathtt{aa}$ and $\sigma(y) = \varepsilon$, then it follows that $\mathtt{abaabaaaaa} \in \mathcal{L}(\alpha)$. By definition $\sigma(\mathtt{a}) = \mathtt{a}$ and $\sigma(\mathtt{b}) = \mathtt{b}$.

Let $\mathsf{PAT} := \{\alpha \mid \alpha \in (\Sigma \cup \Xi)^*\}$ be the set of all patterns and let $\mathcal{L}(\mathsf{PAT})$ be the class of languages definable by a pattern. Note that we always assume some fixed Σ before discussing patterns and the languages they generate.

Typed Patterns. *Typed patterns* extend patterns with a *typing function.* For this paper, we consider regularly typed patterns where the typing function T maps each variable x to a regular language $T(x)$. Then, each variable x must be mapped to a member of $T(x)$. More formally:

Definition 2.2. *A* regularly typed pattern *is a pair $\alpha_T := (\alpha, T)$ consisting of a pattern $\alpha \in (\Sigma \cup \Xi)^*$, and a* typing function *T that maps each $x \in \mathsf{vars}(\alpha)$ to a regular language $T(x) \subseteq \Sigma^*$.*

We denote the language α_T generates as $\mathcal{L}(\alpha_T)$ and this language is defined as $\{\sigma(\alpha) \mid \sigma$ is a substitution where $\sigma(x) \in T(x)$ for all $x \in \mathsf{vars}(\alpha)\}$.

The set of regular typed patterns is denoted by $\mathsf{PAT[REG]}$, and the class of languages definable by a regularly typed pattern is denoted by $\mathcal{L}(\mathsf{PAT[REG]})$.

Example 2.3. Consider $(x\mathtt{b}x, T)$, where $T(x) = \mathtt{a}^+$. Then, for any alphabet Σ where $\mathtt{a}, \mathtt{b} \in \Sigma$, we have that $\mathcal{L}(x\mathtt{b}x, T) = \{\mathtt{a}^n\mathtt{ba}^n \mid n \in \mathbb{N}_+\}$, which is neither a regular language nor a pattern language.

Typed pattern languages have been considered in the context of *learning theory*, for example, see Geilke and Zilles [14] or Koshiba [17]. Schmid [21] compared the expressive power of regularly typed patterns to REGEX and *pattern expressions*.

2.2 FC Conjunctive Queries

A *word equation* is an equation $(\alpha_L \doteq \alpha_R)$, where $\alpha_L, \alpha_R \in (\Sigma \cup \Xi)^*$. For a substitution $\sigma\colon (\Sigma \cup \Xi)^* \to \Sigma^*$, we write $\sigma \models (\alpha_L \doteq \alpha_R)$, if $\sigma(\alpha_L) = \sigma(\alpha_R)$. Let $\mathsf{w} \in \Xi$ be a distinguished variable known as the *universe variable.* For a substitution $\sigma\colon (\Sigma \cup \Xi)^* \to \Sigma^*$, we say that σ is w*-safe* if $\sigma(x) \sqsubseteq \sigma(\mathsf{w})$ for all $x \in \Xi$. Now, we define the syntax and languages of FC-CQ.

Definition 2.4. *An* FC-CQ *is denoted as $\varphi := \bigwedge_{i=1}^n (x_i \doteq \alpha_i)$ where $x_i \in \Xi$, and where $\alpha_i \in (\Sigma \cup \Xi)^*$ for each $i \in [n]$. For a w-safe substitution σ, we write $\sigma \models \varphi$ if $\sigma \models (x_i \doteq \alpha_i)$ for all $i \in [n]$. Then, $\mathcal{L}(\varphi) := \{\sigma(\mathsf{w}) \mid \sigma \models \varphi\}$.*

We use $\mathcal{L}(\mathsf{FC\text{-}CQ})$, to define the class of languages definable in FC-CQ.

Example 2.5. In Example 2.3, we defined a regular typed pattern language that defines the language $\{\mathtt{a}^n\mathtt{b}\mathtt{a}^n \mid n \in \mathbb{N}_+\}$. Now consider:

$$\varphi := (\mathsf{w} \doteq x \cdot \mathtt{b} \cdot x) \wedge (x \doteq \mathtt{a} \cdot y) \wedge (x \doteq y \cdot \mathtt{a}).$$

If $vu = uv$ for $u, v \in \Sigma^*$, then there is some $z \in \Sigma^*$ and $k_1, k_2 \in \mathbb{N}$ such that $u = z^{k_1}$ and $v = z^{k_2}$ (for example, see Proposition 1.3.2 in Lothaire [19]). Thus, if $\sigma \models \varphi$, then $\sigma(\mathsf{w}) = \mathtt{a}^n \cdot \mathtt{b} \cdot \mathtt{a}^n$. Consequently, $\mathcal{L}(\varphi) = \{\mathtt{a}^n\mathtt{b}\mathtt{a}^n \mid n \in \mathbb{N}_+\}$.

In [11], FC was extended to FC[REG] by adding regular constraints. We extend FC-CQ to FC[REG]-CQ in the same way.

Definition 2.6. *Let* FC[REG]-CQ *be the set of* FC-CQ*-formulas extended with regular constraints. That is, formulas of the form* $\varphi := \psi \wedge \bigwedge_{j=1}^{m} (y_i \,\dot{\in}\, \gamma_i)$ *where* ψ *is an* FC-CQ*-formula,* $y_j \in \Xi$ *is a variable for all* $j \in [m]$*, and* γ_j *is a regular expression for all* $j \in [m]$*.*

For a w*-safe substitution* σ*, we write* $\sigma \models \varphi$ *if* $\sigma \models \psi$*, and* $\sigma(y_j) \in \mathcal{L}(\gamma_j)$ *for all* $j \in [m]$*. Let* $\mathcal{L}(\varphi) := \{\sigma(\mathsf{w}) \mid \sigma \models \varphi\}$*.*

Also, we extend FC[REG]-CQ to FC[REG]-UCQ (unions of FC[REG]-CQ formulas) canonically. More formally, if $\varphi_i \in$ FC[REG]-CQ for all $i \in [m]$, then $\varphi := \bigvee_{i=1}^{m} \varphi_i$ is an FC[REG]-UCQ. We define $\mathcal{L}(\varphi) := \bigcup_{i=1}^{m} \mathcal{L}(\varphi_i)$. The class FC-UCQ is defined analogously.

Example 2.7. The language $L_< := \{\mathtt{a}^n\mathtt{b}\mathtt{a}^m \mid n, m \in \mathbb{N} \text{ and } n < m\}$ can be expressed in FC[REG]-CQ using the following formula:

$$\varphi_< := (\mathsf{w} \doteq x \cdot \mathtt{b} \cdot x \cdot y) \wedge (x \,\dot{\in}\, \mathtt{a}^*) \wedge (y \,\dot{\in}\, \mathtt{a}^+).$$

We can represent the language $L_{\neq} := \{\mathtt{a}^n\mathtt{b}\mathtt{a}^m \mid n, m \in \mathbb{N} \text{ and } n \neq m\}$ in FC[REG]-UCQ using a union of two queries analogous to $\varphi_<$. However, as we observe with the following formula, $L_{\neq}$ can also be expressed in FC[REG]-CQ:

$$\psi_{\neq} := (\mathsf{w} \doteq x \cdot y_{\mathtt{b},1} \cdot z \cdot y_{\mathtt{b},2} \cdot x) \wedge (\mathsf{w} \,\dot{\in}\, \mathtt{a}^*\mathtt{b}\mathtt{a}^*) \wedge \\ (z \,\dot{\in}\, \mathtt{a}^+) \wedge (x \,\dot{\in}\, \mathtt{a}^*) \wedge (y_{\mathtt{b}} \doteq \mathtt{b}) \wedge (y_{\mathtt{b}} \doteq y_{\mathtt{b},1} \cdot y_{\mathtt{b},2}).$$

The use of $y_{\mathtt{b}}$ ensures that $y_{\mathtt{b},1} = \mathtt{b}$ and $y_{\mathtt{b},2} = \varepsilon$, or vice versa. Thus, the word $\sigma(z) \in \mathtt{a}^+$, for any substitution σ where $\sigma \models \psi_{\neq}$, appears one side of the $\mathtt{b}$ symbol. This ensures that $\sigma(\mathsf{w}) = \mathtt{a}^m\mathtt{b}\mathtt{a}^n$ and $m \neq n$.

From time to time, we shall make reference to the *existential positive fragment of* FC *and* FC[REG], denoted EP-FC and EP-FC[REG] respectively. This is simply an existential positive first-order logic, where each atom is of the form $x \doteq \alpha$, for some $x \in \Xi$ and $\alpha \in (\Sigma \cup \Xi)^*$. For details, see [11]. While we do not give the formal definitions of EP-FC and EP-FC[REG] in this paper, we note that $\mathcal{L}$(EP-FC) = $\mathcal{L}$(FC-UCQ) and $\mathcal{L}$(EP-FC[REG]) = $\mathcal{L}$(FC[REG]-UCQ). This follows from the fact that FC-UCQ (or FC[REG]-UCQ) is simply a disjunctive normal form of EP-FC (or EP-FC[REG]).

Immediately from the definitions, we know that $\mathcal{L}$(FC-CQ) $\subseteq$ $\mathcal{L}$(EP-FC), and $\mathcal{L}$(FC[REG]-CQ) $\subseteq$ $\mathcal{L}$(EP-FC[REG]). In Sect. 3, we shall look at the expressive power of FC-CQ and FC[REG]-CQ in more detail.

The Theory of Concatenation. Before moving on, let us briefly discuss the theory of concatenation; in particular, the *existential positive fragment of the theory of concatenation*, which we shall denote as EP-C. Informally, EP-C can be seen as a logic that extends word equations with conjunction, disjunction, and existential quantification. While the syntax is similar to EP-FC, the semantics of EP-C allow variables to range over Σ^*. Thus, unlike with EP-FC, there is no separation between the membership problem (does a formula hold for a given word), and the satisfiability problem (does a formula hold for any word).

Regarding the class of languages $\mathcal{L}$(EP-C) generated by EP-C, it is known that this is exactly the class of languages generated by word equations. For more information, see Sect. 5.3 in Chap. 6 of [20]. We use EP-C[REG] to denote the existential theory of concatenation extended with regular constraints. The authors refer the reader to [11] for more information on the theory of concatenation, and its connection to FC.

3 Expressive Power and Closure Properties

This section considers the expressive power and closure properties of FC-CQ and FC[REG]-CQ. The main result is Theorem 3.3, which presents an inclusion diagram of the relative expressive power of FC[REG]-CQ and related models.

Lemma 3.1. *For $\varphi \in$ FC-CQ, we have that $\mathcal{L}(\varphi) = \Sigma^*$ if and only if $\varepsilon, \mathtt{a} \in \mathcal{L}(\varphi)$ for some $\mathtt{a} \in \Sigma$.*

To prove Lemma 3.1, we consider a normal-form for FC-CQ, where each formula can be represented by $\varphi := \bigwedge_{i=1}^{n}(\mathsf{w} \doteq \alpha_i)$. Then, for $\varepsilon \in \mathcal{L}(\varphi)$ to hold, each α_i must solely consist of variables. Furthermore, if $\sigma \models \varphi$ where $\sigma(\mathsf{w}) = \mathtt{a}$, then we can construct the substitution τ, where $\tau \models \varphi$ and $\tau(\mathsf{w}) = w$ for any $w \in \Sigma^*$.

Lemma 3.1 immediately implies that FC-CQ is not closed under union (and thus, $\mathcal{L}$(FC-CQ) $\subset \mathcal{L}$(EP-C)). However, we shall also use Lemma 3.1 in Sect. 4 as a tool to help prove the complexity bounds for the *universality problem.*

Proposition 3.2. *$\mathcal{L}$(FC-CQ) and $\mathcal{L}$(FC[REG]-CQ) are closed under intersection and are not closed complement. $\mathcal{L}$(FC-CQ) is not closed under union.*

The proof that $\mathcal{L}$(FC[REG]-CQ) is closed under intersection follows directly from the fact that if $\varphi, \psi \in$ FC[REG]-CQ, then $\varphi \wedge \psi \in$ FC[REG]-CQ. To show that $\mathcal{L}$(FC[REG]-CQ) is not closed under complement, we show that if $\mathcal{L}$(FC[REG]-CQ) is indeed closed under complement, then there exists $\varphi \in$ FC[REG]-CQ such that $\mathcal{L}(\varphi)$ is the so-called *uniform-0-chunk language*. This language is defined as all words $w \in \{\mathtt{0}, \mathtt{1}\}^*$ such that $w = s_1 \cdot \prod_{i=1}^{n}(t \cdot s_i)$ for any $n \geq 1$, where $t \in \{\mathtt{0}\}^+$ and $s_i \in \{\mathtt{1}\}^+$ for all $i \in [n]$.

The uniform-0-chunk language was shown not to be expressible by core spanners [7]. Immediately from Theorem 5.9 in [11], it follows that the uniform-0-chunk language cannot be expressed by EP-FC[REG], which generates the same class of languages as FC[REG]-UCQ. The fact that $\mathcal{L}$(FC-CQ) is not closed under union or complement follows immediately from Lemma 3.1, using the languages $\{\varepsilon\}$ and $\{\mathtt{a}\}$. Whether FC[REG]-CQ is closed under union is left open.

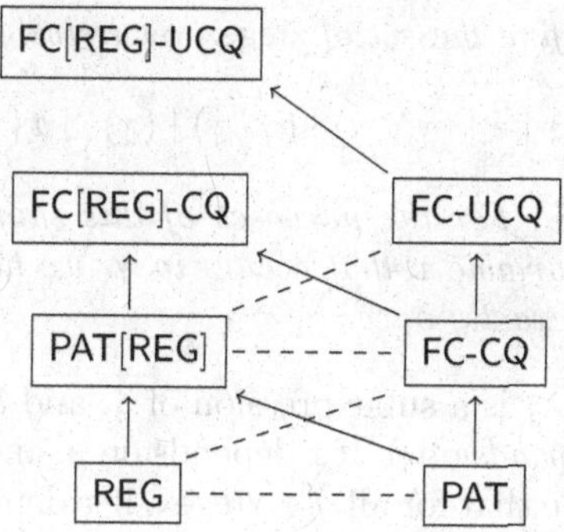

Fig. 1. A directed edge from A to B denotes that $\mathcal{L}(\mathsf{A}) \subset \mathcal{L}(\mathsf{B})$. A dashed, undirected edge between A and B denotes that $\mathcal{L}(\mathsf{A})$ and $\mathcal{L}(\mathsf{B})$ are incomparable.

Theorem 3.3. *Figure 1 describes the relations between language classes.*

The exact relationship between FC-UCQ and FC[REG]-CQ remains open. From [11], we know that FC-UCQ cannot express all regular languages. Therefore, there are languages $L \subseteq \Sigma^*$ such that $L \in$ FC[REG]-CQ \ FC-UCQ. The authors conjecture that $\mathcal{L}$(FC-UCQ) and $\mathcal{L}$(FC[REG]-CQ) are incomparable.

We leave the following problems open:

Open Problem 1. $\mathcal{L}$(FC-UCQ) = $\mathcal{L}$(EP-C)?
Open Problem 2. $\mathcal{L}$(FC[REG]-CQ) = $\mathcal{L}$(FC[REG]-UCQ)?
Open Problem 3. $\mathcal{L}$(FC[REG]-CQ) = $\mathcal{L}$(EP-C[REG])?

The authors conjecture that $\mathcal{L}$(FC[REG]-CQ) $\subset$ $\mathcal{L}$(FC[REG]-UCQ), and therefore conjecture that $\mathcal{L}$(FC[REG]-CQ) $\subset$ $\mathcal{L}$(EP-C[REG]).

3.1 Connection to Related Models

While the left-hand side of an FC-CQ atom is not necessarily w, we can simply re-write formulas of the form $(x \doteq \alpha)$ as $(\mathsf{w} \doteq p_x \cdot \alpha \cdot s_x) \wedge (\mathsf{w} \doteq p_x \cdot x \cdot s_x)$, where $p_x, s_x \in \Xi$ are new and unique variables. Hence, every FC-CQ-formula can be represented as $\varphi := \bigwedge_{i=1}^{n} (\mathsf{w} \doteq \alpha_i)$, and therefore can be understood as a system of word equations $\{(\mathsf{w} \doteq \alpha_1), (\mathsf{w} \doteq \alpha_2), \ldots, (\mathsf{w} \doteq \alpha_n)\}$ that generates the language $\{\sigma(\mathsf{w}) \mid \sigma \models (\mathsf{w} \doteq \alpha_i) \text{ for all } i \in [n]\}$.

Consequently, FC-CQ languages can be seen as a natural extension of pattern languages, or as the languages of restricted cases of systems of word equations. FC[REG]-CQ analogously extends regularly typed pattern languages.

We further explore the connection between FC[REG]-CQ and related language generators by considering a restricted case of xregex. These are regular expressions that are extended with a repetition operator that allows for the definition of non-regular languages, and that is available in most modern regular expression implementation (see for example [12], which also contains a discussion of some peculiarities of these implementations).

Definition 3.4. *We define the set of* xregex *recursively:*

$$\gamma := \emptyset \mid \varepsilon \mid \mathsf{a} \mid (\gamma \vee \gamma) \mid (\gamma \cdot \gamma) \mid (\gamma)^* \mid x\{\gamma\} \mid \&x,$$

where $\mathsf{a} \in \Sigma$ *and* $x \in \Xi$. *For the purposes of this short section on* xregex, *we assume that for every variable* x *that occurs in* γ, *we have that* $x\{\lambda\}$, *for some* $\lambda \in$ xregex, *must occur exactly once.*

Let $\gamma \in$ xregex. If $x\{\lambda\}$ is a subexpression of γ, and $\&y$ is a subexpression of λ, then we say that x *depends* on y. If x depends on y, and y depends on z, then x depends on z. We assume that for all $\gamma \in$ xregex, if x depends on y, then y cannot depend on x. Furthermore, we assume that x cannot depend on x. This avoids defining the language for troublesome expressions such as $x\{\mathsf{a}\cdot\&y\cdot\&x\}\cdot y\{\&x\cdot\mathsf{b}\}$.

Next, we look at the language generated by an xregex. Informally, an xregex of the form $x\{\gamma\}$ matches the same words as γ, and also "stores" the matched word in x. Then, any occurrence of $\&x$ repeats the word stored in x. To define the language generated by an xregex, we use *ref-words*, which were originally introduced by Schmid [22]. We omit the formal definition due to space constraints.

Example 3.5. Let $\gamma_2 := x\{\mathsf{a} \cdot \Sigma^*\} \cdot (\&x)^*$. Then, $\mathcal{L}(\gamma_2)$ is the language of words of the form $(\mathsf{a} \cdot w)^n$ where $w \in \Sigma^*$ and $n \geq 1$.

While the term regular expression is sometimes used to refer to expressions that contain variables, we reserve the term regular expression for its more "classical" definition.

Definition 3.6. *We call* $\gamma \in$ xregex *a* regex path *if for every subexpression of the form* $(\lambda)^*$, *we have that* λ *is a regular expression, and for every subexpression of the form* $(\lambda_1 \vee \lambda_2)$, *we have that* λ_1 *and* λ_2 *are regular expressions. Let* XRP *denote the set of regex paths.*

Regex paths were considered in [10] (in the context of document spanners). Using a proof similar to the proof of Lemma 3.19 in [10] and the proof of Lemma 3.6 in [13], we show the following:

Proposition 3.7. *For every synchronized* $\gamma \in$ XRP, *we can effectively construct a formula* $\varphi \in$ FC[REG]-CQ, *such that* $\mathcal{L}(\varphi) = \mathcal{L}(\gamma)$.

Thus, a large and natural class of xregex can be represented by FC[REG]-CQ, and the conversion is lightweight and straightforward. If one is willing to go beyond FC[REG]-CQs, this could be adapted to larger classes of xregex, although simulating some edge cases in the definition of the latter might make the formulas a bit unwieldy. But a big advantage of logic over xregex is that optimization techniques from relational first-order logic (such as acyclicity [13] or bounded-width [11]) directly lead to tractable fragments.

Furthermore, since we are able to use conjunction directly in an FC[REG]-CQ-formula, FC[REG]-CQ can be used as a more compact representation of XRP.

4 Decision Problems

The first decision problem we shall look at is the membership problem for FC-CQ. This problem is given $w \in \Sigma^*$ and a formula $\varphi \in$ FC-CQ as input, and decides whether $w \in \mathcal{L}(\varphi)$. But first, we define the class of *regular patterns*:

Definition 4.1. *A pattern $\alpha \in (\Sigma \cup \Xi)^*$ is a* regular pattern *if $|\alpha|_x = 1$ for every variable $x \in \mathsf{vars}(\alpha)$. An* FC-CQ *consists only of regular patterns if it is of the form $\bigwedge_{i=1}^{n}(\mathsf{w} \doteq \alpha_i)$ where α_i is a regular pattern for each $i \in [n]$.*

The regular patterns have desirable algorithmic properties (although this comes at the expense of expressive power). For example, the membership problem for regular pattern languages can be solved in linear time [23]. When extending regular patterns to FC-CQ, these desirable algorithmic properties do not carry over.

Theorem 4.2. *The membership problem for* FC-CQ *is* NP*-complete, and remains* NP*-hard even if the formula consists only of regular patterns and the input word is of length one.*

The authors note that the fact that the membership problem for FC-CQ is NP-complete follows from [13], where it was shown that the so-called *model checking problem* for FC-CQ is NP-complete. However, Theorem 4.2 improves upon the result in [13] by showing NP-hardness holds even if the formula consists only of regular patterns and the input word is of length one. Consequently, restricting the size of the input word does not yield a tractable fragment of FC-CQ, even for those FC-CQs that only consist of regular patterns.

Our next focus is on static analysis problems.

Definition 4.3. *For each of* FC-CQ *and* FC[REG]-CQ*, we define the following static analysis problems:*

1. Universality: *Is $\mathcal{L}(\varphi) = \Sigma^*$?*
2. Emptiness: *Is $\mathcal{L}(\varphi) = \emptyset$?*
3. Regularity: *Is $\mathcal{L}(\varphi)$ regular?*
4. Equivalence: *Is $\mathcal{L}(\varphi_1) = \mathcal{L}(\varphi_2)$?*

First, we consider the complexity bounds of the universality problem for FC-CQ.

Corollary 4.4. *Universality is* NP*-complete for* FC-CQ.

This result follows directly from Lemma 3.1 and Theorem 4.2. That is, to determine whether $\mathcal{L}(\varphi) = \Sigma^*$ for a given $\varphi \in$ FC-CQ, we simply need to determine whether $\varepsilon, \mathtt{a} \in \mathcal{L}(\varphi)$ for some $\mathtt{a} \in \Sigma$, which can be done in NP.

Proposition 4.5. *Emptiness is* PSPACE*-complete for* FC[REG]-CQ*, and emptiness is* NP*-hard and in* PSPACE *for* FC-CQ.

The PSPACE upper bound is immediately determined from the fact that the *satisfiability problem* for EP-FC[REG] is in PSPACE [11]; which follows from [4]. Furthermore, the PSPACE lower bounds for FC[REG]-CQ is proven using a reduction from the *intersection problem for regular expressions.*

Showing the exact complexity bounds of the emptiness problem for FC-CQ seems rather difficult. This is because the emptiness problem for FC-CQ is two-way polynomial-time reducible from the satisfiability problem for word equations. More precisely, the word equation $(\alpha_L \doteq \alpha_R)$ is satisfiable if and only if the language generated by $\varphi := (\mathsf{w} \doteq \alpha_L) \wedge (\mathsf{w} \doteq \alpha_R)$ is non-empty. Furthermore, from $\varphi := \bigwedge_{i=1}^{n} (x_i \doteq \alpha_i)$, one can construct – in polynomial time – a word equation $(\alpha_L \doteq \alpha_R)$ such that $\mathcal{L}(\varphi) \neq \emptyset$ if and only if $(\alpha_L \doteq \alpha_R)$ is satisfiable; see Sect. 5.3 in Chap. 6 of [20]. Thus, the emptiness problem for FC-CQ is a reformulation of the big open problem as to whether word equation satisfiability is in NP. See [3,5] for more details on word equation satisfiability.

Next, we consider universality for FC[REG]-CQ and regularity for FC-CQ. To prove that these problems are undecidable, we use *extended Turing machines.* While these extended Turing machines were introduced for the particulars of xregex (see [8]), we observe that they are useful for our purposes. They have also been used to show universality for EP-FC is undecidable (Theorem 4.7 of [11]). Due to space constraints, the formal definitions have been omitted.

Informally, an extended Turing machine is a Turing machine over a binary alphabet $\{0, 1\}$, which has an extra instruction called the CHECK_R-instruction. If $\delta(\mathsf{a}, q) = (\mathsf{CHECK}_R, p)$, then the machine immediately checks – without moving the head's position – whether the tape to the right of the head's current position only contains blank symbols. If the right-side of the tape is blank, then the machine moves to state p. Otherwise, the machine stays in state q. Hence, if the right-side of the tape is not blank, then the machine continuously remains in the same state, and thus we enter an infinite loop.

Without going into details, the CHECK_R-instruction is used in-place of meta-symbols used to mark the start and end of the input word, which allows us to use a truly binary alphabet. See [8] for more details on the CHECK_R-instruction.

For an extended Turing machine $\mathcal{X}$, let us define a language $\mathsf{VALC}(\mathcal{X})$ as:

$$\mathsf{VALC}(\mathcal{X}) := \{\#\#\mathsf{enc}(C_1)\#\# \cdots \#\#\mathsf{enc}(C_n)\#\# \mid (C_i)_{i=1}^{n} \text{ is an accepting run}\},$$

where $\mathsf{enc}(C_i)$ is an encoding of the configuration C_i of $\mathcal{X}$, for each $i \in [n]$.

Thus, $\mathsf{VALC}(\mathcal{X})$ encodes the so-called computational history of every accepting run of $\mathcal{X}$. Let us also define $\mathsf{INVALC}(\mathcal{X}) := \Sigma^* \setminus \mathsf{VALC}(\mathcal{X})$. The language $\mathsf{INVALC}(\mathcal{X})$ can be thought of as the language of computational histories of every erroneous run of $\mathcal{X}$. Similarly to standard Turing machines, deciding whether $\mathsf{INVALC}(\mathcal{X}) = \Sigma^*$ is not semi-decidable, and deciding whether $\mathsf{INVALC}(\mathcal{X})$ is a regular language is neither semi-decidable nor co-semi-decidable.

Recall Corollary 4.4, which states that the universality problem for FC-CQ is NP-complete. Somewhat surprisingly, when we add regular constraints, the universality problem becomes undecidable.

Theorem 4.6. *For* FC[REG]-CQ, *universality is not semi-decidable, and regularity is neither semi-decidable, nor co-semi-decidable.*

To prove Theorem 4.6, we give a construction that takes an extended Turing machine $\mathcal{X}$, and returns $\varphi \in \mathsf{FC[REG]\text{-}CQ}$ such that $\mathcal{L}(\varphi) = \mathsf{INVALC}(\mathcal{X})$.

Observing Corollary 4.4, we know that the universality problem for FC-CQs is NP-complete. Therefore, we cannot effectively construct a formula $\varphi \in \mathsf{FC\text{-}CQ}$ such that $\mathcal{L}(\varphi) = \mathsf{INVALC}(\mathcal{X})$ for a given extended Turing machine $\mathcal{X}$ (otherwise, the emptiness problem for extended Turing machines would be NP-complete). However, using a similar proof idea to Theorem 4.6, we are able to conclude that the regularity problem for FC-CQ is undecidable.

Theorem 4.7. *The regularity problem for* FC-CQ *is neither semi-decidable, nor co-semi-decidable.*

In order to prove Theorem 4.6 and Theorem 4.7, we look at all possible errors that prohibit some $w \in \Sigma^*$ from being in $\mathsf{VALC}(\mathcal{X})$. Then, each of these errors can be encoded as an FC[REG]-CQ or an FC-CQ. However, due to the fact that FC[REG]-CQ and FC-CQ do not have disjunction, we require some encoding gadgets to simulate disjunction using concatenation. Referring back to Example 2.7, we can see a simple example of simulating disjunction in FC[REG]-CQ.

To prove Theorem 4.7, we show how to convert an extended Turing machine $\mathcal{X}$ into $\varphi \in \mathsf{FC\text{-}CQ}$, such that $\mathcal{L}(\varphi) = 0 \cdot \# \cdot 0 \cdot \#^3 \cdot \mathsf{INVALC}(\mathcal{X}) \cdot \#^3$. Thus:

Corollary 4.8. *The equivalence problem for* FC-CQ *is neither semi-decidable, nor co-semi-decidable.*

While these undecidability results are themselves of interest, in the subsequent section, we consider their implications with regards to *minimization* and *non-recursive trade-offs.*

5 Descriptional Complexity

In this section, we consider some of the consequences of the aforementioned undecidability results. First, we look at *minimization.* To examine the problem of minimization, we first must discuss what complexity measure we wish to minimize for. Instead of giving a explicit measure, such as the length, we give the more general definition of a *complexity measure* from Kutrib [18].

Definition 5.1. *Let* $\mathcal{F} \in \{\mathsf{FC\text{-}CQ}, \mathsf{REG}\}$, *where* REG *is the set of regular expressions. A* complexity measure *for* $\mathcal{F}$ *is a recursive function* $c\colon \mathcal{F} \to \mathbb{N}$ *such that the elements of* $\mathcal{F}$ *can be enumerated effectively in increasing order, and for every* $n \in \mathbb{N}$, *there exist finitely many* $\varphi \in \mathcal{F}$ *with* $c(\varphi) = n$.

From Theorem 4.7, we conclude that FC-CQ cannot be minimized effectively:

Corollary 5.2. *Let* c *be a complexity measure for* FC-CQ. *There is no algorithm that given* $\varphi \in \mathsf{FC\text{-}CQ}$, *constructs* $\psi \in \mathsf{FC\text{-}CQ}$ *such that* $\mathcal{L}(\varphi) = \mathcal{L}(\psi)$ *and* ψ *is* c*-minimal.*

Given complexity measures c_1 and c_2 for FC-CQ and REG (respectively), we say that there is a *non-recursive trade-off* from FC-CQ to REG if for every recursive function $f: \mathbb{N} \rightarrow \mathbb{N}$, there exists $\varphi \in \mathsf{FC\text{-}CQ}$ such that $\mathcal{L}(\varphi) \in \mathcal{L}(\mathsf{REG})$, but $c_2(\gamma) > f(c_1(\varphi))$ holds for every $\gamma \in \mathsf{REG}$ with $\mathcal{L}(\varphi) = \mathcal{L}(\gamma)$.

Hartmanis' meta theorem [15] allows us to draw conclusions about the relative succinctness of models from certain undecidability results. Thus, we can conclude the following.

Theorem 5.3. *The trade-off from* FC-CQ *to* REG *is non-recursive.*

Less formally, Theorem 5.3 states that even for those FC-CQs that generate a regular language, the size blowup from the FC-CQ to the regular expression that accepts the same language is not bounded by any recursive function.

While Theorem 5.3 also shows that the trade-off from FC[REG]-CQ to regular expressions is non-recursive, this seems less surprising. Purely from the definition of FC-CQ, it does not seem like FC-CQ should be able to generate many complicated regular languages. Thus, the fact that the size blowup from FC-CQ to an equivalent regular expression is not bounded by any recursive function highlights the deceptive complexity of languages generated by FC-CQ.

6 Conclusions

This paper studies FC-CQ and FC[REG]-CQ, with a particular focus on language theoretic questions. Regarding the expressive power, Fig. 1 gives an inclusion diagram. However, there are still many open problems, such as whether $\mathcal{L}(\mathsf{FC[REG]\text{-}CQ})$ is closed under union. Furthermore, it is not known whether $\mathcal{L}(\mathsf{EP\text{-}C}) = \mathcal{L}(\mathsf{FC\text{-}UCQ})$. As $\mathcal{L}(\mathsf{FC\text{-}UCQ}) = \mathcal{L}(\mathsf{EP\text{-}FC})$, this question is particularly fundamental, as it asks whether the finite-model restriction of FC decreases the expressive power compared to the theory of concatenation (see [11]).

With regards to decision problems, we show that the membership problem for FC-CQ is NP-complete, and remains NP-hard even if the input word is of length one, and the formula only consists of regular patterns. Restrictions like acyclicity (see [13]) and bounded width (see [11]) lead to tractable fragments; however, there is still a lot of research to be done on identifying further tractable fragments and comparing their expressive power. The main technical contribution of this paper is that the universality problem for FC[REG]-CQ, and the regularity problem for FC-CQ is undecidable.

Acknowledgements. This work was funded by EPSRC grant EP/T033762/1. The authors would like to thank Joel D. Day, Anthony W. Lin, Ana Sălăgean, and the anonymous reviewers for their helpful feedback.

Data Availability. Due to the theoretical nature of the research presented in this paper, no data was captured, generated, or analysed for this work.

References

1. Abiteboul, S., Hull, R., Vianu, V.: Foundations of Databases. Addison-Wesley, Boston (1995)
2. Angluin, D.: Finding patterns common to a set of strings. J. Comput. Syst. Sci. **21**(1), 46–62 (1980)
3. Day, J.D., Manea, F., Nowotka, D.: The hardness of solving simple word equations. In: Proceedings of MFCS 2017, pp. 18:1–18:14 (2017)
4. Diekert, V., Gutierrez, C., Hagenah, C.: The existential theory of equations with rational constraints in free groups is PSPACE-complete. Inf. Comput. **202**(2), 105–140 (2005)
5. Diekert, V., Robson, J.M.: Quadratic word equations. Jewels are Forever: Contributions on Theoretical Computer Science in Honor of Arto Salomaa, pp. 314–326 (1999)
6. Durnev, V.G.: Undecidability of the positive ∀∃ 3-theory of a free semigroup. Siberian Math. J. **36**(5), 917–929 (1995)
7. Fagin, R., Kimelfeld, B., Reiss, F., Vansummeren, S.: Document spanners: a formal approach to information extraction. J. ACM **62**(2), 12 (2015)
8. Freydenberger, D.D.: Extended regular expressions: succinctness and decidability. Theory Comput. Syst. **53**(2), 159–193 (2013)
9. Freydenberger, D.D.: A logic for document spanners. Theory Comput. Syst. **63**(7), 1679–1754 (2019)
10. Freydenberger, D.D., Holldack, M.: Document spanners: from expressive power to decision problems. Theory Comput. Syst. **62**(4), 854–898 (2018)
11. Freydenberger, D.D., Peterfreund, L.: The theory of concatenation over finite models. In: Proceedings of ICALP 2021, pp. 130:1–130:17 (2021)
12. Freydenberger, D.D., Schmid, M.L.: Deterministic regular expressions with back-references. J. Comput. Syst. Sci. **105**, 1–39 (2019)
13. Freydenberger, D.D., Thompson, S.M.: Splitting spanner atoms: a tool for acyclic core spanners. In: Proceedings of ICDT 2022, pp. 6:1–6:18 (2022)
14. Geilke, M., Zilles, S.: Polynomial-time algorithms for learning typed pattern languages. In: International Conference on Language and Automata Theory and Applications, pp. 277–288 (2012)
15. Hartmanis, J.: On Gödel speed-up and succinctness of language representations. Theor. Comput. Sci. **26**(3), 335–342 (1983)
16. Karhumäki, J., Mignosi, F., Plandowski, W.: The expressibility of languages and relations by word equations. J. ACM **47**(3), 483–505 (2000)
17. Koshiba, T.: Typed pattern languages and their learnability. In: European Conference on Computational Learning Theory, pp. 367–379 (1995)
18. Kutrib, M.: The phenomenon of non-recursive trade-offs. Int. J. Found. Comput. Sci. **16**(05), 957–973 (2005)
19. Lothaire, M.: Combinatorics on Words, vol. 17. Cambridge University Press, Cambridge (1997)
20. Rozenberg, G., Salomaa, A. (eds.): Handbook of Formal Languages, Volume 1: Word, Language, Grammar. Springer, Cham (1997)
21. Schmid, M.L.: Inside the class of regex languages. In: Proceedings of DLT 2012, pp. 73–84 (2012)
22. Schmid, M.L.: Characterising REGEX languages by regular languages equipped with factor-referencing. Inf. Comput. **249**, 1–17 (2016)
23. Shinohara, T.: Polynomial time inference of extended regular pattern languages. In: RIMS Symposia on Software Science and Engineering, pp. 115–127 (1983)

Groups Whose Word Problems Are Accepted by Abelian G-Automata

Takao Yuyama(✉)

Research Institute for Mathematical Sciences, Kyoto University, Kyoto, Japan
yuyama@kurims.kyoto-u.ac.jp

Abstract. Elder, Kambites, and Ostheimer showed that if a finitely generated group H has word problem accepted by a G-automaton for an abelian group G, then H has an abelian subgroup of finite index. Their proof is, however, non-constructive in the sense that it is by contradiction: they proved that H must have a finite index abelian subgroup without constructing any finite index abelian subgroup of H. In addition, a part of their proof is in terms of geometric group theory, which makes it hard to read without knowledge of the field.

We give a new, elementary, and in some sense more constructive proof of the theorem, in which we construct, from the abelian G-automaton accepting the word problem of H, a group homomorphism from a subgroup of G onto a finite index subgroup of H. Our method is purely combinatorial and contains no geometric arguments.

Keywords: word problem · G-automaton · abelian group

1 Introduction

For a group G, a *G-automaton* is a variant of usual finite state automata, which is augmented with a memory register that stores an element of G. During the computation of a G-automaton, the content of the register may be updated by multiplying on the right by an element of G, but cannot be seen. Such an automaton first initializes the register with the identity element 1_G of G, and the automaton accepts an input word if, by reading this word, it can reach a terminal state, in which the register content is 1_G. (For the precise definition, see Sect. 2.4.) For a positive integer n, $\mathbb{Z}^n$-automata are the same as *blind n-counter automata*, which were defined and studied by Greibach [14,15]. Note that the notion of G-automata is discovered repeatedly by several different authors. The name "G-automaton" is due to Kambites [22]. (In fact, they introduced the notion of M-automata for any monoid M.) Render–Kambites [31] uses *G-valence automata* and Dassow–Mitrana [8] and Mitrana–Stiebe [26] use *extended finite automata* (EFA) over G instead of G-automata.

For a finitely generated group H, the *word problem* of H, with respect to a fixed finite generating set of H, is the set of words over the generating set representing the identity element of H (see Sect. 2.2 for the precise definitions). For several language classes, the class of finitely generated groups whose word

F. Drewes and M. Volkov (Eds.): DLT 2023, LNCS 13911, pp. 246–257, 2023.
https://doi.org/10.1007/978-3-031-33264-7_20

problem is in the class has been determined [1,2,10,17,19,27,28], and many attempts are made for other language classes [3,4,12,13,20,21,24,25,30]. One of the most remarkable theorems about word problems is the well-known result due to Muller and Schupp [27], which states that, with the theorem by Dunwoody [9], a group has a context-free word problem if and only if it is virtually free, i.e., has a free subgroup of finite index. These theorems suggest deep connections between group theory and formal language theory.

Involving both G-automata and word problems, the following broad question was posed implicitly by Elston and Ostheimer [11] and explicitly by Kambites [22].

Question 1. For a given group G, is there any connection between the structural property of G and of the collection of groups whose word problems are accepted by *non-deterministic* G-automata?

Note that by G-automata, we always mean non-deterministic G-automata in this paper. As for *deterministic* G-automata, the following theorem is known.

Theorem 1 (Kambites [22, Theorem 1], 2006). *Let G and H be groups with H finitely generated. Then the word problem of H is accepted by a* deterministic *G-automaton if and only if H has a finite index subgroup which embeds in G.*

For non-deterministic G-automata, several results are known for specific types of groups. For a free group F of rank ≥ 2, it is known that a language is accepted by an F-automaton if and only if it is context-free (essentially by [5, PROPOSITION 2], see also [7, Corollary 4.5] and [23, Theorem 7]). Combining with the Muller–Schupp theorem, the class of groups whose word problems are accepted by F-automata is the class of virtually free groups. The class of groups whose word problems are accepted by $(F \times F)$-automata is exactly the class of recursively presentable groups [7, Corollary 3.5][23, Theorem 8][26, Theorem 10].

For the case where G is (virtually) abelian, the following result was shown by Elder, Kambites, and Ostheimer. Recall that a group G is called virtually abelian if it has an abelian subgroup of finite index.

Theorem 2 (Elder, Kambites, and Ostheimer [10], 2008).

(1) *Let H be a finitely generated group and n be a positive integer. Then the word problem of H is accepted by a $\mathbb{Z}^n$-automaton if and only if H is virtually free abelian of rank at most n [10, Theorem 1].*
(2) *Let G be a virtually abelian group and H be a finitely generated group. Then the word problem of H is accepted by a G-automaton if and only if H has a finite index subgroup which embeds in G [10, Theorem 4].*

However, their proof is non-constructive in the sense that it is by contradiction: they proved that H must have a finite index abelian subgroup without constructing any finite index abelian subgroup of H. In addition, their proof depends on a deep theorem in geometric group theory due to Gromov [16], which states that every finitely generated group with polynomial growth function is virtually nilpotent.

The proof of Theorem 2 in [10] proceeds as follows. Let H be a finitely generated group whose word problem is accepted by a $\mathbb{Z}^n$-automaton. First, some techniques to compute several bounds for linear maps and semilinear sets are developed. Then a map from H to $\mathbb{Z}^n$ with some geometric conditions is constructed to prove that H has polynomial growth function. By Gromov's theorem, H is virtually nilpotent. Finally, it is proved that H is virtually abelian, using some theorems about nilpotent groups and semilinear sets. Theorem 2 (2) is deducible from Theorem 2 (1). Because of the non-constructivity of the proof, the embedding in Theorem 2 (2) is obtained only *a posteriori* and hence has nothing to do with the G-automaton.

To our knowledge, there are almost no attempts so far to obtain explicit algebraic connections between G and H, where H is a finitely generated group with word problem accepted by a G-automaton. The only exception is the result due to Holt, Owens, and Thomas [19, THEOREM 4.2], where they gave a somewhat combinatorial proof to a special case of Theorem 2 (1), for the case where $n = 1$. (In fact, their theorem is slightly stronger than Theorem 2 (1) for $n = 1$ because it is for *non-blind* one-counter automata. See also [10, Section 7].) However, their proof also involves growth functions.

In this paper, we give an elementary, purely combinatorial proof of the following our main theorem, which is equivalent to Theorem 2 (see Sect. 3).

Theorem 3. *Let G be an abelian group and H be a finitely generated group. Suppose that the word problem of H is accepted by a G-automaton. Then there exists a group homomorphism from a subgroup of G onto a finite index subgroup of H.*

Our proof of Theorem 3 proceeds as follows. Suppose that the word problem of a finitely generated group H is accepted by a G-automaton A, where G is an abelian group. First, we prove that there exist only finitely many *minimal accepting paths* in A. Next, for each vertex p of A and each minimal accepting path μ in A, we define a set $M(\mu, p)$ of closed paths that is *pumpable* in μ and starts from p, and prove that each $M(\mu, p)$ forms a monoid with respect to concatenation. Then, we show that each monoid $M(\mu, p)$ induces a group homomorphism $f_{\mu,p}$ from a subgroup $G(\mu, p)$ of G onto a subgroup $H(\mu, p)$ of H. Finally, we show that at least one of the $H(\mu, p)$'s has finite index in H.

In addition to this introduction, this paper comprises four sections. Section 2 provides necessary preliminaries, notations, and conventions. In Sect. 3, we reduce Theorem 2 to Theorem 3 and vice versa. Section 4 is devoted to the proof of Theorem 3. Section 5 concludes the paper.

2 Preliminaries

2.1 Words, Subwords, and Scattered Subwords

For a set Σ, we write Σ^* for the free monoid generated by Σ, i.e., the set of *words* over Σ. For a word $u = a_1 a_2 \cdots a_n \in \Sigma^*$ ($n \geq 0, a_i \in \Sigma$), the number n is called the *length* of u, which is denoted by $|u|$. For two words $u, v \in \Sigma^*$, the *concatenation* of u and v are denoted by $u \cdot v$, or simply uv. The identity element of Σ^* is the *empty word*, denoted by ε, which is the unique word of length zero. For an integer $n \geq 0$, the n-fold concatenation of a word $u \in \Sigma^*$ is denoted by u^n. For an integer $n > 0$, we write $\Sigma^{<n}$ for the set of words of length less than n.

A word $u \in \Sigma^*$ is a *subword* of a word $v \in \Sigma^*$, denoted by $u \sqsubseteq v$, if there exist two words $u_1, u_2 \in \Sigma^*$ such that $u_1 u u_2 = v$. A word $u \in \Sigma^*$ is a *scattered subword* of a word $v \in \Sigma^*$, denoted by $u \sqsubseteq_{\mathrm{sc}} v$, if there exist two finite sequences of words $u_1, u_2, \ldots, u_n \in \Sigma^*$ ($n \geq 0$) and $v_0, v_1, \ldots, v_n \in \Sigma^*$ such that $u = u_1 u_2 \cdots u_n$ and $v = v_0 u_1 v_1 u_2 v_2 \cdots u_n v_n$. That is, v is obtained by inserting some words in u. Note that the two binary relations $\sqsubseteq$ and $\sqsubseteq_{\mathrm{sc}}$ are both partial orderings on Σ^*.

2.2 Word Problem for Groups

Let H be a finitely generated group. A *choice of generators* for H is a surjective monoid homomorphism ρ from the free monoid Σ^*, on a finite alphabet Σ, onto H. The *word problem* of H with respect to ρ, denoted by $\mathrm{WP}_\rho(H)$, is the set of words in Σ^* mapped to the identity element 1_H of H via ρ, i.e., $\mathrm{WP}_\rho(H) = \rho^{-1}(1_H)$.

Although the word problem $\mathrm{WP}_\rho(H)$ depends on the choice of generators ρ, this does not cause problems, at least for our purpose:

Proposition 1 (e.g., [20, Lemma 1]). *Let $\mathcal{C}$ be a class of languages closed under inverse homomorphisms and let H be a finitely generated group. Then $WP_\rho(H) \in \mathcal{C}$ for* some *choice of generators ρ if and only if $WP_\rho(H) \in \mathcal{C}$ for* every *choice of generators ρ.* □

Therefore we usually say "*the* word problem of H" rather than "*a* word problem of H."

2.3 Graphs and Paths

A *graph* is a 4-tuple $(V, E, \mathsf{s}, \mathsf{t})$, where V is the set of vertices, E is the set of (directed) edges, $\mathsf{s}\colon E \to V$ and $\mathsf{t}\colon E \to V$ are functions assigning to every edge $e \in E$ the *source* $\mathsf{s}(e) \in V$ and the *target* $\mathsf{t}(e) \in V$, respectively. A graph is *finite* if it has only finitely many vertices and edges.

A *path* (of *length* n) in a graph $\Gamma = (V, E, \mathsf{s}, \mathsf{t})$ is a word $e_1 e_2 \cdots e_n \in E^*$ ($n \geq 0$) of edges $e_i \in E$ such that $\mathsf{t}(e_i) = \mathsf{s}(e_{i+1})$ for $i = 1, 2, \ldots, n-1$. We usually use Greek letters for paths in a graph. For a non-empty path $\omega = e_1 e_2 \cdots e_n \in E^*$, the source and the target of ω are defined as $\mathsf{s}(\omega) = \mathsf{s}(e_1)$ and $\mathsf{t}(\omega) = \mathsf{t}(e_n)$, respectively. If $\omega = e_1 e_2 \cdots e_n$ and $\omega' = e'_1 e'_2 \cdots e'_k$ are non-empty paths such that $\mathsf{t}(\omega) = \mathsf{s}(\omega')$, or at least one of ω and ω' is empty, then the concatenation of ω and ω', denoted by $\omega \cdot \omega'$ or $\omega\omega'$, is the path $e_1 e_2 \cdots e_n e'_1 e'_2 \cdots e'_k$ of length $n+k$, i.e., the concatenation as words. A path ω in Γ is *closed* if $\mathsf{s}(\omega) = \mathsf{t}(\omega)$, or $\omega = \varepsilon$. For a closed path σ and an integer $n \geq 0$, we write σ^n for the n-fold concatenation of σ.

For a graph $\Gamma = (V, E, \mathsf{s}, \mathsf{t})$, an *edge-labeling function* is a function ℓ from E to a set M. If M is a monoid and $\omega = e_1 e_2 \cdots e_n \in E^*$ is a path in Γ, then the label of ω is defined as $\ell(\omega) = \ell(e_1)\ell(e_2) \cdots \ell(e_n)$ via the multiplication of M.

2.4 G-Automata

For a group G, a (non-deterministic) *G-automaton* over a finite alphabet Σ is defined as a 5-tuple $(\Gamma, \ell_G, \ell_\Sigma, p_{\mathrm{init}}, p_{\mathrm{ter}})$, where $\Gamma = (V, E, \mathsf{s}, \mathsf{t})$ is a finite graph, $\ell_G \colon E \to G$ and $\ell_\Sigma \colon E \to \Sigma^*$ are edge-labeling functions, $p_{\mathrm{init}} \in V$ is the *initial vertex*, and $p_{\mathrm{ter}} \in V$ is the *terminal vertex*. For simplicity, we assume that $\ell_\Sigma(e) \in \Sigma \cup \{\varepsilon\}$ for each $e \in E$. (Note that this assumption does not decrease the accepting power of G-automata. Indeed, if necessary, one can subdivide an edge e with labels $\ell_\Sigma(e) = uv, \ell_G(e) = g$ into two new edges e_1, e_2 with labels $\ell_\Sigma(e_1) = u, \ell_G(e_1) = g$ and $\ell_\Sigma(e_2) = v, \ell_G(e_2) = 1_G$.) An *accepting path* in a G-automaton $A = (\Gamma, \ell_G, \ell_\Sigma, p_{\mathrm{init}}, p_{\mathrm{ter}})$ is a path α in Γ such that $\mathsf{s}(\alpha) = p_{\mathrm{init}}$, $\mathsf{t}(\alpha) = p_{\mathrm{ter}}$, and $\ell_G(\alpha) = 1_G$ (we consider that the empty path $\varepsilon \in E^*$ is accepting if and only if $p_{\mathrm{init}} = p_{\mathrm{ter}}$). We say that a path ω in Γ is *promising* if ω is a subword of some accepting path in A, i.e., there exist two paths $\omega_1, \omega_2 \in E^*$ such that the concatenation $\omega_1 \omega \omega_2 \in E^*$ is an accepting path in A. The *language accepted* by a G-automaton A, denoted by $L(A)$, is the set of all words $u \in \Sigma^*$ such that u is the label of some accepting path in A, i.e.,

$$L(A) = \{\, \ell_\Sigma(\alpha) \in \Sigma^* \mid \alpha \text{ is an accepting path in } A \,\}.$$

We say that a G-automaton A is *abelian* if G is an abelian group.

The class of languages accepted by G-automata satisfies the assumption of Proposition 1:

Proposition 2 (e.g., [23, Proposition 2]). *For a group G, the class of languages accepted by G-automata is closed under inverse homomorphisms.* □

Therefore one can speak of a group H whose word problem is accepted by a G-automaton without any reference to generating set for H.

3 Equivalence of Theorem 3 and Theorem 2

Proposition 3. *Theorem 3 implies Theorem 2.*

Proof. Since Theorem 2 (2) is deducible from Theorem 2 (1) [10, Section 6], it suffices to show Theorem 2 (1). If H is a finitely generated group and $\mathbb{Z}^m$ is a finite index subgroup of H for some $m \leq n$, then one can easily construct a $\mathbb{Z}^n$-automaton that accepts the word problem of H (see e.g., [11, Theorem 7]). Conversely, suppose that the word problem of H is accepted by a $\mathbb{Z}^n$-automaton. By Theorem 3, there exists a group homomorphism f from a subgroup G_0 of $\mathbb{Z}^n$ onto a finite index subgroup H_0 of H. In general, a subgroup S of a free abelian group F is also free abelian, and the rank of S does not exceed that of F (see e.g., [32, 4.2.3]). Thus $G_0 \cong \mathbb{Z}^m$ for some $m \leq n$. Since H_0 is a homomorphic image of $\mathbb{Z}^m$, H_0 is an abelian group generated by at most m elements. Hence, by the fundamental theorem of finitely generated abelian groups (see e.g., [32, 4.2.10]), H_0 has a finite index subgroup isomorphic to $\mathbb{Z}^k$ for some $k \leq m$. Thus H has a finite index subgroup isomorphic to $\mathbb{Z}^k$. □

Proposition 4. *Theorem 2 implies Theorem 3.*

Proof. Suppose that the word problem of a finitely generated group H is accepted by a G-automaton A, where G is an abelian group. By Theorem 2 (2), there exist a finite index subgroup H_0 of H and an embedding $f\colon H_0 \to G$. Since f is injective, the homomorphism $f^{-1}\colon f(H_0) \to H_0$ is the desired one.

4 Proof of Theorem 3

Throughout this section, we fix an abelian group G, a finitely generated group H, a choice of generators $\rho\colon \Sigma^* \to H$, and an abelian G-automaton $A = (\Gamma = (V, E, \mathsf{s}, \mathsf{t}), \ell_G, \ell_\Sigma, p_{\text{init}}, p_{\text{ter}})$ such that $\mathrm{WP}_\rho(H) = L(A)$. We write the group operation of G additively and write 0_G for the identity element of G.

4.1 Minimal Accepting Paths

The minimal accepting paths, defined below, play a central role in this paper.

Definition 1. *An accepting path α in A is* minimal *if it is minimal with respect to the scattered subword relation $\sqsubseteq_{\text{sc}}$ on E^* among the accepting paths. An accepting path α in A* dominates *a minimal accepting path μ in A if $\mu \sqsubseteq_{\text{sc}} \alpha$.*

A similar notion of minimal accepting paths can be found in [6, Section 4].

Proposition 5 (Higman's lemma [18, Theorem 4.4]). *For any finite alphabet Σ, the scattered subword relation $\sqsubseteq_{\text{sc}}$ on Σ^* is a* well-quasi-order*, i.e., for any infinite sequence $u_1, u_2, \ldots \in \Sigma^*$, there exist some $i < j$ such that $u_i \sqsubseteq_{\text{sc}} u_j$.* □

Corollary 1. *There are only finitely many minimal accepting paths in A, and every accepting path on A dominates some minimal accepting path in A.*

Proof. Suppose the contrary that there are infinitely many distinct minimal accepting paths $\mu_1, \mu_2, \ldots \in E^*$. Then we have $\mu_i \not\sqsubseteq_{\text{sc}} \mu_j$ for any $i < j$ because of the minimality of μ_j, a contradiction. The second half of the lemma is also true since a well-quasi-ordered set admits no infinite descending sequence. □

Note that if $p_{\text{init}} = p_{\text{ter}}$ then the only minimal accepting path is the empty path ε.

4.2 Pumpable Paths and the Monoids $M(\mu, p)$

Definition 2. *Let $\mu = e_1 e_2 \cdots e_n \in E^*$ $(e_i \in E)$ be a minimal accepting path in A. A closed path $\sigma \in E^*$ in Γ is* pumpable *in μ if there exists an accepting path α in A dominating μ such that $\alpha = \alpha_0 e_1 \alpha_1 e_2 \cdots e_n \alpha_n$ for some paths $\alpha_0, \alpha_1, \ldots, \alpha_n \in E^*$ in Γ and $\sigma \sqsubseteq \alpha_j$ for some $j \in \{0, 1, \ldots, n\}$.*

Remark 1.

(1) In Definition 2, each α_i inserted to μ is a closed path in Γ. In addition, $\ell_G(\alpha_0) + \ell_G(\alpha_1) + \cdots + \ell_G(\alpha_n) = 0_G$ since G is abelian and $\ell_G(\alpha) = \ell_G(\mu) = 0_G$.
(2) Every closed path σ pumpable in a minimal accepting path μ is promising since σ is a subword (not a scattered subword) of an accepting path α dominating μ.

Definition 3. *For a minimal accepting path μ in A and a vertex $p \in V$, define*

$$M(\mu, p) = \left\{ \sigma \in E^* \;\middle|\; \begin{array}{c} \sigma \textit{ is a closed path in } \Gamma \textit{ pumpable in } \mu \\ \textit{such that } \mathsf{s}(\sigma) = p, \textit{ or } \sigma = \varepsilon \end{array} \right\}.$$

Note that there are only finitely many $M(\mu, p)$'s by Corollary 1.

Lemma 1. *Each $M(\mu, p)$ is a monoid with respect to the concatenation operation, i.e., $\sigma_1, \sigma_2 \in M(\mu, p)$ implies $\sigma_1 \sigma_2 \in M(\mu, p)$.*

Proof. Since both σ_1 and σ_2 are pumpable in $\mu = e_1 e_2 \cdots e_n \in E^*$ $(e_i \in E)$, there exist two accepting paths $\alpha = \alpha_0 e_1 \alpha_1 e_2 \cdots e_n \alpha_n$ $(\alpha_i \in E^*)$ and $\beta = \beta_0 e_1 \beta_1 e_2 \cdots e_n \beta_n$ $(\beta_i \in E^*)$ such that $\sigma_1 \sqsubseteq \alpha_i$ and $\sigma_2 \sqsubseteq \beta_j$ for some $i, j \in \{0, 1, \ldots, n\}$. Then we have $\alpha_i = \alpha'_i \sigma_1 \alpha''_i$ for some $\alpha'_i, \alpha''_i \in E^*$ and $\beta_j = \beta'_j \sigma_2 \beta''_j$ for some $\beta'_j, \beta''_j \in E^*$. We may assume that $i \leq j$. Since G is abelian, the merged path $\gamma = (\alpha_0 \beta_0) e_1 (\alpha_1 \beta_1) e_2 \cdots e_n (\alpha_n \beta_n)$ and its permutation

$$\gamma' = (\alpha_0 \beta_0) e_1 (\alpha_1 \beta_1) e_2 \cdots e_i (\alpha'_i \sigma_1 \sigma_2 \alpha''_i \beta_i) e_{i+1} \cdots e_j (\alpha_j \beta'_j \beta''_j) e_{j+1} \cdots e_n (\alpha_n \beta_n) \tag{1}$$

are accepting paths in A by Remark 1 (1) (Fig. 1). □

Lemma 2. *Let σ and τ be closed paths in Γ such that $\mathsf{s}(\tau) = p$ (or $\tau = \varepsilon$) and $\tau \sqsubseteq_{\mathrm{sc}} \sigma \in M(\mu, p)$. Then $\tau \in M(\mu, p)$.*

Proof. Suppose that $\tau = e'_1 e'_2 \cdots e'_k$ $(k \geq 0, e'_i \in E)$ and $\sigma = \sigma_0 e'_1 \sigma_1 e'_2 \cdots e'_k \sigma_k$ $(\sigma_i \in E^*)$. Note that each σ_i is a closed path in Γ. Since, by Lemma 1, σ^2 is pumpable in $\mu = e_1 e_2 \cdots e_n$ $(n \geq 0, e_i \in E)$, there exists an accepting path $\alpha = \alpha_0 e_1 \alpha_1 e_2 \cdots e_n \alpha_n$ dominating μ such that $\sigma^2 \sqsubseteq_{\mathrm{sc}} \alpha_i$ for some $i \in \{0, 1, \ldots, n\}$. If $\alpha_i = \alpha'_i \sigma^2 \alpha''_i$, then the path

$$\alpha_0 e_1 \alpha_1 e_2 \cdots e_i (\alpha'_i \cdot \tau \cdot (\sigma_0^2 e'_1 \sigma_1^2 e'_2 \cdots e'_k \sigma_k^2) \cdot \alpha''_i) e_{i+1} \cdots e_n \alpha_n \tag{2}$$

is an accepting path in A by Remark 1 (1) (Fig. 2). □

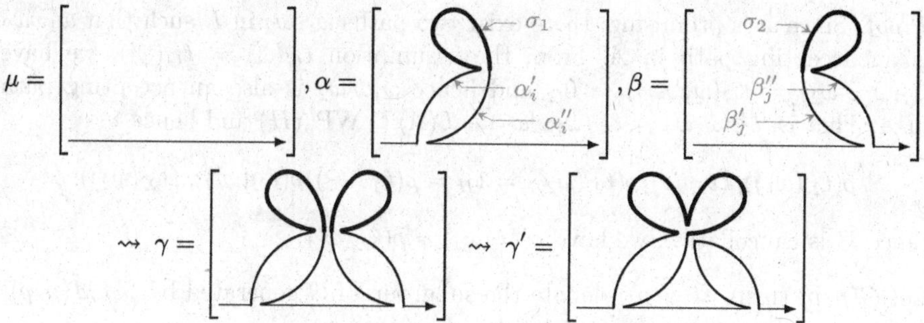

Fig. 1. Construction of the accepting path γ' in (1)

Fig. 2. Construction of the path $\tau \cdot (\sigma_0^2 e_1' \sigma_1^2 e_2' \cdots e_k' \sigma_k^2)$ in (2)

Lemma 3. *Let $\sigma \in M(\mu, p)$ and $\omega \sqsubseteq \sigma$ be a path. Then there exist two paths $\omega_1, \omega_2 \in E^{<|V|}$ such that $\omega_1 \omega \omega_2 \in M(\mu, p)$.*

Proof. Let $(\omega_1, \omega_2) \in E^* \times E^*$ be a pair of two paths such that $\omega_1 \omega \omega_2 \in M(\mu, p)$ and $\max\{|\omega_1|, |\omega_2|\}$ is minimum. Such a pair exists since $\omega \sqsubseteq \sigma \in M(\mu, p)$. Suppose the contrary that $\max\{|\omega_1|, |\omega_2|\} \geq |V|$, say $|\omega_1| \geq |V|$. By the pigeonhole principle, ω_1 must visit some vertex $p \in V$ at least twice. That is, there exist three paths α, β, γ such that $\omega_1 = \alpha\beta\gamma$ and β is a non-empty closed path. Now we have $\alpha\gamma\omega\omega_2 \sqsubseteq_{\mathrm{sc}} \omega_1 \omega \omega_2 \in M(\mu, p)$, and Lemma 2 implies $\alpha\gamma\omega\omega_2 \in M(\mu, p)$, which contradicts the minimality of (ω_1, ω_2). □

4.3 Group Homomorphisms $f_{\mu,p}$ from $G(\mu, p)$ onto $H(\mu, p)$

For each $M(\mu, p)$, Lemma 1 allows us to define a surjective monoid homomorphism $\varphi_{\mu,p} \colon M(\mu, p) \to \rho(\ell_\Sigma(M(\mu, p)))$ as the composition function $\rho \circ \ell_\Sigma$. Then each $\varphi_{\mu,p}$ induces a well-defined surjective monoid homomorphism $\bar{\varphi}_{\mu,p} \colon \ell_G(M(\mu, p)) \to \rho(\ell_\Sigma(M(\mu, p)))$ thanks to the following Lemma 4.

Lemma 4. *Let ω and ω' be paths in Γ such that $\mathsf{s}(\omega) = \mathsf{s}(\omega')$ and $\mathsf{t}(\omega) = \mathsf{t}(\omega')$, and suppose that ω is promising. Then $\ell_G(\omega) = \ell_G(\omega')$ implies $\rho(\ell_\Sigma(\omega)) = \rho(\ell_\Sigma(\omega'))$.*

Proof. Since ω is promising, there exist two paths ω_1, ω_2 in Γ such that $\omega_1\omega\omega_2$ is an accepting path in A. From the assumption $\ell_G(\omega) = \ell_G(\omega')$, we have $\ell_G(\omega_1\omega'\omega_2) = \ell_G(\omega_1\omega\omega_2) = 0_G$ and hence $\omega_1\omega'\omega_2$ is also an accepting path in A. That is, $\ell_\Sigma(\omega_1\omega\omega_2), \ell_\Sigma(\omega_1\omega'\omega_2) \in L(A) \subseteq \mathrm{WP}_\rho(H)$ and hence

$$\rho(\ell_\Sigma(\omega_1))\rho(\ell_\Sigma(\omega))\rho(\ell_\Sigma(\omega_2)) = 1_H = \rho(\ell_\Sigma(\omega_1))\rho(\ell_\Sigma(\omega'))\rho(\ell_\Sigma(\omega_2)).$$

Since H is cancellative, we have $\rho(\ell_\Sigma(\omega)) = \rho(\ell_\Sigma(\omega'))$. □

Let $G(\mu, p)$ (resp. $H(\mu, p)$) denote the subgroup of G generated by $\ell_G(M(\mu, p))$ (resp. the subgroup of H generated by $\rho(\ell_\Sigma(M(\mu, p)))$).

Lemma 5. *One can extend $\bar{\varphi}_{\mu,p}\colon \ell_G(M(\mu, p)) \to \rho(\ell_\Sigma(M(\mu, p)))$ to a unique surjective group homomorphism $f_{\mu,p}\colon G(\mu, p) \to H(\mu, p)$.*

Proof. Since G is an abelian group, every element $g \in G(\mu, p)$ can be written as $g = \ell_G(\sigma_1) - \ell_G(\sigma_2)$ for some $\sigma_1, \sigma_2 \in M(\mu, p)$. Defining $f_{\mu,p}(g) = \rho(\ell_\Sigma(\sigma_1)) - \rho(\ell_\Sigma(\sigma_2))$, one can easily check the well-definedness, the uniqueness, and the surjectivity. □

4.4 One of the $H(\mu, p)$'s has Finite Index in H

The remaining task is to prove that at least one of the $H(\mu, p)$'s has finite index in H. To do this, we use B. H. Neumann's lemma in the following form.

Proposition 6 (B. H. Neumann's lemma [29, (4.1) LEMMA and (4.2)]). *Let H be a group, $H_1, H_2, \ldots, H_n$ be subgroups of H, and $a_1, b_1, a_2, b_2, \ldots, a_n, b_n$ be elements of H. If $H = \bigcup_{i=1}^{n} a_i H_i b_i$, then at least one of the H_i's is of index at most n in H.* □

Lemma 6. *The following holds.*

$$H = \bigcup \left\{ h_1^{-1} H(\mu, p) h_2^{-1} \;\middle|\; \begin{array}{c} \mu \text{ is a minimal accepting path in } A, \\ p \in V, \text{ and } h_1, h_2 \in \rho(\Sigma^{<|V|}) \end{array} \right\}. \tag{3}$$

Proof. Let $h \in H$ and fix a word $v \in \Sigma^*$ such that $\rho(v) = h$. Since ρ is surjective, there exists a word $\bar{v} \in \Sigma^*$ such that $\rho(\bar{v}) = \rho(v)^{-1}$. Define

$$N = 1 + \max\{\, |\mu| \mid \mu \in E^* \text{ is a minimal accepting path in } A \,\},$$

and we have $N < \infty$ by Corollary 1. Since $(v\bar{v})^N \in \mathrm{WP}_\rho(H) \subseteq L(A)$, there exists an accepting path

$$\alpha = \omega_1\bar{\omega}_1\omega_2\bar{\omega}_2 \cdots \omega_N\bar{\omega}_N \tag{4}$$

in A such that $\ell_\Sigma(\omega_i) = v$ and $\ell_\Sigma(\bar{\omega}_i) = \bar{v}$ for $i = 1, 2, \ldots, N$. Let $\mu = e_1 e_2 \cdots e_n$ ($e_i \in E$) be a minimal accepting path such that α dominates μ. Then we have another decomposition

$$\alpha = \alpha_0 e_1 \alpha_1 e_2 \cdots e_n \alpha_n \tag{5}$$

for some closed paths $\alpha_0, \alpha_1, \ldots, \alpha_n \in E^*$. Since $N > |\mu| = n$ and each e_i in the decomposition (5) is contained in at most one ω_i in the decomposition (4), at least one of the ω_i's is "disjoint" from all e_i's, i.e., there exist $i \in \{1, 2, \ldots, N\}$ and $j \in \{0, 1, \ldots, n\}$ such that $\omega_i \sqsubseteq \alpha_j$. Since α_j is a pumpable closed path in μ, we have $\alpha_j \in M(\mu, p)$, where $p = \mathsf{s}(\alpha_j)$. By Lemma 3, there exist $\alpha'_j, \alpha''_j \in E^{<|V|}$ such that $\alpha'_j \omega_i \alpha''_j \in M(\mu, p)$. Then we have $|\ell_\Sigma(\alpha'_j)|, |\ell_\Sigma(\alpha''_j)| < |V|$ and $\rho(\ell_\Sigma(\alpha'_j))\rho(\ell_\Sigma(\omega_i))\rho(\ell_\Sigma(\alpha''_j)) \in \rho(\ell_\Sigma(M(\mu, p))) \subseteq H(\mu, p)$, hence

$$h = \rho(v) = \rho(\ell_\Sigma(\omega_i)) \in \rho(\ell_\Sigma(\alpha'_j))^{-1} H(\mu, p) \rho(\ell_\Sigma(\alpha''_j))^{-1}.$$

Thus (3) holds. □

Proof of Theorem 3. Since G is an abelian group and $L(A) \subseteq \mathrm{WP}_\rho(H)$, we have surjective group homomorphisms $f_{\mu,p}\colon G(\mu, p) \to H(\mu, p)$ by Lemmata 4 and 5. Since $\mathrm{WP}_\rho(H) \subseteq L(A)$, the equation (3) holds by Lemma 6, and the right-hand side of (3) is a finite union of cosets of H by Corollary 1. Thus, by B. H. Neumann's lemma (Proposition 6), at least one of the $H(\mu, p)$'s is of index at most $|\{\, \mu \mid \mu \text{ is a minimal accepting path in } A \}| \times |V| \times |\Sigma^{<|V|}|^2$ in H. □

5 Conclusion

We gave a new, elementary, purely combinatorial proof of the theorem due to Elder, Kambites, and Ostheimer, which states that if a finitely generated group H has word problem accepted by an abelian G-automaton A, then H is virtually abelian. In contrast to their original geometric argument, we stuck to the combinatorial link—the abelian G-automaton A—between the two groups G and H, and we obtained explicit algebraic connections between them as finitely many group homomorphisms $f_{\mu,p}\colon G(\mu, p) \to H(\mu, p)$.

We leave the following question as our future work.

Question 2. Let G and H be (not necessarily abelian) groups with H finitely generated. Suppose that the word problem of H is accepted by a G-automaton. Can one combinatorially obtain a group homomorphism from a subgroup of G onto a finite index subgroup of H?

For example, how about the case where G is free or nilpotent? Our Theorem 3 is the very first step for approaching Question 2.

Acknowledgements. The author would like to thank Ryoma Sin'ya (Akita University) for encouragement and helpful comments. They would also like to thank anonymous reviewers for many valuable comments. This work was supported by JSPS KAKENHI Grant Numbers JP20J23039 and JP20H05961.

References

1. Anīsīmov, A.V.: Certain algorithmic questions for groups and context-free languages. Kibernetika (Kiev) **2**, 4–11 (1972)

2. Boone, W.W., Higman, G.: An algebraic characterization of groups with soluble word problem. J. Aust. Math. Soc. **18**, 41–53 (1974). Collection of articles dedicated to the memory of Hanna Neumann, IX
3. Brough, T.: Groups with poly-context-free word problem. Groups Complex. Cryptol. **6**(1), 9–29 (2014). https://doi.org/10.1515/gcc-2014-0002
4. Ceccherini-Silberstein, T., Coornaert, M., Fiorenzi, F., Schupp, P.E., Touikan, N.W.M.: Multipass automata and group word problems. Theor. Comput. Sci. **600**, 19–33 (2015). https://doi.org/10.1016/j.tcs.2015.06.054
5. Chomsky, N., Schützenberger, M.P.: The algebraic theory of context-free languages. In: Computer Programming and Formal Systems, pp. 118–161. North-Holland, Amsterdam (1963)
6. Cleary, S., Elder, M., Ostheimer, G.: The word problem distinguishes counter languages (2006). https://arxiv.org/abs/math/0606415
7. Corson, J.M.: Extended finite automata and word problems. Int. J. Algebra Comput. **15**(3), 455–466 (2005). https://doi.org/10.1142/S0218196705002360
8. Dassow, J., Mitrana, V.: Finite automata over free groups. Int. J. Algebra Comput. **10**(6), 725–737 (2000). https://doi.org/10.1142/S0218196700000315
9. Dunwoody, M.J.: The accessibility of finitely presented groups. Invent. Math. **81**(3), 449–457 (1985). https://doi.org/10.1007/BF01388581
10. Elder, M., Kambites, M., Ostheimer, G.: On groups and counter automata. Int. J. Algebra Comput. **18**(8), 1345–1364 (2008). https://doi.org/10.1142/S0218196708004901
11. Elston, G.Z., Ostheimer, G.: On groups whose word problem is solved by a counter automaton. Theor. Comput. Sci. **320**(2–3), 175–185 (2004). https://doi.org/10.1016/j.tcs.2003.09.007
12. Gebhardt, K., Meunier, F., Salvati, S.: O_n is an n-MCFL. J. Comput. Syst. Sci. **127**, 41–52 (2022). https://doi.org/10.1016/j.jcss.2022.02.003
13. Gilman, R.H., Kropholler, R.P., Schleimer, S.: Groups whose word problems are not semilinear. Groups Complex. Cryptol. **10**(2), 53–62 (2018). https://doi.org/10.1515/gcc-2018-0010
14. Greibach, S.A.: Remarks on the complexity of nondeterministic counter languages. Theor. Comput. Sci. **1**(4), 269–288 (1975/76). https://doi.org/10.1016/0304-3975(76)90072-4
15. Greibach, S.A.: Remarks on blind and partially blind one-way multicounter machines. Theor. Comput. Sci. **7**(3), 311–324 (1978). https://doi.org/10.1016/0304-3975(78)90020-8
16. Gromov, M.: Groups of polynomial growth and expanding maps. Inst. Hautes Études Sci. Publ. Math. (53), 53–73 (1981). https://www.numdam.org/item?id=PMIHES_1981__53__53_0
17. Higman, G.: Subgroups of finitely presented groups. Proc. Roy. Soc. Lond. Ser. A **262**, 455–475 (1961). https://doi.org/10.1098/rspa.1961.0132
18. Higman, G.: Ordering by divisibility in abstract algebras. Proc. Lond. Math. Soc. **3**(1), 326–336 (1952). https://doi.org/10.1112/plms/s3-2.1.326
19. Holt, D.F., Owens, M.D., Thomas, R.M.: Groups and semigroups with a one-counter word problem. J. Aust. Math. Soc. **85**(2), 197–209 (2008). https://doi.org/10.1017/S1446788708000864
20. Holt, D.F., Rees, S., Röver, C.E., Thomas, R.M.: Groups with context-free co-word problem. J. Lond. Math. Soc. **71**(3), 643–657 (2005). https://doi.org/10.1112/S002461070500654X
21. Holt, D.F., Röver, C.E.: Groups with indexed co-word problem. Int. J. Algebra Comput. **16**(5), 985–1014 (2006). https://doi.org/10.1142/S0218196706003359

22. Kambites, M.: Word problems recognisable by deterministic blind monoid automata. Theor. Comput. Sci. **362**(1–3), 232–237 (2006). https://doi.org/10.1016/j.tcs.2006.06.026
23. Kambites, M.: Formal languages and groups as memory. Commun. Algebra **37**(1), 193–208 (2009). https://doi.org/10.1080/00927870802243580
24. Kanazawa, M., Salvati, S.: MIX is not a tree-adjoining language. In: Proceedings of the 50th Annual Meeting of the Association for Computational Linguistics (Volume 1: Long Papers), Jeju Island, Korea, pp. 666–674. Association for Computational Linguistics, July 2012. https://clanthology.org/P12-1070
25. Kropholler, R.P., Spriano, D.: Closure properties in the class of multiple context-free groups. Groups Complex. Cryptol. **11**(1), 1–15 (2019). https://doi.org/10.1515/gcc-2019-2004
26. Mitrana, V., Stiebe, R.: Extended finite automata over groups. Discrete Appl. Math. **108**(3), 287–300 (2001). https://doi.org/10.1016/S0166-218X(00)00200-6
27. Muller, D.E., Schupp, P.E.: Groups, the theory of ends, and context-free languages. J. Comput. Syst. Sci. **26**(3), 295–310 (1983). https://doi.org/10.1016/0022-0000(83)90003-X
28. Rino Nesin, G.A., Thomas, R.M.: Groups whose word problem is a Petri net language. In: Shallit, J., Okhotin, A. (eds.) DCFS 2015. LNCS, vol. 9118, pp. 243–255. Springer, Cham (2015). https://doi.org/10.1007/978-3-319-19225-3_21
29. Neumann, B.H.: Groups covered by permutable subsets. J. Lond. Math. Soc. **29**, 236–248 (1954). https://doi.org/10.1112/jlms/s1-29.2.236
30. Nyberg-Brodda, C.F.: On the word problem for free products of semigroups and monoids. J. Algebra **622**, 721–741 (2023). https://doi.org/10.1016/j.jalgebra.2023.02.007
31. Render, E., Kambites, M.: Polycyclic and bicyclic valence automata. In: Martín-Vide, C., Otto, F., Fernau, H. (eds.) LATA 2008. LNCS, vol. 5196, pp. 464–475. Springer, Heidelberg (2008). https://doi.org/10.1007/978-3-540-88282-4_42
32. Robinson, D.J.S.: A Course in the Theory of Groups. Graduate Texts in Mathematics, vol. 80, 2nd edn. Springer, New York (1996). https://doi.org/10.1007/978-1-4419-8594-1

Author Index

F. Drewes and M. Volkov (Eds.): DLT 2023, LNCS 13911, p. 259, 2023.
https://doi.org/10.1007/978-3-031-33264-7

Printed in the United States
by Baker & Taylor Publisher Services

Printed in the United States
Baker & Taylor Publisher Services